T0245226

CAMBRIDGE LIBRARY COLLECTION

Books of enduring scholarly value

Physical Sciences

From ancient times, humans have tried to understand the workings of the world around them. The roots of modern physical science go back to the very earliest mechanical devices such as levers and rollers, the mixing of paints and dyes, and the importance of the heavenly bodies in early religious observance and navigation. The physical sciences as we know them today began to emerge as independent academic subjects during the early modern period, in the work of Newton and other 'natural philosophers', and numerous sub-disciplines developed during the centuries that followed. This part of the Cambridge Library Collection is devoted to landmark publications in this area which will be of interest to historians of science concerned with individual scientists, particular discoveries, and advances in scientific method, or with the establishment and development of scientific institutions around the world.

Elements of Natural Philosophy

In 1867, Sir William Thomson (later Lord Kelvin) and Peter Guthrie Tate revolutionised physics with the publication of their *Treatise on Natural Philosophy* (also reissued in this series), in which they demonstrated the centrality of energy conservation to systems of dynamic movement. Popularly known as 'T&T' for its authors' initials, the *Treatise* became the standard textbook on natural philosophy, introducing generations of mathematicians to the 'new energy-based dynamics'. In *Elements of Natural Philosophy* (1873), they distil the portions of the *Treatise* not requiring higher calculus into a primer suitable for use in university courses. The first half covers the basic principles of kinematics and dynamics, including the motion of points, lines, and volumes, while the second half concerns questions of 'abstract dynamics', including particle attraction. The result of one of the most important collaborations in modern physics, this book remains a thorough introduction to the major principles of Thomson and Tait's larger work.

Cambridge University Press has long been a pioneer in the reissuing of out-of-print titles from its own backlist, producing digital reprints of books that are still sought after by scholars and students but could not be reprinted economically using traditional technology. The Cambridge Library Collection extends this activity to a wider range of books which are still of importance to researchers and professionals, either for the source material they contain, or as landmarks in the history of their academic discipline.

Drawing from the world-renowned collections in the Cambridge University Library, and guided by the advice of experts in each subject area, Cambridge University Press is using state-of-the-art scanning machines in its own Printing House to capture the content of each book selected for inclusion. The files are processed to give a consistently clear, crisp image, and the books finished to the high quality standard for which the Press is recognised around the world. The latest print-on-demand technology ensures that the books will remain available indefinitely, and that orders for single or multiple copies can quickly be supplied.

The Cambridge Library Collection will bring back to life books of enduring scholarly value (including out-of-copyright works originally issued by other publishers) across a wide range of disciplines in the humanities and social sciences and in science and technology.

Elements of Natural Philosophy

William Thomson, Baron Kelvin
P.G. Tait

CAMBRIDGE UNIVERSITY PRESS

Cambridge, New York, Melbourne, Madrid, Cape Town, Singapore,
São Paolo, Delhi, Dubai, Tokyo

Published in the United States of America by Cambridge University Press, New York

www.cambridge.org
Information on this title: www.cambridge.org/9781108014489

This edition first published 1873
This digitally printed version 2010

ISBN 978-1-108-01448-9 Paperback

NATURAL PHILOSOPHY

THOMSON AND TAIT

ELEMENTS

OF

NATURAL PHILOSOPHY

BY

PROFESSORS SIR W. THOMSON
AND P. G. TAIT

PART I

Cambridge
AT THE UNIVERSITY PRESS

PREFACE.

THE following work consists, in great part, of the large-type, or non-mathematical, portion of our Treatise on Natural Philosophy.

As it is designed more especially for use in Schools and in the junior classes in Universities, the mathematical methods employed are, almost without exception, limited to those of the most elementary geometry, algebra, and trigonometry. Where higher methods are required for an investigation, the reader is, in general, simply referred to our larger work.

It is particularly interesting to note how many theorems, even among those not ordinarily attacked without the help of the Differential Calculus, have here been found to yield easily to geometrical methods of the most elementary character.

Simplification of modes of proof is not merely an indication of advance in our knowledge of a subject, but is also the surest guarantee of readiness for farther progress.

A large part of Chapter VII is reprinted from a series of notes of a part of the Glasgow course, drawn up for Sir W. Thomson by John Ferguson, Esq., and printed for the use of his students.

We have had considerable difficulty in compiling this treatise from the larger work—arising from the necessity for condensation to a degree almost incompatible with the design to omit nothing of importance: and we feel that it would have given us much less trouble and anxiety, and would probably have ensured a better result, had we written the volume anew without keeping the larger book constantly before us. The sole justification of the course we have pursued is that wherever, in the present volume, the student may feel further information to be desirable, he will have no difficulty in finding it in the corresponding pages of the larger work.

A great portion of the present volume has been in type since the autumn of 1863, and has been printed for the use of our classes each autumn since that date.

W. THOMSON.
P. G. TAIT.

November, 1872.

CONTENTS.

DIVISION I. PRELIMINARY.

DIVISION II. ABSTRACT DYNAMICS.

DIVISION I.

PRELIMINARY.

CHAPTER I.—KINEMATICS.

1. THE science which investigates the action of Force is called, by the most logical writers, DYNAMICS. It is commonly, but erroneously, called MECHANICS; a term employed by Newton in its true sense, the Science of Machines, and the art of making them.

2. Force is recognized as acting in two ways:

1° so as to compel rest or to prevent change of motion, and
2° so as to produce or to change motion.

Dynamics, therefore, is divided into two parts, which are conveniently called STATICS and KINETICS.

3. In Statics the action of force in maintaining rest, or preventing change of motion, the 'balancing of forces,' or Equilibrium, is investigated; in Kinetics, the action of force in producing or in changing motion.

4. In Kinetics it is not mere *motion* which is investigated, but the relation of *forces* to motion. The circumstances of mere motion, considered without reference to the bodies moved, or to the forces producing the motion, or to the forces called into action by the motion, constitute the subject of a branch of Pure Mathematics, which is called KINEMATICS, or, in its more practical branches, MECHANISM.

5. Observation and experiment have afforded us the means of translating, as it were, from Kinematics into Dynamics, and *vice versâ*. This is merely mentioned now in order to show the necessity for, and the value of, the preliminary matter we are about to introduce.

6. Thus it appears that there are many properties of motion, displacement, and deformation, which may be considered altogether independently of force, mass, chemical constitution, elasticity, temperature, magnetism, electricity; and that the preliminary consideration of such properties in the abstract is of very great use for Natural Philosophy. We devote to it, accordingly, the whole of this chapter;

which will form, as it were, the Geometry of the subject, embracing what can be observed or concluded with regard to actual motions, as long as the *cause* is not sought. In this category we shall first take up the free motion of a point, then the motion of a point attached to an inextensible cord, then the motions and displacements of rigid systems—and finally, the deformations of solid and fluid masses.

7. When a point moves from one position to another it must evidently describe a *continuous* line, which may be curved or straight, or even made up of portions of curved and straight lines meeting each other at any angles. If the motion be that of a *material particle*, however, there can be no abrupt change of velocity, nor of direction unless where the velocity is zero, since (as we shall afterwards see) such would imply the action of an *infinite* force. It is useful to consider at the outset various theorems connected with the geometrical notion of the path described by a moving point; and these we shall now take up, deferring the consideration of Velocity to a future section, as being more closely connected with physical ideas.

8. The *direction* of motion of a moving point is at each instant the tangent drawn to its path, if the path be a curve; or the path itself if a straight line. This is evident from the definition of the tangent to a curve.

9. If the path be not straight the direction of motion changes from point to point, and the *rate* of this change, per unit of length of the curve, is called the *Curvature.* To exemplify this, suppose two tangents PT, QU, drawn to a circle, and radii OP, OQ, to the points of contact. The angle between the tangents is the change of direction between P and Q, and the rate of change is to be measured by the relation between this angle and the length of the circular arc PQ. Now, if θ be the angle, s the arc, and r the radius, we see at once that (as the angle between the radii is equal to the angle between the tangents, and as the measure of an angle is the ratio of the arc to the radius, § 54)

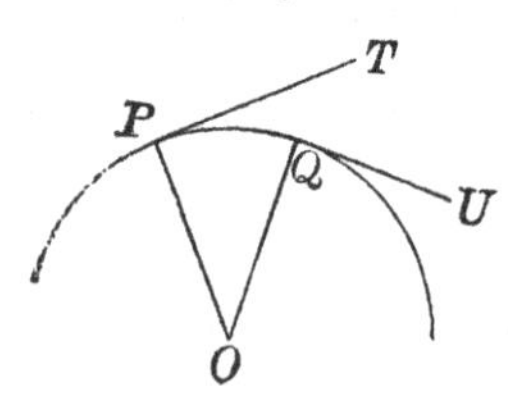

$$r\theta = s,$$

and therefore $\frac{\theta}{s} = \frac{1}{r}$ is the measure of the curvature. Hence the curvature of a circle is inversely as its radius, and is measured, in terms of the proper unit of curvature, simply by the reciprocal of the radius.

10. Any small portion of a curve may be approximately taken as a circular arc, the approximation being closer and closer to the truth, as the assumed arc is smaller. The curvature at any point is the reciprocal of the radius of this circle for a small arc on each side of the point.

11. If all the points of the curve lie in one plane, it is called a *plane* *curve*, and if it be made up of portions of straight or curved lines it is called a *plane polygon*. If the line do not lie in one plane, we have in one case what is called a *curve of double curvature*, in the other a *gauche polygon*. The term 'curve of double curvature' is a very bad one, and, though in very general use, is, we hope, not ineradicable. The fact is, that there are not two curvatures, but only a curvature (as above defined) of which the plane is continuously changing, or twisting, round the tangent line. The course of such a curve is, in common language, well called 'tortuous;' and the measure of the corresponding property is conveniently called *Tortuosity*.

12. The nature of this will be best understood by considering the curve as a polygon whose sides are indefinitely small. Any two consecutive sides, of course, lie in a plane—and in that plane the curvature is measured as above; but in a curve which is not plane the third side of the polygon will not be in the same plane with the first two, and therefore the new plane in which the curvature is to be measured is different from the old one. The plane of the curvature on each side of any point of a tortuous curve is sometimes called the *Osculating Plane* of the curve at that point. As two successive positions of it contain the second side of the polygon above mentioned, it is evident that the osculating plane passes from one position to the next by revolving about the tangent to the curve.

13. Thus, as we proceed along such a curve, the curvature in general varies; and, at the same time, the plane in which the curvature lies is turning about the tangent to the curve. The rate of torsion, or the tortuosity, is therefore to be measured by the rate at which the osculating plane turns about the tangent, per unit length of the curve. The simplest illustration of a tortuous curve is the thread of a screw. Compare § 41 (*d*).

14. The *Integral Curvature*, or *whole change of direction*, of an arc of a plane curve, is the angle through which the tangent has turned as we pass from one extremity to the other. The *average curvature* of any portion is its whole curvature divided by its length. Suppose a line, drawn through any fixed point, to turn so as always to be parallel to the direction of motion of a point describing the curve: the angle through which this turns during the motion of the point exhibits what we have defined as the integral curvature. In estimating this, we must of course take the enlarged modern meaning of an angle, including angles greater than two right angles, and also negative angles. Thus the integral curvature of any closed curve or broken line, whether everywhere concave to the interior or not, is four right angles, provided it does not cut itself. That of a Lemniscate, 8, is *zero*. That of the Epicycloid ⑥ is eight right angles; and so on.

15. The definition in last section may evidently be extended to a plane polygon, and the integral change of direction, or the angle between the first and last sides, is then the sum of its exterior angles, all the sides being produced each in the direction in which the moving point describes it while passing round the figure. This is true whether the polygon be closed or not. If closed, then, as long as it is not crossed, this sum is four right angles,—an extension of the result in Euclid, where all *reëntrant* polygons are excluded. In the star-shaped figure , it is ten right angles, wanting the sum of the five acute angles of the figure; i.e. it is eight right angles.

16. A chain, cord, or fine wire, or a fine fibre, filament, or hair, may suggest, what is not to be found among natural or artificial productions, a perfectly *flexible and inextensible line.* The elementary kinematics of this subject require no investigation. The mathematical condition to be expressed in any case of it is simply that the distance measured along the line from any one point to any other, remains constant, however the line be bent.

17. The use of a cord in mechanism presents us with many practical applications of this theory, which are in general extremely simple; although curious, and not always very easy, geometrical problems occur in connexion with it. We shall say nothing here about such cases as knots, knitting, weaving, etc., as being excessively difficult in their general development, and too simple in the ordinary cases to require explanation.

18. The simplest and most useful applications are to the *Pulley* and its combinations. In *theory* a pulley is simply a smooth body which *changes the direction* of a flexible and inextensible cord stretched across part of its surface; in *practice* (to escape as much as possible of the inevitable friction) it is a wheel, on part of whose circumference the cord is wrapped.

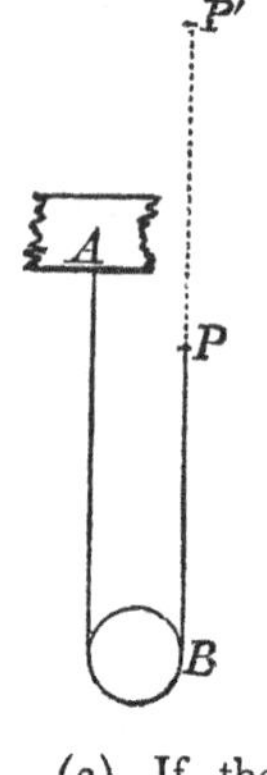

(1) Suppose we have a single pulley B, about which the flexible and inextensible cord ABP is wrapped, and suppose its free portions to be parallel. If (A being fixed) a point P of the cord be moved to P', it is evident that each of the portions AB and PB will be shortened by one-half of PP'. Hence, when P moves through any space in the direction of the cord, the pulley B moves in the same direction, through half the space.

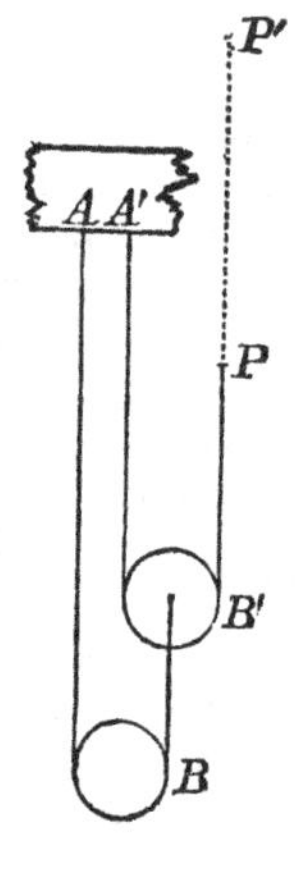

(2) If there be two cords and two pulleys, the ends AA' being fixed, and the other end of AB being attached to the pulley B'—then, if all free parts of the cord are parallel, when P is moved to P', B' moves in the same direction through *half* the space, and carries with it one end of the cord AB. Hence B moves through half the space B' did, that is, *one fourth* of PP'.

(3) And so on for any number of pulleys, if they be arranged in the above manner. Similar considerations enable us to determine the relative motions of all parts of other systems of pulleys and cords as long as all the free parts of the cords are parallel.

Of course, if a pulley be *fixed*, the motion of a point of one end of the cord *to* or *from* it involves an equal motion of the other end *from* or *to* it.

If the strings be not parallel, the relations of a single pulley or of a system of pulleys are a little complex, but present no difficulty.

19. In the mechanical tracing of curves, a flexible and inextensible cord is often supposed. Thus, in drawing an ellipse, the focal property of the curve shows us that if we fix the ends of such a cord to the foci and keep it stretched by a pencil, the pencil will trace the curve.

By a ruler moveable about one focus, and a string attached to a point in the ruler and to the other focus, and kept tight by a pencil sliding along the edge of the ruler, the hyperbola may be described by the help of its analogous focal property; and so on.

20. But the consideration of evolutes is of some importance in Natural Philosophy, especially in certain mechanical and optical questions, and we shall therefore devote a section or two to this application of Kinematics.

Def. If a flexible and inextensible string be fixed at one point of a plane curve, and stretched along the curve, and be then unwound in the plane of the curve, its extremity will describe an *Involute* of the curve. The original curve, considered with reference to the other, is called the *Evolute.*

21. It will be observed that we speak of *an* involute, and of *the* evolute, of a curve. In fact, as will be easily seen, a curve can have but one evolute, but it has an infinite number of involutes. For all that we have to do to vary an involute, is to change the point of the curve from which the tracing-point starts, or consider the involutes described by different points of the string; and these will, in general, be different curves. But the following section shows that there is but one evolute.

22. Let AB be any curve, PQ a portion of an involute, pP, qQ positions of the free part of the string. It will be seen at once that these must be tangents to the arc AB at p and q. Also the string at any stage, as pP, ultimately revolves about p. Hence pP is *normal* (or perpendicular to the tangent) to the curve PQ. And thus the evolute of PQ is a definite curve, viz. the envelop of (or line which is touched by) the normals drawn at every point of PQ, or, which is the same thing, the locus of the centres of the circles which have at each point the same tangent and curvature as the curve PQ. And we may merely mention, as an obvious result of the

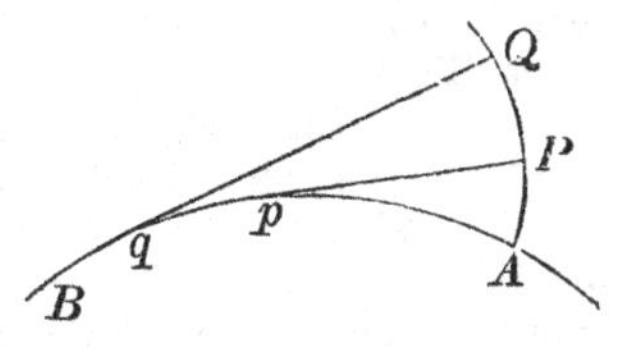

mode of tracing, that the arc qp is equal to the difference of qQ and pP, or that the arc pA is equal to pP. Compare § 104.

23. The rate of motion of a point, or its *rate of change of position*, is called its *Velocity*. It is greater or less as the space passed over in a given time is greater or less: and it may be *uniform*, i.e. the same at every instant; or it may be *variable*.

Uniform velocity is measured by the space passed over in unit of time, and is, in general, expressed in feet or in metres per second; if very great, as in the case of light, it may be measured in miles per second. It is to be observed that *Time* is here used in the abstract sense of a uniformly-increasing quantity—what in the differential calculus is called an independent variable. Its physical definition is given in the next chapter.

24. Thus a point, which moves uniformly with velocity v, describes a space of v feet each second, and therefore vt feet in t seconds, t being any number whatever. Putting s for the space described in t seconds, we have $s=vt$.

Thus with unit velocity a point describes unit of space in unit of time.

25. It is well to observe here, that since, by our formula, we have generally

$$v=\frac{s}{t},$$

and since nothing has been said as to the magnitudes of s and t, we may take these as small as we choose. Thus *we get the same result whether we derive v from the space described in a million seconds, or from that described in a millionth of a second.* This idea is very useful, as it makes our results intelligible when a variable velocity has to be measured, and we find ourselves obliged to approximate to its value (as in § 28) by considering the space described in an interval so short, that during its lapse the velocity does not sensibly alter in value.

26. When the point does not move uniformly, the velocity is variable, or different at different successive instants: but we define the *average* velocity during any time as the space described in that time, divided by the time; and, the less the interval is, the more nearly does the average velocity coincide with the actual velocity at any instant of the interval. Or again, we define the exact velocity at any instant as the space which the point would have described in one second, if for such a period it kept its velocity unchanged.

27. That there is at every instant a definite velocity for any moving point, is evident to all, and is matter of everyday conversation. Thus, a railway train, after starting, gradually increases its speed, and every one understands what is meant by saying that at a particular instant it moves at the rate of ten or of fifty miles an hour,—although, in the course of an hour, it may not have moved a mile altogether. We may suppose that, at any instant during the motion, the steam is so adjusted as to keep the train running for some time at a uniform velocity. This is the velocity which the train had at the instant in question. Without supposing any such definite adjustment of the driving-power to be made, we can evidently obtain an approximation to the velocity at a particular instant, by considering (§ 25) the motion for so short a time, that during that time the actual variation of speed may be small enough to be neglected.

28. In fact, if v be the velocity at either beginning or end, or at any instant, of an interval t, and s the space actually described in that interval; the equation $v = \frac{s}{t}$ (which expresses the definition of the average velocity, § 26) is more and more nearly true, as the velocity is more nearly uniform during the interval t; so that if we take the interval small enough the equation may be made as nearly exact as we choose. Thus the set of values—

Space described in one second,
Ten times the space described in the first tenth of a second,
A hundred „ „ „ hundredth „

and so on, give nearer and nearer approximations to the velocity at the beginning of the first second.

The whole foundation of Newton's differential calculus is, in fact, contained in the simple question, 'What is the rate at which the space described by a moving point increases?' i. e. What is the velocity of the moving point? Newton's notation for the velocity, i. e. the rate at which s increases, or the *fluxion* of s, is $\dot{s}$. This notation is very convenient, as it saves the introduction of a second letter.

29. The preceding definition of velocity is equally applicable whether the point move in a straight or a curved line; but, since, in the latter case, the direction of motion continually changes, the mere amount of the velocity is not sufficient completely to describe the motion, and we must have in every such case additional data to thoroughly specify the motion.

In such cases as this the method most commonly employed, whether we deal with velocities, or (as we shall do farther on) with accelerations and forces, consists in studying, not the velocity, acceleration, or force, *directly*, but its resolved parts parallel to any three assumed directions at right angles to each other. Thus, for a train moving up an incline in a N.E. direction, we may have the whole velocity and the steepness of the incline given; or we may express the same ideas thus—the train is moving simultaneously northward, eastward, and upward—and the motion as to amount and direction will be completely known if we know separately the northward, eastward, and upward velocities—these being called the *components* of the whole velocity in the three mutually perpendicular directions N., E., and up.

30. A velocity in any direction may be resolved in, and perpendicular to, any other direction. The first component is found by multiplying the velocity by the cosine of the angle between the two directions; the second by using as factor the sine of the same angle.

Thus a point moving with velocity V up an *Inclined Plane*, making an angle α with the horizon, has a vertical velocity $V \sin \alpha$ and a horizontal velocity $V \cos \alpha$.

Or it may be resolved into components in any three rectangular directions, each component being found by multiplying the whole velocity by the cosine of the angle between its direction and that of the component. The velocity resolved in any direction is the sum of the resolved parts (in that direction) of the three rectangular components of the whole velocity. And if we consider motion in one plane, this is still true, only we have but *two* rectangular components.

31. These propositions are virtually equivalent to the following obvious geometrical construction:—

To compound any two velocities as OA, OB in the figure; where OA, for instance, represents in magnitude and direction the space which would be described in one second by a point moving with the first of the given velocities—and similarly OB for the second; from A draw AC parallel and equal to OB. Join OC: then OC is the resultant velocity in magnitude and direction.

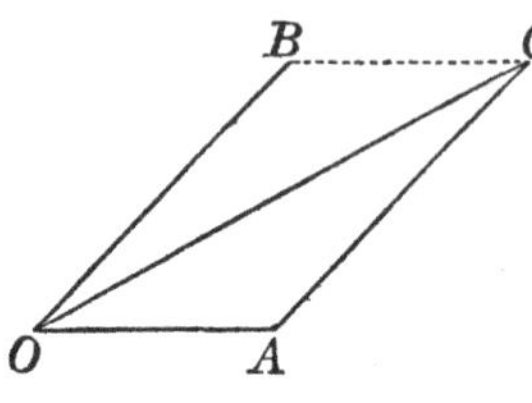

OC is evidently the diagonal of the parallelogram two of whose sides are OA, OB.

Hence the resultant of any two velocities as OA, AC, in the figure, is a velocity represented by the third side, OC, of the triangle OAC.

Hence if a point have, at the same time, velocities represented by OA, AC, and CO, the sides of a triangle *taken in the same order*, it is at rest.

Hence the resultant of velocities represented by the sides of any closed polygon whatever, whether in one plane or not, taken all in the same order, is zero.

Hence also the resultant of velocities represented by all the sides of a polygon but one, taken in order, is represented by that one taken in the opposite direction.

When there are two velocities, or three velocities, in two or in three rectangular directions, the resultant is the square root of the sum of their squares; and the cosines of its inclination to the given directions are the ratios of the components to the resultant.

32. The velocity of a point is said to be accelerated or retarded according as it increases or diminishes, but the word *acceleration* is generally used in either sense, on the understanding that we may regard its quantity as either positive or negative: and (§ 34) is farther generalized so as to include change of direction as well as change of speed. Acceleration of velocity may of course be either uniform or variable. It is said to be uniform when the point receives equal increments of velocity in equal times, and is then measured by the actual increase of velocity per unit of time. If we choose as the unit of acceleration that which adds a unit of velocity per unit of time to the velocity of a point, an acceleration measured by a will add a units of velocity in unit of time—and, therefore, $a t$ units of velocity in t units of time. Hence if v be the change in the velocity during the interval t,

$$v = at, \text{ or } a = \frac{v}{t}.$$

33. Acceleration is variable when the point's velocity does not receive equal increments in successive equal periods of time. It is then measured by the increment of velocity, which would have been generated in a unit of time had the acceleration remained throughout that unit the same as at its commencement. The *average* acceleration during any time is the whole velocity gained during that time, divided by the time. In Newton's notation $\dot{v}$ is used to express the acceleration in the direction of motion; and, if $v = \dot{s}$ as in § 28, we have

$$a = \dot{v} = \ddot{s}.$$

34. But there is *another* form in which acceleration may manifest itself. Even if a point's velocity remain unchanged, yet if its *direction* of motion change, the resolved parts of its velocity in fixed directions will, in general, be accelerated. And as acceleration is merely a change of the component velocity in a stated direction, it is evident that its laws of composition and resolution are the same as those of velocity.

We therefore *expand* the definition just given, thus:—Acceleration is the *rate of change of velocity whether that change take place in the direction of motion or not.*

35. What is meant by change of velocity is evident from § 31. For if a velocity OA become OC, its change is AC, or OB.

Hence, just as the direction of motion of a point is the tangent to its path, so the direction of acceleration of a moving point is to be found by the following construction:—

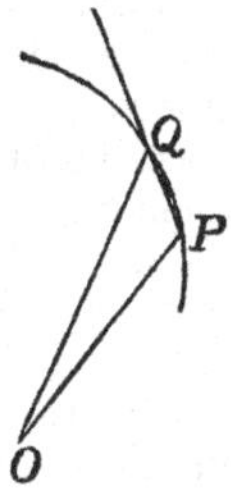

From any point O draw lines OP, OQ, etc., representing in magnitude and direction the velocity of the moving point at every instant. (Compare § 49.) The points, P, Q, etc., must form a continuous curve, for (§ 7) OP cannot change *abruptly* in direction. Now if Q be a point near to P, OP and OQ represent two successive values of the velocity. Hence PQ is the whole change of velocity during the interval. As the interval becomes smaller, the direction PQ more and more nearly becomes the tangent at P. Hence the direction of acceleration is that of the tangent to the curve thus described.

The amount of acceleration is the rate of change of velocity, and is therefore measured by the velocity of P in the curve PQ.

36. Let a point describe a circle, ABD, radius R, with uniform velocity V. Then, to determine the direction of acceleration, we must draw, as below, from a fixed point O, lines OP, OQ, etc., representing the velocity at A, B, etc., in direction and magnitude. Since the velocity in ABD is constant, all the lines OP, OQ, etc., will be equal (to V), and therefore PQS is a circle whose centre is O. The direction of acceleration at A is parallel to the tangent at P, that is, is perpendicular to OP, i.e. to Aa, and is therefore that of the radius AC.

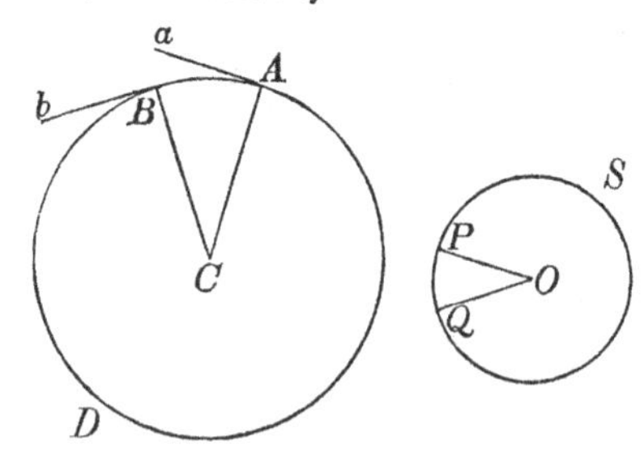

Now P describes the circle PQS, while A describes ABD. Hence the velocity of P is to that of A as OP to CA, i.e. as V to R; and is therefore equal to

$$\frac{V}{R}.V \text{ or } \frac{V^2}{R},$$

and this (§ 35) is the amount of the acceleration in the circular path ABD.

37. The whole acceleration in any direction is the sum of the components (in that direction) of the accelerations parallel to any three rectangular axes—each component acceleration being found by the same rule as component velocities, that is, by multiplying by the cosine of the angle between the direction of the acceleration and the line along which it is to be resolved.

38. When a point moves in a curve the whole acceleration may be resolved into two parts, one in the direction of the motion and equal to the acceleration of the velocity; the other towards the centre of curvature (perpendicular therefore to the direction of motion), whose magnitude is proportional to the square of the velocity and also to the curvature of the path. The former of these changes the velocity, the other affects only the form of the path, or the direction of motion. Hence if a moving point be subject to an acceleration, constant or not, whose direction is continually perpendicular to the direction of motion, the velocity will not be altered—and the only effect of the acceleration will be to make the point move in a curve whose curvature is proportional to the acceleration at each instant, and inversely as the square of the velocity.

39. In other words, if a point move in a curve, whether with a uniform or a varying velocity, its change of direction is to be regarded as constituting an acceleration towards the centre of curvature, equal in amount to the square of the velocity divided by the radius of curvature. The whole acceleration will, in every case, be the resultant of the acceleration thus measuring change of direction and the acceleration of actual velocity along the curve.

40. If for any case of motion of a point we have given the whole velocity and its direction, or simply the components of the velocity in three rectangular directions, at any *time*, or, as is most commonly the case, for any *position*; the determination of the form of the path described, and of other circumstances of the motion, is a question of pure mathematics, and in all cases is capable (if not of an exact solution, at all events) of a solution to any degree of approximation that may be desired.

This is true also if the total acceleration and its direction at every instant, or simply its rectangular components, be given, provided the velocity and its direction, as well as the position of the point, at any one instant be given. But these are in general questions requiring for their solution a knowledge of the integral calculus.

41. From the principles already laid down, a great many interesting results may be deduced, of which we enunciate a few of the simpler and more important.

(*a*) If the velocity of a moving point be uniform, and if its direction revolve uniformly in a plane, the path described is a circle.

(*b*) If a point moves in a plane, and its component velocity parallel to each of two rectangular axes is proportional to its distance from that axis, the path is an ellipse or hyperbola whose principal diameters coincide with those axes; and the acceleration is directed to or from the centre of the curve at every instant (§§ 66, 78).

(*c*) If the components of the velocity parallel to each axis be equimultiples of the distances from the other axis, the path is a straight line passing through the origin.

(*d*) When the velocity is uniform, but in a direction revolving uniformly in a right circular cone, the motion of the point is in a circular helix whose axis is parallel to that of the cone.

42. When a point moves uniformly in a circle of radius R, with velocity V, the whole acceleration is directed towards the centre, and has the constant value $\frac{V^2}{R}$. See § 36.

43. With uniform acceleration in the direction of motion, a point describes spaces proportional to the squares of the times elapsed since the commencement of the motion. This is the case of a body falling vertically in vacuo under the action of gravity.

In this case the space described in any interval is that which would be described in the same time by a point moving uniformly with a velocity equal to that at the middle of the interval. In other words, the average velocity (when the acceleration is uniform) is, during any interval, the arithmetical mean of the initial and final velocities. For, since the velocity increases uniformly, its value at any time *before* the middle of the interval is as much *less* than this mean as its value at the same time after the middle of the interval is *greater* than the mean: and hence its value at the middle of the interval must be the mean of its first and last values.

In symbols; if at time $t=0$ the velocity was V, then at time t it is

$$v=V+at.$$

Also the space (x) described is equal to the product of the time by the average velocity. But we have just shown that the average velocity is

$$=\tfrac{1}{2}(V+\overline{V+at})=V+\tfrac{1}{2}at,$$

and therefore $$x=Vt+\tfrac{1}{2}at^2.$$

Hence, by algebra,

$$V^2+2ax=V^2+2Vat+a^2t^2=(V+at)^2=v^2,$$

or $$\tfrac{1}{2}v^2-\tfrac{1}{2}V^2=ax.$$

If there be no initial velocity our equations become

$$v=at,\quad x=\tfrac{1}{2}at^2,\quad \tfrac{1}{2}v^2=ax.$$

Of course the preceding formulae apply to a constant retardation, as in the case of a projectile moving vertically upwards, by simply giving a a negative sign.

44. When there is uniform acceleration in a constant direction, the path described is a parabola, whose axis is parallel to that direction. This is the case of a projectile moving in vacuo.

For the velocity (V) in the original direction of motion remains unchanged; and therefore, in time t, a space Vt is described parallel to this line. But in the same interval, by the above reasoning, we see that a space $\tfrac{1}{2}at^2$ is described parallel to the direction of acceleration.

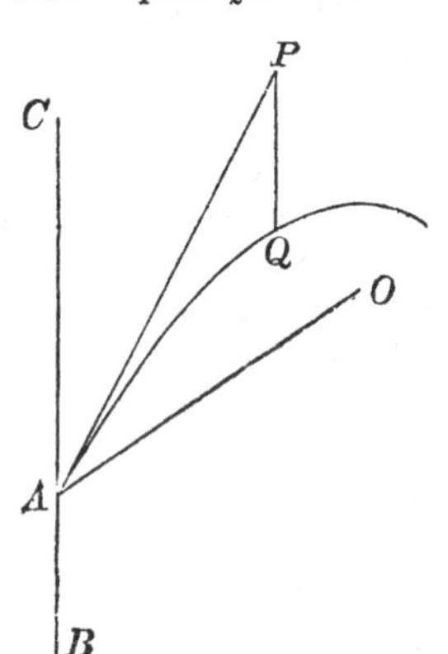

Hence, if AP be the direction of motion at A, AB the direction of acceleration, and Q the position of the point at time t; draw QP parallel to BA, meeting AP in P: then

$$AP=Vt,\quad PQ=\tfrac{1}{2}at^2.$$

Hence $$AP^2=\frac{2V^2}{a}PQ.$$

This is a property of a parabola, of which the axis is parallel to AB; AB being a diameter, and AP a tangent. If O be the focus of this curve, we know that

$$AP^2=4OA\cdot PQ.$$

Hence $$OA=\frac{V^2}{2a},$$

and is therefore known. Also OA is known in direction, for AP bisects the angle, OAC, between the focal distance of a point and the diameter through it.

45. When the acceleration is directed to a fixed point, the path is in a plane passing through that point; and in this plane the areas traced out by the radius-vector are proportional to the times employed. This includes the case of a satellite or planet revolving about its primary, according to Kepler's first law.

Evidently there is no acceleration perpendicular to the plane containing the fixed point and the line of motion of the moving point at any instant; and there being no velocity perpendicular to this plane at starting, there is therefore none throughout the motion;

thus the point moves in the plane. For the proof of the second part of the proposition we must make a slight digression.

46. The *Moment* of a velocity or of a force about any point is the product of its magnitude into the perpendicular from the point upon its direction. The moment of the resultant velocity of a particle about any point in the plane of the components is equal to the algebraic sum of the moments of the components, the proper sign of each moment depending on the *direction* of motion about the point. The same is true of moments of forces and of moments of momentum, as defined in Chapter II.

First, consider two component motions, AB and AC, and let AD be their resultant (§ 31). Their half-moments round the point O are respectively the areas OAB, OCA. Now OCA, together with half the area of the parallelogram $CABD$, is equal to OBD. Hence the sum of the two half-moments together with half the area of the parallelogram is equal to AOB together with BOD, that is to say, to the area of the whole figure $OABD$. But ABD, a part of this figure, is equal to half the area of the parallelogram; and therefore the remainder, OAD, is equal to the sum of the two half-moments. But OAD is half the moment of the resultant velocity round the point O. Hence the moment of the resultant is equal to the sum of the moments of the two components. By attending to the *signs* of the moments, we see that the proposition holds when O is within the angle CAB.

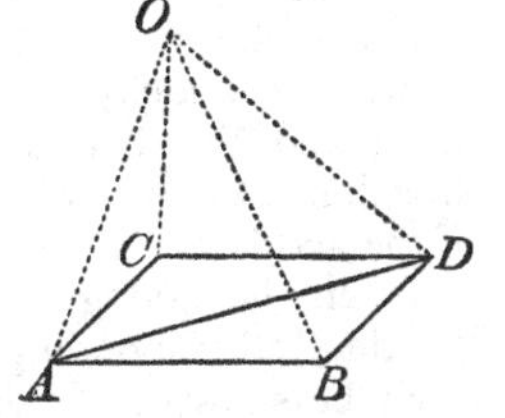

If there be any number of component rectilineal motions, we may compound them in order, any two taken together first, then a third, and so on; and it follows that the sum of their moments is equal to the moment of their resultant. It follows, of course, that the sum of the moments of any number of component velocities, all in one plane, into which the velocity of any point may be resolved, is equal to the moment of their resultant, round any point in their plane. It follows also, that if velocities, in different directions all in one plane, be successively given to a moving point, so that at any time its velocity is their resultant, the moment of its velocity at any time is the sum of the moments of all the velocities which have been successively given to it.

47. Thus if one of the components always passes through the point, its moment vanishes. This is the case of a motion in which the acceleration is directed to a fixed point, and we thus prove the second theorem of § 45, that in the case supposed the areas described by the radius-vector are proportional to the times; for, as we have seen, the moment of the velocity is double the area traced out by the radius-vector in unit of time.

48. Hence in this case the velocity at any point is inversely as the perpendicular from the fixed point to the tangent to the path or the momentary direction of motion.

For the product of this perpendicular and the velocity at any instant gives double the area described in one second about the fixed point, which has just been shown to be a constant quantity.

Other examples of these principles will be met with in the chapters on Kinetics.

49. If, as in § 35, from any fixed point, lines be drawn at every instant representing in magnitude and direction the velocity of a point describing any path in any manner, the extremities of these lines form a curve which is called the *Hodograph*. The fixed point from which these lines are drawn is called the hodographic origin. The invention of this construction is due to Sir W. R. Hamilton; and one of the most beautiful of the many remarkable theorems to which it leads is this: *The Hodograph for the motion of a planet or comet is always a circle, whatever be the form and dimensions of the orbit.* The proof will be given immediately.

It was shown (§ 35) that an arc of the hodograph represents the change of velocity of the moving point during the corresponding time; and also that the tangent to the hodograph is parallel to the direction, and the velocity in the hodograph is equal to the amount of the acceleration of the moving point.

When the hodograph and its origin, and the velocity along it, or the time corresponding to each point of it, are given, the orbit may easily be shown to be determinate.

[An important improvement in nautical charts has been suggested by Archibald Smith[1]. It consists in drawing a curve, which may be called the tidal hodograph with reference to any point of a chart for which the tidal currents are to be specified throughout the chief tidal period (twelve lunar hours). Numbers from I. to XII. are placed at marked points along the curve, corresponding to the lunar hours. Smith's curve is precisely the Hamiltonian hodograph for an imaginary particle moving at each instant with the same velocity and the same direction as the particle of fluid passing, at the same instant, through the point referred to.]

50. In the case of a projectile (§ 44), the horizontal velocity is unchanged, and the vertical velocity increases uniformly. Hence the hodograph is a vertical straight line, whose distance from the origin is the horizontal velocity, and which is described uniformly.

51. To prove Hamilton's proposition (§ 49), let APB be a portion of a conic section and S one focus. Let P move so that SP describes equal areas in equal times, that is (§ 48), let the velocity be inversely as the perpendicular SY from S to the tangent to the orbit. If ABP be an ellipse or hyperbola, the intersection of the perpendicular with the tangent lies in the circle YAZ, whose diameter is the major axis. Produce YS to cut the circle again in Z. Then $YS \cdot SZ$ is constant, and therefore SZ is inversely as SY, that is, SZ is proportional to the velocity at P. Also

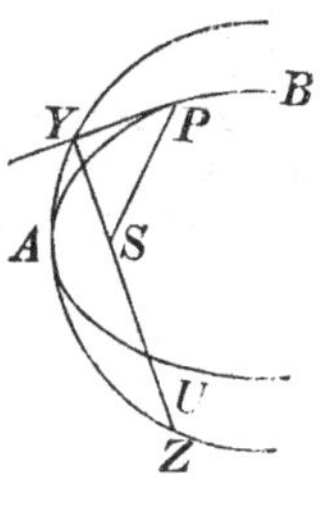

B Y P A S U Z

[1] *Proc. R.S.* 1865.

SZ is perpendicular to the direction of motion PY, and thus the circular locus of Z is the hodograph turned through a right angle about S in the plane of the orbit. If APB be a parabola, AY is a straight line. But if another point U be taken in YS produced, so that $YS \cdot SU$ is constant, the locus of U is easily seen to be a circle. Hence the proposition is generally true for all conic sections. The hodograph surrounds its origin if the orbit be an ellipse, passes through it when the orbit is a parabola, and the origin is without the hodograph if the orbit is a hyperbola.

52. A reversal of the demonstration of § 51 shows that, if the acceleration be towards a fixed point, and if the hodograph be a circle, the orbit must be a conic section of which the fixed point is a focus.

But we may also prove this important proposition as follows: Let A be the centre of the circle, and O the hodographic origin. Join OA and draw the perpendiculars PM to OA and ON to PA. Then OP is the velocity in the orbit: and ON, being parallel to the tangent at P, is the direction of acceleration in the orbit; and is therefore parallel to the radius-vector to the fixed point about which there is equable description of areas. The velocity parallel to the radius-vector is therefore ON, and the velocity perpendicular to the fixed line OA is PM. But $\dfrac{ON}{PM} = \dfrac{OA}{AP} = \text{constant}.$

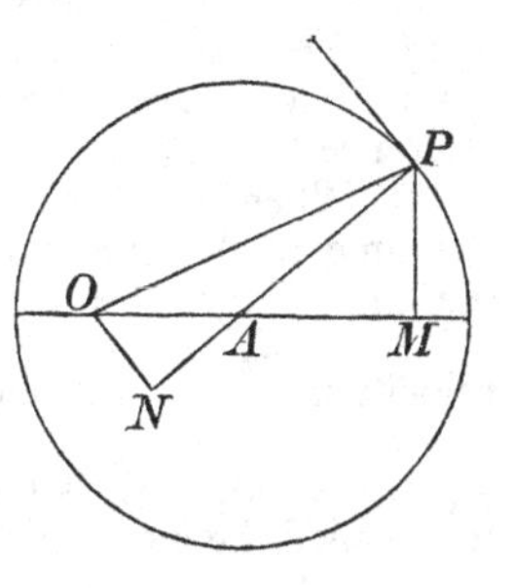

Hence, in the orbit, the velocity along the radius-vector is proportional to that perpendicular to a fixed line: and therefore the radius-vector of any point is proportional to the distance of that point from a fixed line — a property belonging exclusively to the conic sections referred to their focus and directrix.

53. The path which, in consequence of *Aberration*, a fixed star seems to describe, is the hodograph of the earth's orbit, and is therefore a circle whose plane is parallel to the plane of the ecliptic.

54. When a point moves in any manner, the line joining it with a fixed point generally changes its direction. If, for simplicity, we consider the motion to be confined to a plane passing through the fixed point, the angle which the joining line makes with a fixed line in the plane is continually altering, and its rate of alteration at any instant is called the *Angular Velocity* of the first point about the second. If uniform, it is of course measured by the angle described in unit of time; if variable, by the angle which would have been described in unit of time if the angular velocity at the instant in question were maintained constant for so long. In this respect the process is precisely similar to that which we have already explained for the measurement of velocity and acceleration.

We may also speak of the angular velocity of a moving plane with respect to a fixed one, as the rate of increase of the angle contained by them; but unless their line of intersection remain fixed, or at all events parallel to itself, a somewhat more laboured statement is required to give a complete specification of the motion.

55. The unit angular velocity is that of a point which describes, or would describe, unit angle about a fixed point in unit of time. The usual unit angle is (as explained in treatises on plane trigonometry) that which subtends at the centre of a circle an arc whose length is equal to the radius; being an angle of $\frac{180^\circ}{\pi} = 57^\circ \cdot 29578 \ldots = 57^\circ 17' 44'' \cdot 8$ nearly.

56. The angular velocity of a point in a plane is evidently to be found by dividing the velocity perpendicular to the radius-vector by the length of the radius-vector.

57. When the angular velocity is variable its rate of increase or diminution is called the *Angular Acceleration*, and is measured with reference to the same unit angle.

58. When one point describes uniformly a circle about another, the time of describing a complete circumference being T, we have the angle 2π described uniformly in T; and, therefore, the angular velocity is $\frac{2\pi}{T}$. Even when the angular velocity is not uniform, as in a planet's motion, it is useful to introduce the quantity $\frac{2\pi}{T}$, which is then called the *mean* angular velocity.

59. When a point moves uniformly in a straight line its angular velocity evidently diminishes as it recedes from the point about which the angles are measured, and it may easily be shown that it varies inversely as the square of the distance from this point. The same proposition is true for *any* path, when the acceleration is towards the point about which the angles are measured: being merely a different mode of stating the result of § 48.

60. The intensity of heat and light emanating from a point, or from a uniformly radiating spherical surface, diminishes according to the inverse square of the distance from the centre. Hence the rate at which a planet receives heat and light from the sun varies in simple proportion to the angular velocity of the radius-vector. Hence the whole heat and light received by the planet in any time is proportional to the whole angle turned through by its radius-vector in the same time.

61. A further instance of this use of the idea of angular velocity may now be given, to solve the problem of finding the hodograph (§ 35) for any case of motion in which the acceleration is directed to a fixed point, and varies inversely as the square of the distance from that point. The velocity of P, in the hodograph PQ, being the acceleration in the orbit, varies inversely as the square of the radius-vector; and therefore (§ 59) directly as the angular velocity. Hence the arc of PQ, described in any time, is proportional to the corresponding angle-vector in the orbit, i.e. to the angle through which the tangent to PQ has turned. Hence (§ 9) the curvature of PQ is constant, or PQ is a circle.

This demonstration, reversed, proves that if the hodograph be a circle, and the acceleration be towards a fixed point, the acceleration varies inversely as the square of the distance of the moving point from the fixed point.

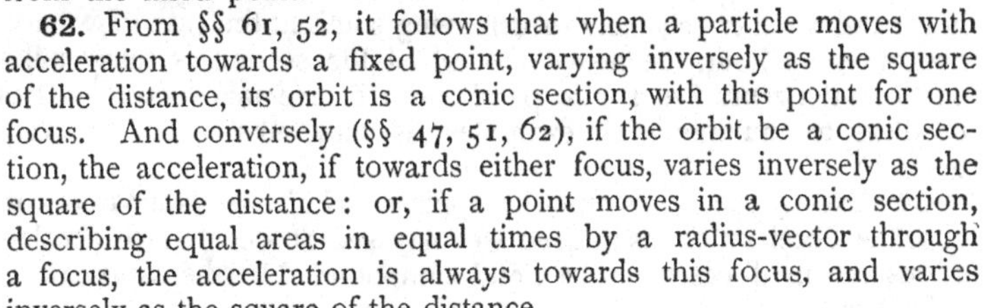

62. From §§ 61, 52, it follows that when a particle moves with acceleration towards a fixed point, varying inversely as the square of the distance, its orbit is a conic section, with this point for one focus. And conversely (§§ 47, 51, 62), if the orbit be a conic section, the acceleration, if towards either focus, varies inversely as the square of the distance: or, if a point moves in a conic section, describing equal areas in equal times by a radius-vector through a focus, the acceleration is always towards this focus, and varies inversely as the square of the distance.

63. All motion that we are, or can be, acquainted with, is *Relative* merely. We can calculate from astronomical data for any instant the direction in which, and the velocity with which, we are moving on account of the earth's diurnal rotation. We may compound this with the (equally calculable) velocity of the earth in its orbit. This resultant again we may compound with the (roughly-known) velocity of the sun relatively to the so-called fixed stars; but, even if all these elements were accurately known, it could not be said that we had attained any idea of an *absolute* velocity; for it is only the sun's relative motion among the stars that we can observe; and, in all probability, sun and stars are moving on (it may be with inconceivable rapidity) relatively to other bodies in space. We must therefore consider how, from the actual motions of a set of bodies, we may find their relative motions with regard to any one of them; and how, having given the relative motions of all but one with regard to the latter, and the actual motion of the latter, we may find the actual motions of all. The question is very easily answered. Consider for a moment a number of passengers walking on the deck of a steamer. Their relative motions with regard to the deck are what we immediately observe, but if we compound with these the velocity of the steamer itself we get evidently their actual motion relatively to the earth. Again, in order to get the relative motion of all with regard to the deck, we eliminate the motion of the steamer altogether; that is, we alter the velocity of each relatively to the earth by compounding with it the actual velocity of the vessel taken in a reversed direction.

Hence to find the relative motions of any set of bodies with regard to one of their number, imagine, impressed upon each in composition with its own motion, a motion equal and opposite to the motion of that one, which will thus be reduced to rest, while the motions of the others will remain the same with regard to it as before.

Thus, to take a very simple example, two trains are running in opposite directions, say north and south, one with a velocity of fifty, the other of thirty, miles an hour. The relative velocity of the second with regard to the first is to be found by imagining impressed on both a southward velocity of fifty miles an hour; the effect of this being to bring the first to rest, and to give the second a southward velocity of eighty miles an hour, which is the required relative motion.

Or, given one train moving north at the rate of thirty miles an hour, and another relatively to it moving south at the rate of twenty-five miles an hour, the actual motion of the second is thirty miles north, and twenty-five south, per hour, i. e. five miles north. It is needless to multiply such examples, as they must occur to every one.

64. Exactly the same remarks apply to relative as compared with absolute acceleration, as indeed we may see at once, since accelerations are in all cases resolved and compounded by the same law as velocities.

65. The following proposition in relative motion is of considerable importance:—

Any two moving points describe similar paths relatively to each other and relatively to any point which divides in a constant ratio the line joining them.

Let A and B be any simultaneous positions of the points. Take G or G' in AB such that the ratio $\frac{GA}{GB}$ or $\frac{G'A}{G'B}$ has a constant value. Then, as the form of the relative path depends only upon the *length* and *direction* of the line joining the two points at any instant, it is obvious that these will be the same for A with regard to B, as for B with regard to A, saving only the inversion of the direction of the joining line. Hence B's path about A is A's about B turned through two right angles. And with regard to G and G' it is evident that the directions remain the same, while the lengths are altered in a given ratio; but this is the definition of similar curves.

G' A G B

66. An excellent example of the transformation of relative into absolute motion is afforded by the family of *Cycloids*. We shall in a future section consider their mechanical description, by the *rolling* of a circle on a fixed straight line or circle. In the meantime, we take a different form of enunciation, which however leads to precisely the same result.

The actual path of a point which revolves uniformly in a circle about another point—the latter moving uniformly in a straight line or circle in the same plane—belongs to the family of Cycloids.

67. As an additional illustration of this part of our subject, we may define as follows:—

If one point A executes any motion whatever with reference to a second point B; if B executes any other motion with reference to a third point C; and so on—the first is said to execute, with reference to the last, a movement which is the resultant of these several movements.

The relative position, velocity, and acceleration are in such a case the geometrical resultants of the various components combined according to preceding rules.

68. The following practical methods of effecting such a combination in the simple case of the movements of two points are useful in scientific illustrations and in certain mechanical arrangements. Let two moving points be joined by a uniform elastic string; the middle point of this string will evidently execute a movement which is *half* the resultant of the motions of the two points. But for drawing, or engraving, or for other mechanical applications, the following method is preferable:—

CF and ED are rods of equal length moving freely round a pivot at P, which passes through the middle point of each—CA, AD, EB, and BF are rods of half the length of the two former, and so pivotted to them as to form a pair of equal rhombi CD, EF, whose angles can be altered at will. Whatever motions, whether in a plane, or in space of three dimensions, be given to A and B, P will evidently be subjected to half their resultant.

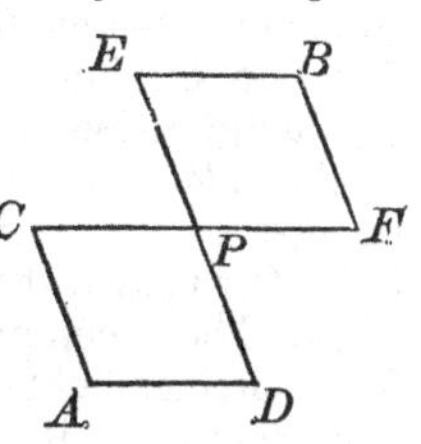

69. Amongst the most important classes of motions which we have to consider in Natural Philosophy, there is one, namely, *Harmonic Motion*, which is of such immense use, not only in ordinary kinetics, but in the theories of sound, light, heat, etc., that we make no apology for entering here into some little detail regarding it.

70. *Def.* When a point Q moves uniformly in a circle, the perpendicular QP drawn from its position at any instant to a fixed diameter AA' of the circle, intersects the diameter in a point P, whose position changes by a *simple harmonic motion.*

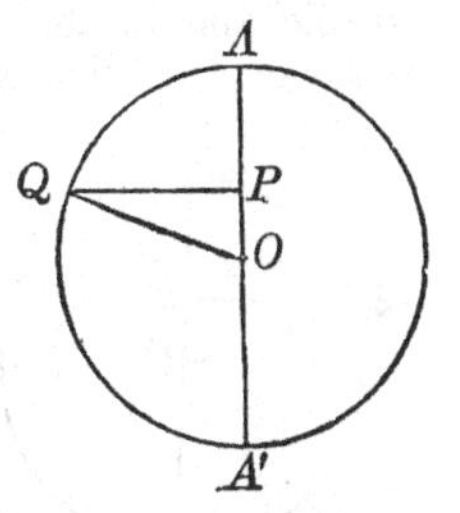

Thus, if a planet or satellite, or one of the constituents of a double star, be supposed to move uniformly in a circular orbit about its primary, and be viewed from a very distant position in the plane of its orbit, it will appear to move backwards and forwards in a straight line with a simple harmonic motion. This is nearly the case with such bodies as the satellites of Jupiter when seen from the earth.

Physically, the interest of such motions consists in the fact of their being approximately those of the simplest vibrations of sounding bodies such as a tuning-fork or pianoforte-wire; whence their name; and of the various media in which waves of sound, light, heat, etc., are propagated.

71. The *Amplitude* of a simple harmonic motion is the range on one side or the other of the middle point of the course, i.e. OA or OA' in the figure.

An arc of the circle referred to, measured from any fixed point to the uniformly moving point Q, is the *Argument* of the harmonic motion.

[The distance of a point, performing a simple harmonic motion, from the middle of its course or range, is a *simple harmonic function of the time*. The *argument* of this function is what we have defined as the argument of the motion.]

The *Epoch* in a simple harmonic motion is the interval of time which elapses from the era of reckoning till the moving point first comes to its greatest elongation in the direction reckoned as positive, from its mean position or the middle of its range. Epoch in angular measure is the angle described on the circle of reference in the period of time defined as the epoch.

The *Period* of a simple harmonic motion is the time which elapses from any instant until the moving point again moves in the same direction through the same position, and is evidently the time of revolution in the auxiliary circle.

The *Phase* of a simple harmonic motion at any instant is the fraction of the whole period which has elapsed since the moving point last passed through its middle position in the positive direction.

72. Those common kinds of mechanism, for producing rectilineal from circular motion, or *vice versâ*, in which a crank moving in a circle works in a straight slot belonging to a body which can only move in a straight line, fulfil strictly the definition of a simple harmonic motion in the part of which the motion is rectilineal, if the motion of the rotating part is uniform.

The motion of the treadle in a spinning-wheel approximates to the same condition when the wheel moves uniformly; the approximation being the closer, the smaller is the angular motion of the treadle and of the connecting string. It is also approximated to more or less closely in the motion of the piston of a steam-engine connected, by any of the several methods in use, with the crank, provided always the rotatory motion of the crank be uniform.

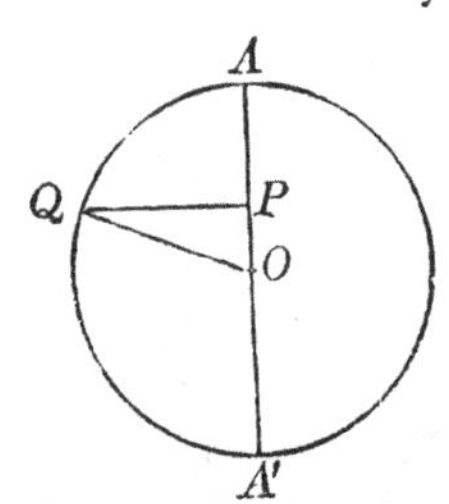

73. The velocity of a point executing a simple harmonic motion is a simple harmonic function of the time, a quarter of a period earlier in phase than the displacement, and having its maximum value equal to the velocity in the circular motion by which the given function is defined.

For, in the fig., if V be the velocity in the circle, it may be represented by OQ in a direction perpendicular to its own, and therefore by OP and PQ in directions perpendicular to those lines. That is, the velocity of P in the simple harmonic motion is $\frac{PQ}{OQ}V$ or $\frac{V}{OQ}PQ$; which, when P passes through O, becomes V.

74. The acceleration of a point executing a simple harmonic motion is at any time simply proportional to the displacement from the middle point, but in opposite direction, or always towards the middle point. Its maximum value is that with which a velocity equal to that of the circular motion would be acquired in the time in which an arc equal to the radius is described.

For in the fig., the acceleration of Q (by § 36) is $\frac{V^2}{QO}$ along QO. Supposing, for a moment, QO to represent the magnitude of this acceleration, we may resolve it into QP, PO. The acceleration of P is therefore represented on the same scale by PO. Its magnitude is therefore $\frac{V^2}{QO} \cdot \frac{PO}{QO} = \frac{V^2}{QO^2} PO$, which is proportional to PO, and has at A its maximum value, $\frac{V^2}{QO}$, an acceleration under which the velocity V would be acquired in the time $\frac{QO}{V}$ as stated. Thus we have in simple harmonic motion

$$\frac{\text{Acceleration}}{\text{Displacement}} = \frac{V^2}{QO^2} = \frac{4\pi^2}{T^2}$$

where T is the time of describing the circle, or the period of the harmonic motion.

75. Any two simple harmonic motions in one line, and of one period, give, when compounded, a single simple harmonic motion; of the same period; of amplitude equal to the diagonal of a parallelogram described on lengths equal to their amplitudes measured on lines meeting at an angle equal to their difference of epochs; and of epoch differing from their epochs by angles equal to those which this diagonal makes with the two sides of the parallelogram. Let P and P' be two points executing simple harmonic motions of one period, and in one line $B'BCAA'$. Let Q and Q' be the uniformly moving points in the relative circles. On CQ and CQ' describe a parallelogram $SQCQ'$; and through S draw SR perpendicular to $B'A'$ produced. We have obviously $P'R = CP$ (being projections of the equal and parallel lines $Q'S$, CQ, on CR). Hence $CR = CP + CP'$; and therefore the point R executes the

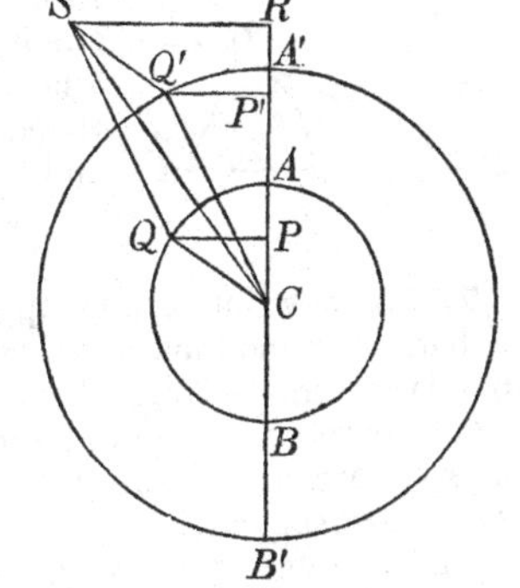

resultant of the motions P and P'. But CS, the diagonal of the parallelogram, is constant (since the angular velocities of CQ and CQ' are equal, and therefore the angle QCQ' is constant), and revolves with the same angular velocity as CQ or CQ'; and therefore the resultant motion is simple harmonic, of amplitude CS, and of epoch exceeding that of the motion of P, and falling short of that of the motion of P', by the angles QCS and SCQ' respectively.

76. [The construction described in the preceding section exhibits the resultant of two simple harmonic motions, whether of the same period or not. Only, if they are not of the same period, the diagonal of the parallelogram will not be constant, but will diminish from a maximum value, the sum of the component amplitudes, which it has at the instant when the phases of the component motions agree; to a minimum, the difference of those amplitudes, which is its value when the phases differ by half a period. Its direction, which always must be nearer to the greater than to the less of the two radii constituting the sides of the parallelogram, will oscillate on each side of the greater radius to a maximum deviation amounting on either side to the angle whose sine is the less radius divided by the greater, and reached when the less radius deviates by this together with a quarter circumference, from the greater. The full period of this oscillation is the time in which either radius gains a full turn on the other. The resultant motion is therefore not simple harmonic, but is, as it were, simple harmonic with periodically increasing and diminishing amplitude, and with periodical acceleration and retardation of phase. This view is most appropriate for the case in which the periods of the two component motions are nearly equal, but the amplitude of one of them much greater than that of the other.

To find the time and the amount of the maximum acceleration or retardation of phase, let CA be equal to the greater half-amplitude. From A as centre, with AB the less half-amplitude as radius, describe a circle. CB touching this circle represents the most deviated resultant. Hence CBA is a right angle; and

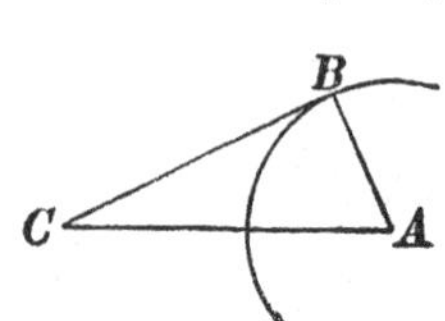

$$\sin BCA = \frac{AB}{CA}.]$$

77. A most interesting application of this case of the composition of harmonic motions is to the lunar and solar tides; which, except in tidal rivers, or long channels or bays, follow each very nearly the simple harmonic law, and produce, as the actual result, a variation of level equal to the sum of variations that would be produced by the two causes separately.

The amount of the lunar tide in the equilibrium theory is about 2·1 times that of the solar. Hence the spring tides of this theory are 3·1, and the neap tides only 1·1, each reckoned in terms of the solar tide; and at spring and neap tides the hour of high water is that of the lunar tide alone. The greatest deviation of the actual tide from the phases (high, low, or mean water) of the lunar tide alone, is about ·95 of a lunar hour, that is, ·98 of a solar hour (being the same part of 12 lunar hours that 28° 26′, or the angle whose sine is $\frac{1}{2\cdot1}$, is of 360°). This maximum deviation will be in advance or in arrear according as the crown of the solar tide precedes or follows the crown of the lunar tide; and it will be exactly reached when the interval of phase between the two component tides is 3·95 lunar hours. That is to say, there will be maximum advance of the time of high water 4½ days after, and maximum retardation the same number of days before, spring tides.

78. We may consider next the case of equal amplitudes in the two given motions. If their periods are equal, their resultant is a simple harmonic motion, whose phase is at every instant the mean of their phases, and whose amplitude is equal to twice the amplitude of either multiplied by the cosine of half the difference of their phase. The resultant is of course nothing when their phases differ by half the period, and is a motion of double amplitude and of phase the same as theirs when they are of the same phase.

When their periods are very nearly, but not quite, equal (their amplitudes being still supposed equal), the motion passes very slowly from the former (zero, or no motion at all) to the latter, and back, in a time equal to that in which the faster has gone once oftener through its period than the slower has.

In practice we meet with many excellent examples of this case, which will, however, be more conveniently treated of when we come to apply kinetic principles to various subjects in practical mechanics, acoustics, and physical optics; such as the marching of troops over a suspension bridge, the sympathy of pendulums or tuning-forks, etc.

79. We may exhibit, graphically, the various preceding cases of single or compound simple harmonic motions in one line by curves in which the abscissae represent intervals of time, and the ordinates the corresponding distances of the moving point from its mean position. In the case of a single simple harmonic motion, the corresponding curve would be that described by the point P in § 66, if, while Q maintained its uniform circular motion, the circle were to move with uniform velocity in any direction perpendicular to OA. This construction gives the harmonic curve, or curve of sines, in which the ordinates are proportional to the sines of the abscissae, the straight line in which O moves being the axis of abscissae. It is the simplest possible form assumed by a vibrating string; and when it is assumed that at each instant the motion of every particle of the string is simple harmonic. When the harmonic motion is complex, but in one line, as is the case for any point in a violin-, harp-, or pianoforte-string (differing, as these do, from one another in their motions on account of the different modes of excitation used), a similar construction may be made. Investigation regarding complex harmonic functions has led to results of the highest importance, having their most general expression in *Fourier's Theorem*, to be presently enunciated. We give below a graphic representation of the composition of two simple harmonic motions in one line, of *equal* amplitudes and of periods which are as 1 : 2 and as 2 : 3, the epochs being each a quarter circumference. The horizontal line is the axis of abscissae of the curves; the vertical line to the left of each being the axis of ordinates. In the first case the slower motion goes through one complete period, in the second it goes through two periods.

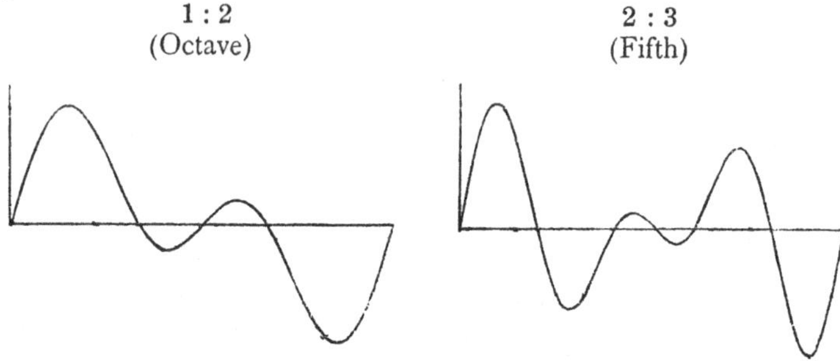

These and similar cases when the periodic times are not commensurable, will be again treated of under Acoustics.

80. We have next to consider the composition of simple harmonic motions in different directions. In the first place, we see that any number of simple harmonic motions of one period, and of the same phase, superimposed, produce a single simple harmonic motion of the same phase. For, the displacement at any instant being, according to the principle of the composition of motions, the geometrical resultant of the displacements due to the component motions separately, these component displacements in the case supposed, all vary in simple proportion to one another, and are in constant directions. Hence the resultant displacement will vary in simple proportion to each of them, and will be in a constant direction.

But if, while their periods are the same, the phases of the several component motions do not agree, the resultant motion will generally be elliptic, with equal areas described in equal times by the radius-vector from the centre; although in particular cases it may be uniform circular, or, on the other hand, rectilineal and simple harmonic.

81. To prove this, we may first consider the case, in which we have two equal simple harmonic motions given, and these in perpendicular lines, and differing in phase by a quarter period. Their resultant is a uniform circular motion. For, let BA, $B'A'$ be their ranges; and from O, their common middle point as centre, describe a circle through AA' BB'. The given motion of P in BA will be (§ 67) defined by the motion of a point Q round the circumference of this circle; and the same point, if moving in the direction indicated by the arrow, will give a simple harmonic motion of P', in $B'A'$, a quarter of a period behind that of the motion of P in BA. But, since $A'OA$, QPO, and $QP'O$ are right angles, the figure $QP'OP$ is a parallelogram, and therefore Q is in the position of the displacement compounded of OP and OP'. Hence two equal simple harmonic motions in perpendicular lines, of phases differing by a quarter period, are equivalent to a uniform circular motion of radius equal to the maximum displacement of either singly, and in the direction from the positive end of the range of the component in advance of the other towards the positive end of the range of this latter.

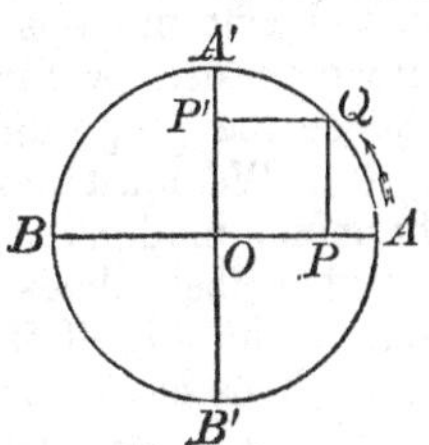

82. Now, orthogonal projections of simple harmonic motions are clearly simple harmonic with unchanged phase. Hence, if we project the case of § 81 on any plane, we get motion in an ellipse, of which the projections of the two component ranges are conjugate diameters, and in which the radius-vector from the centre describes equal areas (being the projections of the areas described by the radius of the circle) in equal times. But the plane and position of the circle of which this projection is taken may clearly be found so as to fulfil the condition of having the projections of the ranges coincident with any two given mutually bisecting lines. Hence any two given simple harmonic motions, equal or unequal in range, and oblique or at right angles to one another in direction, provided only they differ by a quarter period in phase, produce elliptic motion, having their ranges for conjugate axes, and describing, by the radius-vector from the centre, equal areas in equal times.

83. Returning to the composition of any number of equal simple harmonic motions in lines in all directions and of all phases: each component simple harmonic motion may be determinately resolved into two in the same line, differing in phase by a quarter period, and one of them having any given epoch. We may therefore reduce the given motions to two sets, differing in phase by a quarter period, those of one set agreeing in phase with any one of the given, or with any other simple harmonic motion we please to choose (i. e. having their epoch anything we please).

All of each set may (§ 75) be compounded into one simple harmonic motion of the same phase, of determinate amplitude, in a determinate line; and thus the whole system is reduced to two simple fully-determined harmonic motions differing from one another in phase by a quarter period.

Now the resultant of two simple harmonic motions, one a quarter of a period in advance of the other, in different lines, has been proved (§ 82) to be motion in an ellipse of which the ranges of the component motions are conjugate axes, and in which equal areas are described by the radius-vector from the centre in equal times. Hence the proposition of § 80.

84. We must next take the case of the composition of simple harmonic motions of *different* kinds and in different lines. In general, whether these lines be in one plane or not, the line of motion returns into itself if the periods are commensurable; and if not, not. This is evident without proof.

Also we see generally that the composition of any number of simple harmonic motions in any directions and of any periods, may be effected by compounding, according to previously explained methods, their resolved parts in each of any three rectangular directions, and then compounding the final resultants in these directions.

85. By far the most interesting case, and by far the simplest, is that of *two* simple harmonic motions of any periods, whose directions must of course be in one plane.

Mechanical methods of obtaining such combinations will be afterwards described, as well as cases of their occurrence in Optics and Acoustics.

We may suppose, for simplicity, the two component motions to take place in perpendicular directions. Also, it is easy to see that we can only have a reëntering curve when their periods are commensurable.

The following figures represent the paths produced by the combination of simple harmonic motions of *equal* amplitude in two rectangular directions, the periods of the components being as 1 : 2, and the epochs differing successively by 0, 1, 2, etc., sixteenths of a circumference.

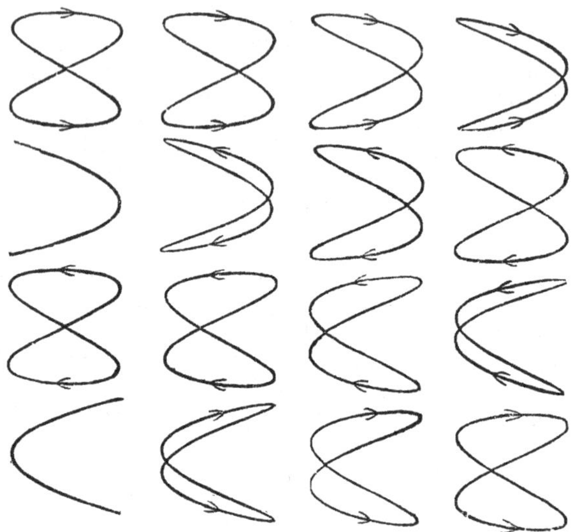

In the case of epochs equal, or differing by a multiple of π, the curve is a portion of a parabola, and is gone over twice in opposite directions by the moving point in each complete period.

If the periods be not exactly as 1 : 2 the form of the path produced by the combination changes gradually from one to another of the series above figured; and goes through all its changes in the time in which one of the components gains a complete vibration on the other.

86. Another very important case is that of two groups of two simple harmonic motions in one plane, such that the resultant of each group is uniform circular motion.

If their periods are equal, we have a case belonging to those already treated (§ 80), and conclude that the resultant is, in general, motion in an ellipse, equal areas being described in equal times about the centre. As particular cases we may have simple harmonic, or uniform circular, motion.

If the circular motions are in the *same* direction, the resultant is evidently circular motion in the same direction. This is the case of the motion of S in § 75, and requires no further comment, as its amplitude, epoch, etc., are seen at once from the figure.

87. If the radii of the component motions are equal, and the periods very nearly equal, but the motions in *opposite* directions, we have cases of great importance in modern physics, one of which is figured below (in general, a non-reëntrant curve).

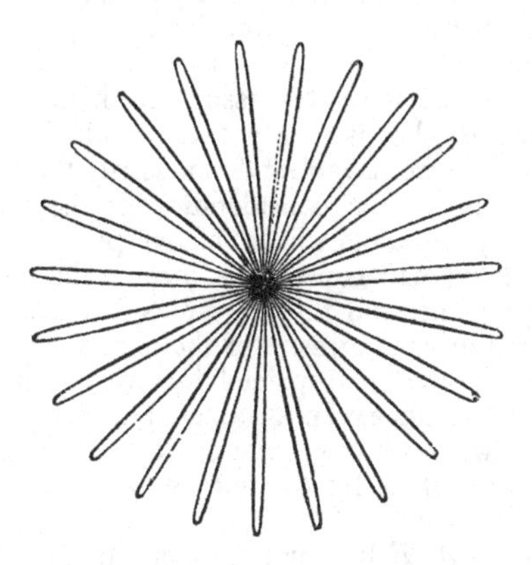

This is intimately connected with the explanation of two sets of important phenomena,—the rotation of the plane of polarization of light, by quartz and certain fluids on the one hand, and by transparent bodies under magnetic forces on the other. It is a case of the hypotrochoid, and its corresponding mode of description will be described in § 104. It may be exhibited experimentally as the path of a pendulum, hung so as to be free to move in any vertical plane through its point of suspension, and containing in its bob a fly-wheel in rapid rotation.

88. [Before leaving for a time the subject of the composition of harmonic motions, we must enunciate Fourier's Theorem, which is not only one of the most beautiful results of modern analysis, but may be said to furnish an indispensable instrument in the treatment of nearly every recondite question in modern physics. To mention only sonorous vibrations, the propagation of electric signals along a telegraph wire, and the conduction of heat by the earth's crust, as subjects in their generality intractable without it, is to give but a feeble idea of its importance. Unfortunately it is impossible to give a satisfactory proof of it without introducing some rather troublesome analysis, which is foreign to the purpose of so elementary a treatise as the present.

The following seems to be the most intelligible form in which it can be presented to the general reader:—

THEOREM.—*A complex harmonic function, with a constant term added, is the proper expression, in mathematical language, for any arbitrary periodic function; and consequently can express any function whatever between definite values of the variable.*

89. Any arbitrary periodic function whatever being given, the amplitudes and epochs of the terms of a complex harmonic function, which shall be equal to it for every value of the independent variable, may be investigated by the 'method of indeterminate co-efficients.' Such an investigation is sufficient as a solution of the problem,—to find a complex harmonic function expressing a given arbitrary periodic function,—when once we are assured that the problem is possible; and when we have this assurance, it proves that the resolution is determinate; that is to say, that no other complex harmonic function than the one we have found can satisfy the conditions.]

90. We now pass to the consideration of the displacement of a rigid body or group of points whose relative positions are unalterable. The simplest case we can consider is that of the motion of a plane figure in its own plane, and this, as far as kinematics is concerned, is entirely summed up in the result of the next section.

91. If a plane figure be displaced in any way in its own plane, there is always (with an exception treated in § 93) one point of it common to any two positions; that is, it may be moved from any one position to any other by rotation in its own plane about one point held fixed.

To prove this, let A, B be any two points of the plane figure in a first position, A', B' the position of the same two after a displacement. The lines AA', BB' will not be parallel, except in one case to be presently considered. Hence the line equidistant from A and A' will meet that equidistant from B and B' in some point O. Join OA, OB, OA', OB'. Then, evidently, because $OA'=OA$, $OB'=OB$, and $A'B'=AB$, the triangles $OA'B'$ and OAB are equal and similar. Hence O is similarly situated with regard to $A'B'$ and AB, and is therefore one and the same point of the plane figure in its two positions. If, for the sake of illustration, we actually trace the angle OAB upon the plane, it becomes $OA'B'$ in the second position of the figure.

92. If from the equal angles $A'OB'$, AOB of these similar triangles we take the common part $A'OB$, we have the remaining angles AOA', BOB' equal, and each of them is clearly equal to the angle through which the figure must have turned round the point O to bring it from the first to the second position.

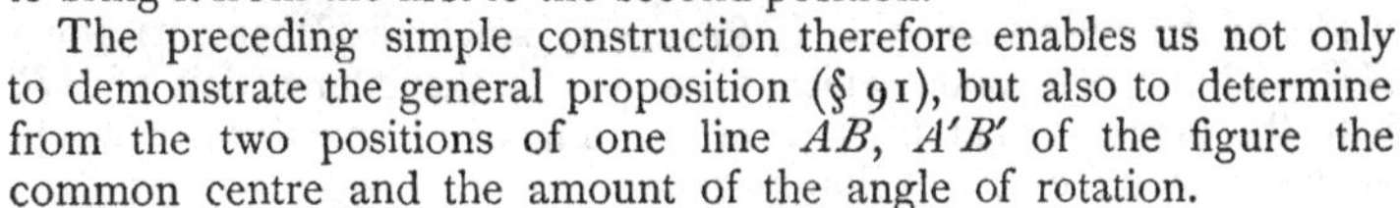

The preceding simple construction therefore enables us not only to demonstrate the general proposition (§ 91), but also to determine from the two positions of one line AB, $A'B'$ of the figure the common centre and the amount of the angle of rotation.

93. The lines equidistant from A and A', and from B and B', are parallel if AB is parallel to $A'B'$; and therefore the construction fails, the point O being infinitely distant, and the theorem becomes nugatory. In this case the motion is in fact a simple translation of the figure in its own plane without rotation—since as AB is parallel and equal to $A'B'$, we have AA' parallel and equal to BB'; and instead of there being one point of the figure common to both positions, the lines joining the successive positions of every point in the figure are equal and parallel.

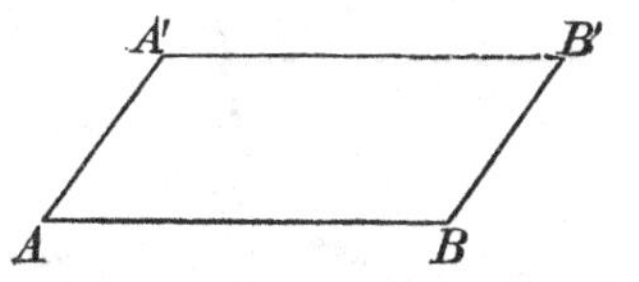

94. It is not necessary to suppose the figure to be a mere flat disc or plane—for the preceding statements apply to any one of a set of parallel planes in a rigid body, moving in any way subject to the condition that the points of any one plane in it remain always in a fixed plane in space.

95. There is yet a case in which the construction in § 91 is nugatory—that is when AA' is parallel to BB', but AB intersects $A'B'$. In this case, however, it is easy to see at once that this point of intersection is the point O required, although the former method would not have enabled us to find it.

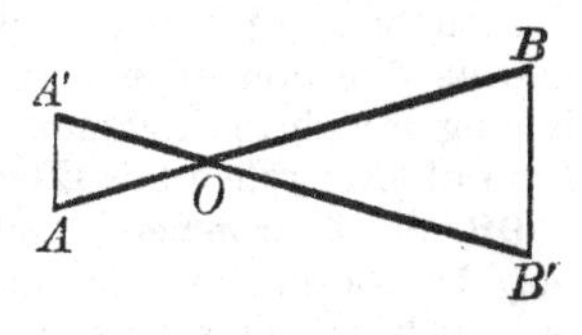

96. Very many interesting applications of this principle may be made, of which, however, few belong strictly to our subject, and we shall therefore give only an example or two. Thus we know that if a line of given length AB move with its extremities always in two fixed lines OA, OB, any point in it as P describes an ellipse. (This is proved in § 101 below.) It is required to find the direction of motion of P at any instant, i.e. to draw a tangent to the ellipse.

BA will pass to its next position by rotating about the point Q; found by the method of § 91 by drawing perpendiculars to OA and OB at A and B. Hence P for the instant revolves about Q, and thus its direction of motion, or the tangent to the ellipse, is perpendicular to QP. Also AB in its motion always touches a curve (called in geometry its envelop); and the same principle enables us to find the point of the envelop which lies in AB, for the motion of that point must evidently be ultimately (that is for a very small displacement) along AB, and the only point which so moves is the intersection of AB, with the perpendicular to it from Q. Thus our construction would enable us to trace the envelop by points.

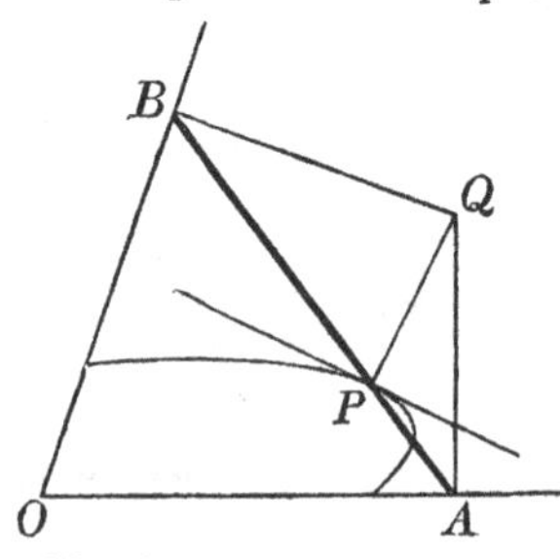

97. Again, suppose $ABDC$ to be a jointed frame, AB having a reciprocating motion about A, and by a link BD turning CD in the same plane about C. Determine the relation between the angular velocities of AB and CD in any position. Evidently the instantaneous direction of motion of B is transverse to AB, and of D transverse to CD—hence if AB, CD produced meet in O, the motion of BD is for an instant as if it turned about O. From this it may easily be seen that if the angular velocity of AB be ω, that of CD is $\frac{AB}{OB}\frac{OD}{CD}\omega$. A similar process is of course applicable to any combination of machinery, and we shall find it very convenient when we come to apply the principle of work in various problems of Mechanics.

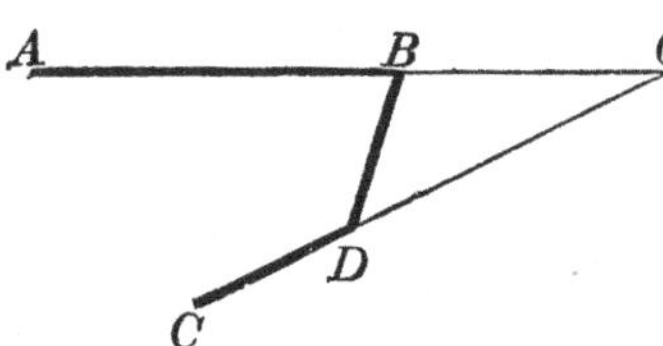

Thus in any *Lever*, turning in the plane of its arms—the rate of motion of any point is proportional to its distance from the fulcrum, and its direction of motion at any instant perpendicular to the line joining it with the fulcrum. This is of course true of the particular form of lever called the *Wheel and Axle*.

98. Since, in general, any movement of a plane figure in its plane may be considered as a rotation about one point, it is evident that two such rotations may, in general, be compounded into one; and therefore, of course, the same may be done with any number of rotations. Thus let A and B be the points of the figure about which in succession the rotations are to take place. By rotation about A, B is brought say to B', and by a rotation about B', A is brought to A'. The construction of § 91 gives us at once the point O and the amount of rotation about it which singly gives the same effect as those about A and B in succession. But there is one case of exception, viz. when the rotations about A and B are of equal amount and in opposite directions. In this case $A'B'$ is evidently parallel to AB, and therefore the compound result is a *translation* only. That is, if a body revolve in succession through equal angles, but in opposite directions, about two parallel axes, it finally takes a position to which it could have been brought by a simple translation perpendicular to the lines of the body in its initial or final position, which were successively made axes of rotation; and inclined to their plane at an angle equal to half the supplement of the common angle of rotation.

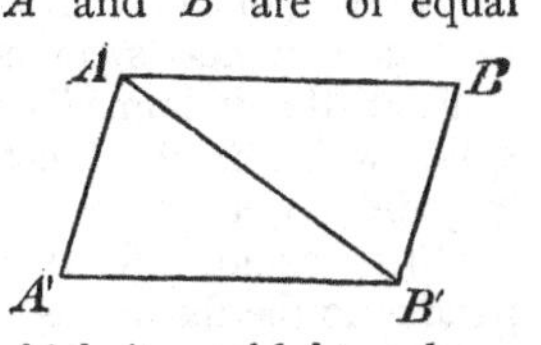

99. Hence to compound into an equivalent rotation a rotation and a translation, the latter being effected parallel to the plane of the former, we may decompose the translation into two rotations of equal amounts and opposite directions, compound one of them with the given rotation by § 98, and then compound the other with the resultant rotation by the same process. Or we may adopt the following far simpler method:—Let OA be the translation common to all points in the plane, and let BOC be the angle of rotation about O, BO being drawn so that OA bisects the exterior angle COB'. Evidently there is a point B' in BO produced, such that $B'C'$, the space through which the rotation carries it, is equal and opposite to OA. This point retains its former position after the performance of the compound operation; so that a rotation and a translation in one plane can be compounded into an equal rotation about a different axis.

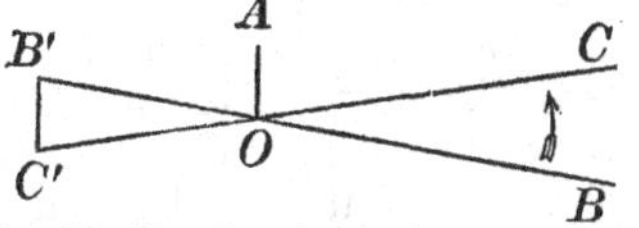

100. Any motion whatever of a plane figure in its own plane might be produced by the rolling of a curve fixed to the figure upon a curve fixed in the plane.

For we may consider the whole motion as made up of successive elementary displacements, each of which corresponds, as we have seen, to an elementary rotation about some point in the plane. Let O_1, O_2, O_3, etc., be the successive points of the *figure* about which the rotations take place, o_1, o_2, o_3, etc., the positions of these points on the *plane* when each is the instantaneous centre of rotation. Then the figure rotates about O_1 (or o_1, which coincides with it) till O_2 coincides with o_2, then about the latter till O_3 coincides with o_3, and so on. Hence, if we join O_1, O_2, O_3, etc., in the plane of the figure, and o_1, o_2, o_3, etc., in the fixed plane, the motion will be the same as if the polygon $O_1 O_2 O_3$, etc., rolled upon the fixed polygon $o_1 o_2 o_3$, etc. By supposing the successive displacements small enough, the sides

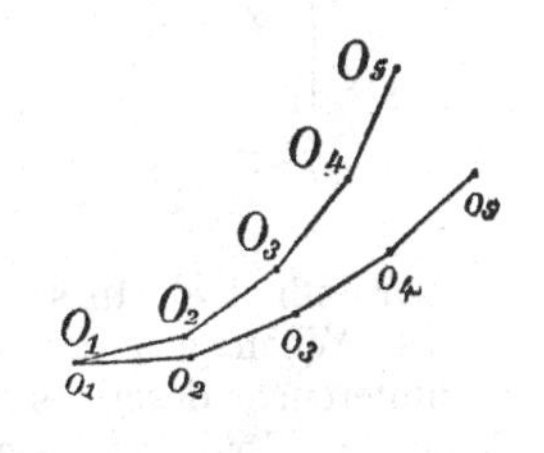

of these polygons gradually diminish, and the polygons finally become continuous curves. Hence the theorem.

From this it immediately follows, that any displacement of a rigid solid, which is in directions wholly perpendicular to a fixed line, may be produced by the rolling of a cylinder fixed in the solid on another cylinder fixed in space, the axes of the cylinders being parallel to the fixed line.

101. As an interesting example of this theorem, let us recur to the case of § 96:—A circle may evidently be circumscribed about $OBQA$; and it must be of invariable magnitude, since in it a chord of given length AB subtends a given angle O at the circumference. Also OQ is a diameter of this circle, and is therefore constant. Hence, as Q is momentarily at rest, the motion of the circle circumscribing $OBQA$ is one of internal rolling on a circle of double its diameter. Hence if a circle roll internally on another of twice its diameter any point in its circumference describes a diameter of the fixed circle, any other point in its plane an ellipse. This is precisely the same proposition as that of § 86, although the ways of arriving at it are very different.

102. We may easily employ this result, to give the proof, promised in § 96, that the point P of AB describes an ellipse. Thus let OA, OB be the fixed lines, in which the extremities of AB move. Draw the circle $AOBD$, circumscribing AOB, and let CD be the diameter of this circle which passes through P. While the two points A and B of this circle move along OA and OB, the points C and D must, because of the invariability of the angles BOD, AOC, move along straight lines OC, OD, and these are evidently at right angles. Hence the path of P may be considered as that of a point in a line whose ends move on two *mutually perpendicular* lines. Let E be the centre of the circle; join OE, and produce it to meet, in F, the line FPG drawn through P parallel to DO. Then evidently $EF=EP$, hence F describes a circle about O. Also $FP:FG::2FE:FO$, or PG is a constant submultiple of FG; and therefore the locus of P is an ellipse whose major axis is a diameter of the circular path of F. Its semi-axes are DP along OC, and PC along OD.

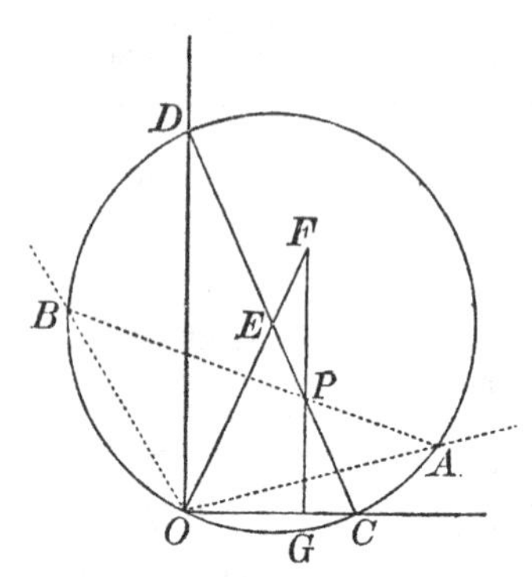

103. When a circle rolls upon a straight line, a point in its circumference describes a Cycloid, an internal point describes a Prolate Cycloid, an external point a Curtate Cycloid. The two latter varieties are sometimes called Trochoids.

The general form of these curves will be seen in the succeeding figures; and in what follows we shall confine our remarks to the cycloid itself, as it is of greater consequence than the others. The next section contains a simple investigation of those properties of the cycloid which are most useful in our subject.

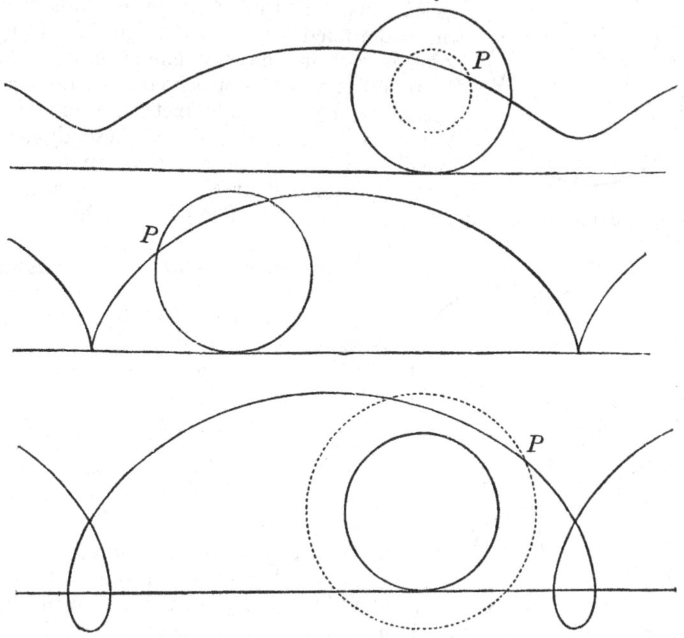

104. Let AB be a diameter of the generating (or rolling) circle, BC the line on which it rolls. The points A and B describe similar and equal cycloids, of which AQC and BS are portions. If PQR be any subsequent position of the generating circle, Q and S the new positions of A and B, QPS is of course a right angle. If, therefore, QR be drawn parallel to PS, PR is a diameter of the rolling circle, and R lies in a straight line AH drawn parallel to BC. Thus $AR = BP$. Produce QR to T, making $RT = QR = PS$. Evidently the curve AT, which is the locus of T, is similar and equal to BS, and is therefore a cycloid similar and equal to AC. But QR is perpendicular to PQ, and is therefore the instantaneous direction of motion of Q, or is the tangent to the cycloid AQC. Similarly, PS is perpendicular to the cycloid BS at S, and therefore TQ is perpendicular to AT at T. Hence (§ 22) AQC is the evolute of AT, and arc $AQ = QT = 2QR$.

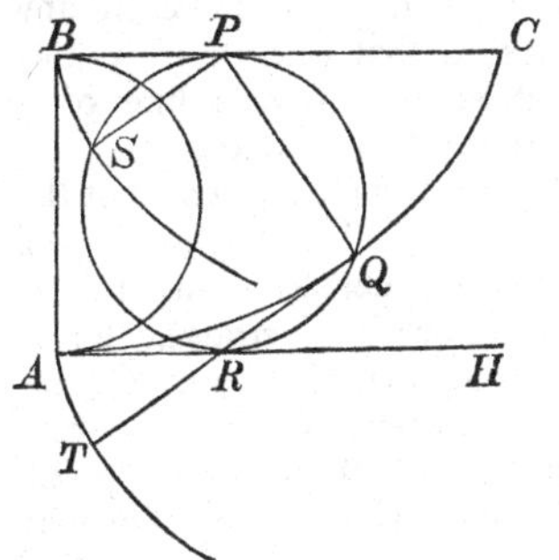

105. When a circle rolls upon another circle, the curve described by a point in its circumference is called an Epicycloid, or a Hypocycloid, as the rolling circle is without or within the fixed circle; and when the tracing-point is not in the circumference, we have Epitrochoids and Hypotrochoids. Of the latter classes we have already met with examples (§§ 87, 101), and others will be presently mentioned. Of the former we have, in the first of the appended figures, the case of a circle rolling externally on another of equal size. The curve in this case is called the Cardioid.

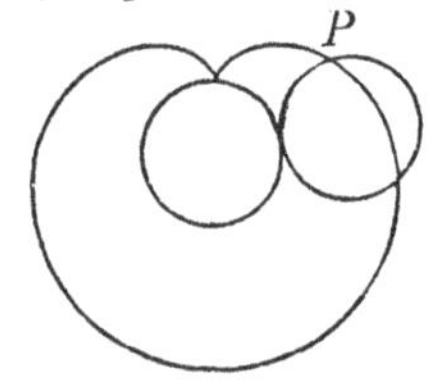

In the second figure a circle rolls externally on another of twice its radius. The epicycloid so described is of importance in optics, and will, with others, be referred to when we consider the subject of Caustics by reflexion.

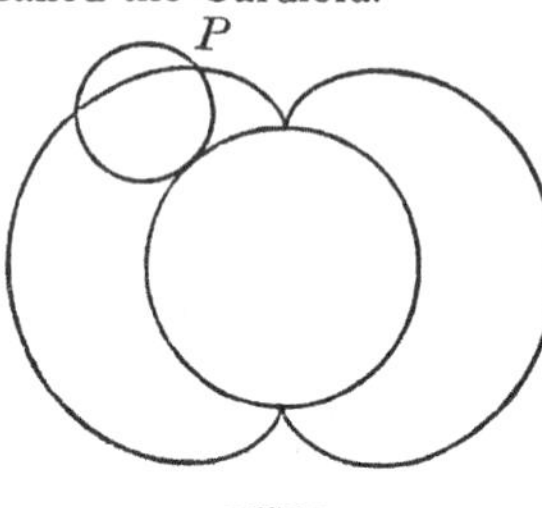

In the third figure we have a hypocycloid traced by the rolling of one circle internally on another of four times its radius.

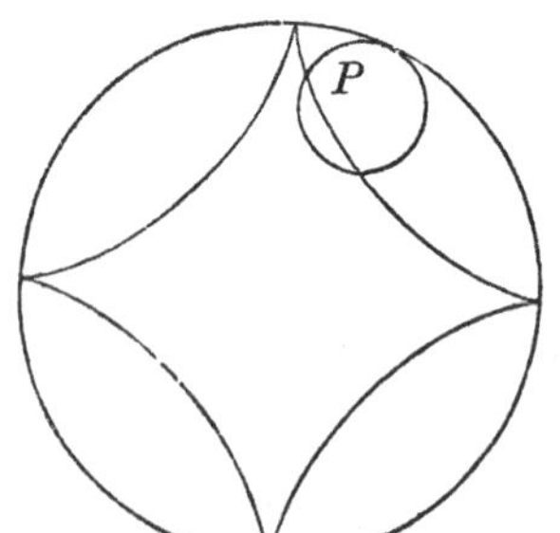

The curve of § 87 is a hypotrochoid described by a point in the plane of a circle which rolls internally on another of rather more than twice its diameter, the tracing-point passing through the centre of the fixed circle. Had the diameters of the circles been exactly as 1 : 2, § 101 shows us that this curve would have been reduced to a single straight line.

106. If a rigid body move in any way whatever, subject only to the condition that one of its points remains fixed, there is always (without exception) one line of it through this point common to the body in any two positions.

Consider a spherical surface within the body, with its centre at the fixed point *C*. All points of this sphere attached to the body will move on a sphere fixed in space. Hence the construction of § 91 may be made, only with great circles instead of straight lines; and the same reasoning will apply to prove that the point *O* thus obtained is common to the body in its two positions. Hence every point of the body in the line *OC*, joining *O* with the fixed point, must be common to it in the two positions. Hence the body may pass from any one position to any other by a definite amount of rotation about a definite axis. And hence, also, successive or simultaneous rotations about any number of axes through the fixed point may be compounded into one such rotation.

107. Let OA, OB be two axes about which a body revolves with angular velocities ω, ω_1 respectively.

With radius unity describe the arc AB, and in it take any point I. Draw Ia, $I\beta$ perpendicular to OA, OB respectively. Let the rotations about the two axes be such that that about OB tends to *raise* I above the plane of the paper, and that about OA to depress it. In an infinitely short interval of time τ, the amounts of these displacements will be $\omega_1 I\beta.\tau$ and $-\omega Ia.\tau$. The point I, and therefore every point in the line OI, will be at rest during the interval τ if the sum of these displacements is zero—i. e. if

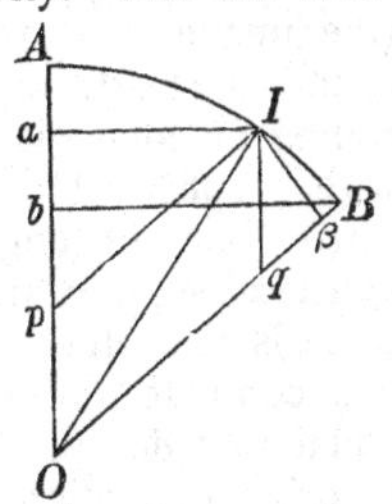

$$\omega_1.I\beta = \omega.Ia.$$

Hence the line OI is instantaneously at rest, or the *two* rotations about OA and OB may be compounded into *one* about OI. Draw Ip, Iq, parallel to OB, OA respectively. Then, expressing in two ways the area of the parallelogram $IpOq$, we have

$$Oq.I\beta = Op.Ia.$$

Hence $$Oq : Op :: \omega_1 : \omega.$$

In words, if on the axes OA, OB, we measure off from O lines Op, Oq, proportional respectively to the angular velocities about these axes—the diagonal of the parallelogram of which these are contiguous sides is the resultant axis.

Again, if Bb be drawn perpendicular to OA, and if Ω be the angular velocity about OI, the whole displacement of B may evidently be represented either by

$$\omega.Bb \text{ or } \Omega.IB.$$

Hence $$\Omega : \omega :: Bb : I\beta$$
$$:: OI : Op.$$

And thus on the scale on which Op, Oq represent the component angular velocities, the diagonal OI represents their resultant.

108. Hence rotations are to be compounded according to the same law as velocities, and therefore the single angular velocity, equivalent to three co-existent angular velocities about three mutually perpendicular axes, is determined in magnitude, and the direction of its axis is found, as follows:—The square of the resultant angular velocity is the sum of the squares of its components, and the ratios of the three components to the resultant are the direction-cosines of the axis.

Hence also, an angular velocity about any line may be resolved into three about any set of rectangular lines, the resolution in each case being (like that of simple velocities) effected by multiplying by the cosine of the angle between the directions.

Hence, just as in § 38 a uniform acceleration, acting perpendicularly to the direction of motion of a point, produces a change in the *direction* of motion, but does not influence the *velocity;* so, if a body be rotating about an axis, and be subjected to an action tending to produce rotation about a perpendicular axis, the result will be a change of *direction* of the axis about which the body revolves, but no change in the *angular velocity*.

109. If a pyramid or cone of any form roll on a similar pyramid (the image in a plane mirror of the first position of the first) all round, it clearly comes back to its primitive position. This (as all rolling of cones) is exhibited best by taking the intersection of each with a spherical surface. Thus we see that if a spherical polygon turns about its angular points in succession, always keeping on the spherical surface, and if the angle through which it turns about each point is twice the supplement of the angle of the polygon, or, which will come to the same thing, if it be in the other direction, but equal to twice the angle itself of the polygon, it will be brought to its original position.

110. The method of § 100 also applies to the case of § 106; and it is thus easy to show that the most general motion of a spherical figure on a fixed spherical surface is obtained by the rolling of a curve fixed in the figure on a curve fixed on the sphere. Hence as at each instant the line joining C and O contains a set of points of the body which are momentarily at rest, the most general motion of a rigid body of which one point is fixed consists in the rolling of a cone fixed in the body upon a cone fixed in space—the vertices of both being at the fixed point.

111. To complete our kinematical investigation of the motion of a body of which one point is fixed, we require a solution of the following problem:—From the given angular velocities of the body about three rectangular axes attached to it to determine the position of the body in space after a given time. But the general solution of this problem demands higher analysis than can be admitted into the present treatise.

112. We shall next consider the most general possible motion of a rigid body of which no point is fixed—and first we must prove the following theorem. There is one set of parallel planes in a rigid body which are parallel to each other in any two positions of the body. The parallel lines of the body perpendicular to these planes are of course parallel to each other in the two positions.

Let C and C' be any point of the body in its first and second positions. Move the body without rotation from its second position to a third in which the point at C' in the second position shall occupy its original position C. The preceding demonstration shows that there is a line CO common to the body in its first and third positions. Hence a line $C'O'$ of the body in its second position is parallel to the same line CO in the first position. This of course clearly applies to every line of the body parallel to CO, and the planes perpendicular to these lines also remain parallel.

113. Let S denote a plane of the body, the two positions of which are parallel. Move the body from its first position, without rotation, in a direction perpendicular to S, till S comes into the plane of its second position. Then to get the body into its actual position, such a motion as is treated in § 91 is farther required. But by § 91 this may be effected by rotation about a certain axis perpendicular to the plane S, unless the motion required belongs to the exceptional case of pure translation. Hence (this case excepted), the body may be brought from the first position to the second by translation through a determinate distance perpendicular to a given plane, and rotation through a determinate angle about a determinate axis perpendicular to that plane. This is precisely the motion of a *screw* in its nut.

114. To understand the nature of this motion we may commence with the sliding of one straight-edged board on another.

Thus let $GDEF$ be a plane board whose edge, DE, slides on the edge, AB, of another board, ABC, of which for convenience we suppose the edge, AC, to be horizontal. By § 30, if the upper board move horizontally to the right, the constraint will give it, in addition, a vertically upward motion, and the rates of these motions are in the constant ratio of AC to CB. Now, if both planes be bent so as to form portions of the surface of a vertical right cylinder, the motion of DF parallel to AC will become a rotation about the axis of the cylinder, and the necessary accompaniment of vertical motion will remain unchanged. As it is evident that all portions of AB will be equally inclined to the axis of the cylinder, it is obvious that the thread of the screw, which corresponds to the edge, DE, of the upper board, must be traced on the cylinder so as always to make a constant angle with its generating lines (§ 128). A hollow mould taken from the screw itself forms what is called the nut—the representative of the board, ABC—and it is obvious that the screw cannot move without rotating about its axis, if the nut be fixed. If a be the radius of the cylinder, ω the angular velocity, α the inclination of the screw thread to a generating line, u the linear velocity of the axis of the screw, we see at once from the above construction that

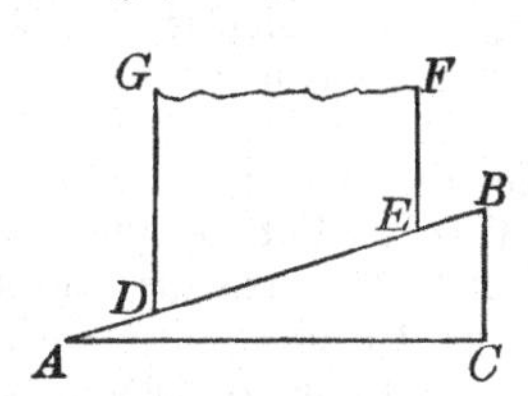

$$a\omega : u :: AC : CB :: \sin \alpha : \cos \alpha,$$

which gives the requisite relation between ω and u.

115. In the excepted case of § 113, the whole motion consists of two translations, which can of course be compounded into a single one: and thus, in this case, there is no rotation at all, or every plane of it fulfils the specified condition for S of § 113.

116. We may now briefly consider the case in which the guiding cones (§ 110) are both circular, as it has important applications to the motion of the earth, the evolutions of long or flattened projectiles, the spinning of tops and gyroscopes, etc. The motion in this case may be called *Precessional Rotation.* The plane through the instantaneous axis and the axis of the fixed cone passes through the axis of the rolling cone. This plane turns round the axis of the fixed cone with an angular velocity Ω, which must clearly bear a constant ratio to the angular velocity ω of the rigid body about its instantaneous axis.

117. The motion of the plane containing these axes is called the *precession* in any such case. What we have denoted by Ω is the angular velocity of the precession, or, as it is sometimes called, the rate of precession.

The angular motions ω, Ω are to one another inversely as the distances of a point in the axis of the rolling cone from the instantaneous axis and from the axis of the fixed cone.

For, let OA be the axis of the fixed cone, OB that of the rolling cone, and OI the instantaneous axis. From any point P in OB draw PN perpendicular to OI, and PQ perpendicular to OA. Then we perceive that P moves always in the circle whose centre is Q, radius PQ, and plane perpendicular to OA. Hence the actual velocity of the point P is $\Omega.QP$. But, by the principles explained above (§ 110) the velocity of P is the same as that of a point moving in a circle whose centre is N, plane perpendicular to ON, and radius NP, which, as this radius revolves with angular velocity ω, is $\omega.NP$. Hence

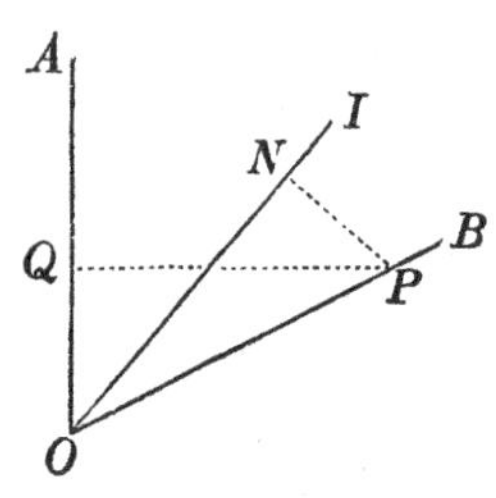

$$\Omega.QP = \omega.NP,$$

or
$$\omega : \Omega :: QP : NP.$$

118. Suppose a rigid body bounded by any curved surface to be touched at any point by another such body. Any motion of one on the other must be of one or more of the forms *sliding*, *rolling*, or *spinning*. The consideration of the first is so simple as to require no comment.

Any motion in which the bodies have no relative velocity at the point of contact, must be rolling or spinning, separately or combined.

Let one of the bodies rotate about successive instantaneous axes, all lying in the common tangent plane at the point of instantaneous contact, and each passing through this point—the other body being fixed. This motion is what we call rolling, or simple rolling, of the movable body on the fixed.

On the other hand, let the instantaneous axis of the moving body be the common normal at the point of contact. This is pure spinning, and does not change the point of contact.

Let it move, so that the instantaneous axis, still passing through the point of contact, is neither in, nor perpendicular to, the tangent plane. This motion is combined rolling and spinning.

119. As an example of pure *rolling*, we may take that of one cylinder on another, the axes being parallel.

Let ρ be the radius of curvature of the rolling, σ of the fixed, cylinder; ω the angular velocity of the former, V the linear velocity of the point of contact. We have

$$\omega = \left(\frac{1}{\rho} + \frac{1}{\sigma}\right) V.$$

For, in the figure, suppose P to be at any time the point of contact, and Q and p the points which are to be in contact after a very small interval τ; O, O' the centres of curvature; $POp = \theta$, $PO'Q = \phi$.

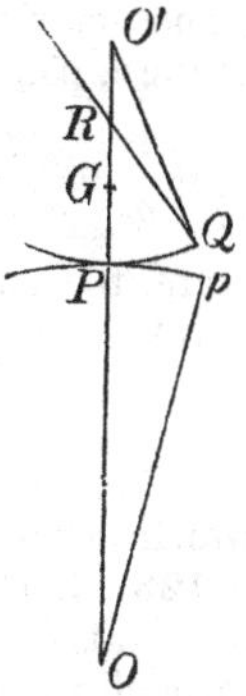

Then $PQ = Pp =$ space described by point of contact. In symbols

$$\rho\phi = \sigma\theta = V\tau.$$

Also, before $O'Q$ and OP can coincide in direction, the former must evidently turn through an angle $\theta + \phi$.

Therefore $$\omega\tau = \theta + \phi;$$
and by eliminating θ and ϕ, and dividing by τ, we get the above result.

It is to be understood here, that as the radii of curvature have been considered positive when both surfaces are convex, the negative sign must be introduced for either radius when the corresponding surface is concave.

Hence the angular velocity of the rolling curve is in this case equal to the product of the linear velocity of the point of contact into the sum or difference of the curvatures, according as the curves are both convex, or one concave and the other convex.

120. We may now take up a few points connected with the curvature of surfaces, which are useful in various parts of our subject.

The tangent plane at any point of a surface may or may not cut it at that point. In the former case, the surface bends away from the tangent plane partly towards one side of it, and partly towards the other, and has thus, in some of its normal sections, curvatures oppositely directed to those in others. In the latter case, the surface on every side of the points bends away from the same side of its tangent plane, and the curvatures of all normal sections are similarly directed. Thus we may divide curved surfaces into *Anticlastic* and *Synclastic*. A saddle gives a good example of the former class; a ball of the latter. Curvatures in opposite directions, with reference to the tangent plane, have of course different signs. The outer portion of the surface of an anchor-ring is synclastic, the inner anticlastic.

121. *Meunier's Theorem.*—The curvature of an oblique section of a surface is equal to that of the normal section through the same tangent line multiplied by the secant of the inclination of the planes of the sections. This is evident from the most elementary considerations regarding projections.

122. *Euler's Theorem.*—There are at every point of a synclastic surface two normal sections, in one of which the curvature is a maximum, in the other a minimum; and these are at right angles to each other.

In an anticlastic surface there is maximum curvature (but in opposite directions) in the two normal sections whose planes bisect the angles between the lines in which the surface cuts its tangent plane. On account of the difference of sign, these may be considered as a maximum and a minimum.

Generally the sum of the curvatures at a point, in any two normal planes at right angles to each other, is independent of the position of these planes.

If $\frac{1}{\rho}$ and $\frac{1}{\sigma}$ be the maximum and minimum curvatures at any point, the curvature of a normal section making an angle θ with the normal section of maximum curvature is

$$\frac{1}{\rho}\cos^2\theta + \frac{1}{\sigma}\sin^2\theta,$$

which includes the above statements as particular cases.

123. Let P, p be two points of a surface indefinitely near to each other, and let r be the radius of curvature of a normal section passing through them. Then the radius of curvature of an oblique section through the same points, inclined to the former at an angle a, is $r \cos a$ (§ 121). Also the length along the normal section, from P to p, is less than that along the oblique section—since a given chord cuts off an arc from a circle, longer the less is the radius of that circle.

124. Hence, if the shortest possible line be drawn from one point of a surface to another, its osculating plane, or plane of curvature, is everywhere perpendicular to the surface.

Such a curve is called a *Geodetic* line. And it is easy to see that it is the line in which a flexible and inextensible string would touch the surface if stretched between those points, the surface being supposed smooth.

125. A perfectly flexible but inextensible surface is suggested, although not realized, by paper, thin sheet-metal, or cloth, when the surface is plane; and by sheaths of pods, seed-vessels, or the like, when not capable of being stretched flat without tearing. The process of changing the form of a surface by bending is called '*developing*.' But the term '*Developable Surface*' is commonly restricted to such inextensible surfaces as can be developed into a plane, or, in common language, 'smoothed flat.'

126. The geometry or kinematics of this subject is a great contrast to that of the flexible line (§ 16), and, in its merest elements, presents ideas not very easily apprehended, and subjects of investigation that have exercised, and perhaps even overtasked, the powers of some of the greatest mathematicians.

127. Some care is required to form a correct conception of what is a perfectly flexible inextensible surface. First let us consider a plane sheet of paper. It is very flexible, and we can easily form the conception from it of a sheet of ideal matter perfectly flexible. It is very inextensible; that is to say, it yields very little to any application of force tending to pull or stretch it in any direction, up to the strongest it can bear without tearing. It does, of course, stretch a little. It is easy to test that it stretches when under the influence of force, and that it contracts again when the force is removed, although not always to its original dimensions, as it may and generally does remain to some sensible extent permanently stretched. Also, flexure stretches one side and condenses the other temporarily; and, to a less extent, permanently. Under elasticity we may return to this. In the meantime, in considering illustrations of our kinematical propositions, it is necessary to anticipate such physical circumstances.

128. The flexure of an inextensible surface which can be plane, is a subject which has been well worked by geometrical investigators and writers, and, in its elements at least, presents little difficulty. The first elementary conception to be formed is, that such a surface (if perfectly flexible), taken plane in the first place, may be bent about any straight line ruled on it, so that the two plane parts may make any angle with one another.

Such a line is called a 'generating line' of the surface to be formed.

Next, we may bend one of these plane parts about any other line which does not (within the limits of the sheet) intersect the former; and so on. If these lines are infinite in number, and the angles of bending infinitely small, but such that their sum may be finite, we have our plane surface bent into a curved surface, which is of course 'developable' (§ 125).

129. Lift a square of paper, free from folds, creases, or ragged edges, gently by one corner, or otherwise, without crushing or forcing it, or very gently by two points. It will hang in a form which is very rigorously a developable surface; for although it is not absolutely inextensible, yet the forces which tend to stretch or tear it, when it is treated as above described, are small enough to produce absolutely no sensible stretching. Indeed the greatest stretching it can experience without tearing, in any direction, is not such as can affect the form of the surface much when sharp flexures, singular points, etc., are kept clear off.

130. Prisms and cylinders (when the lines of bending, § 128, are parallel, and finite in number with finite angles, or infinite in number with infinitely small angles), and pyramids and cones (the lines of bending meeting in a point if produced), are clearly included.

131. If the generating lines, or line-edges of the angles of bending, are not parallel, they must meet, since they are in a plane when the surface is plane. If they do not meet all in one point, they must meet in several points: in general, let each one meet its predecessor and its successor in different points.

132. There is still no difficulty in understanding the form of, say a square, or circle, of the plane surface when bent as explained above, provided it does not include any of these points of intersection. When the number is infinite, and the surface finitely curved, the developable lines will, in general, be tangents to a curve (the locus of the points of intersection when the number is infinite). This curve is called the *edge of regression.* The surface must clearly, when *complete* (according to mathematical ideas), consist of two sheets meeting in this edge of regression (just as a cone consists of two sheets meeting in the vertex), because each tangent may be produced beyond the point of contact, instead of stopping at it, as in the preceding diagram.

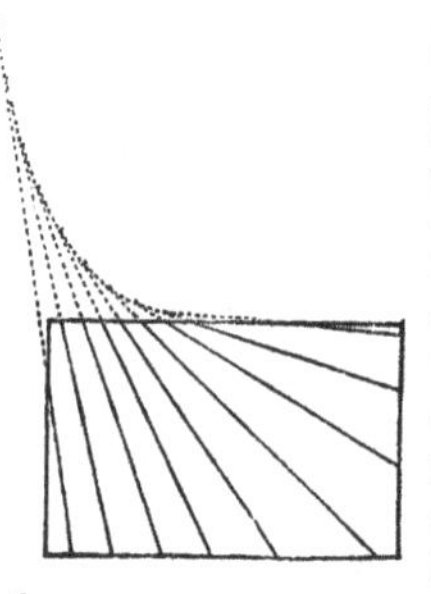

133. To construct a complete developable surface in two sheets from its edge of regression—

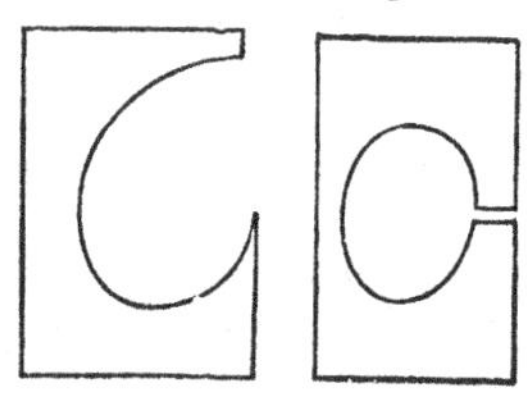

Lay one piece of perfectly flat, unwrinkled, smooth-cut paper on the top of another. Trace any curve on the other, and let it have no point of inflection, but everywhere finite curvature. Cut the paper quite away on the concave side. If the curve traced is closed, it must be cut open (see second diagram). The limits to the extent that may be left uncut away, are the tangents drawn outwards from the two ends, so that, in short, no portion of the paper through which a real tangent does not pass is to be left.

Attach the two sheets together by very slight paper or muslin clamps gummed to them along the common curved edge. These must be so slight as not to interfere sensibly with the flexure of the two sheets. Take hold of one corner of one sheet and lift the whole. The two will open out into two sheets of a developable surface, of which the curve, bending into a curve of double curvature, is the edge of regression. The tangent to the curve drawn in one direction from the point of contact, will always lie in one of the sheets, and its continuation on the other side in the other sheet. Of course a double-sheeted developable polyhedron can be constructed by this process, by starting from a polygon instead of a curve.

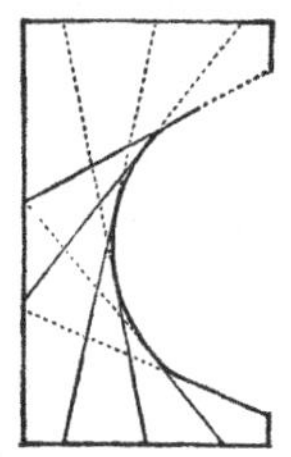

134. A flexible but perfectly inextensible surface, altered in form in any way possible for it, must keep any line traced on it unchanged in length; and hence any two intersecting lines unchanged in mutual inclination. Hence, also, geodetic lines must remain geodetic lines.

135. We have now to consider the very important kinematical conditions presented by the changes of volume or figure experienced by a solid or liquid mass, or by a group of points whose positions with regard to each other are subject to known conditions.

Any such definite alteration of form or dimensions is called a *Strain*.

Thus a rod which becomes longer or shorter is strained. Water, when compressed, is strained. A stone, beam, or mass of metal, in a building or in a piece of framework, if condensed or dilated in any direction, or bent, twisted, or distorted in any way, is said to experience a strain. A ship is said to 'strain' if, in launching, or when working in a heavy sea, the different parts of it experience relative motions.

136. If, when the matter occupying any space is strained in any way, all pairs of points of its substance which are initially at equal distances from one another in parallel lines remain equidistant, it may be at an altered distance; and in parallel lines, altered, it may be, from their initial direction; the strain is said to be homogeneous.

137. Hence if any straight line be drawn through the body in its initial state, the portion of the body cut by it will continue to be a straight line when the body is homogeneously strained. For, if ABC be any such line, AB and BC, being parallel to one line in the initial, remain parallel to one line in the altered, state; and therefore remain in the same straight line with one another. Thus it follows that a plane remains a plane, a parallelogram a parallelogram, and a parallelepiped a parallelepiped.

138. Hence, also, similar figures, whether constituted by actual portions of the substance, or mere geometrical surfaces, or straight or curved lines passing through or joining certain portions or points of the substance, similarly situated (i.e. having corresponding parameters parallel) when altered according to the altered condition of the body, remain similar and similarly situated among one another.

139. The lengths of parallel lines of the body remain in the same proportion to one another, and hence all are altered in the same proportion. Hence, and from § 137, we infer that any plane figure becomes altered to another plane figure which is a diminished or magnified orthographic projection of the first on some plane.

The elongation of the body along any line is the proportion which the addition to the distance between any two points in that line bears to their primitive distance.

140. Every orthogonal projection of an ellipse is an ellipse (the case of a circle being included). Hence, and from § 139, we see that an ellipse remains an ellipse; and an ellipsoid remains a surface of which every plane section is an ellipse; that is, remains an ellipsoid.

141. The ellipsoid which any surface of the body initially spherical becomes in the altered condition, may, to avoid circumlocutions, be called the *Strain Ellipsoid.*

142. In any absolutely unrestricted homogeneous strain there are three directions (the three principal axes of the strain ellipsoid), at right angles to one another, which remain at right angles to one another in the altered condition of the body. Along one of these the elongation is greater, and along another less, than along any other direction in the body. Along the remaining one the elongation is less than in any other line in the plane of itself and the first mentioned, and greater than along any other line in the plane of itself and the second.

Note.—Contraction is to be reckoned as a negative elongation: the maximum elongation of the preceding enunciation may be a minimum contraction: the minimum elongation may be a maximum contraction.

143. The ellipsoid into which a sphere becomes altered may be an ellipsoid of revolution, or, as it is called, a spheroid, prolate, or oblate. There is thus a maximum or minimum elongation along the axis, and equal minimum or maximum elongation along all lines perpendicular to the axis.

Or it may be a sphere; in which case the elongations are equal in all directions. The effect is, in this case, merely an alteration of dimensions without change of figure of any part.

144. The principal axes of a strain are the principal axes of the ellipsoid into which it converts a sphere. The principal elongations of a strain are the elongations in the direction of its principal axes.

145. When the positions of the principal axes, and the magnitudes of the principal elongations of a strain are given, the elongation of any line of the body, and the alteration of angle between any two lines, may be obviously determined by a simple geometrical construction.

146. With the same data the alteration of angle between any two planes of the body may also be easily determined, geometrically.

147. Let the ellipse of the annexed diagram represent the section of the strain ellipsoid through the greatest and least principal axes.

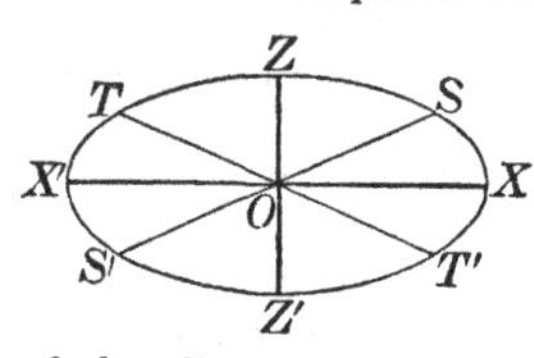

Let $S'OS$, $T'OT$ be the two diameters of this ellipse, which are equal to the mean principal axis of the ellipsoid. Every plane through O, perpendicular to the plane of the diagram, cuts the ellipsoid in an ellipse of which one principal axis is the diameter in which it cuts the ellipse of the diagram, and the other, the mean principal diameter of the ellipsoid. Hence a plane through either SS' or TT', perpendicular to the plane of the diagram, cuts the ellipsoid in an ellipse of which the two principal axes are equal, that is to say, in a circle. Hence the elongations along all lines in either of these planes are equal to the elongation along the mean principal axis of the strain ellipsoid.

148. The consideration of the circular sections of the strain ellipsoid is highly instructive, and leads to important views with reference to the analysis of the most general character of a strain. First let us suppose there to be no alteration of volume on the whole, and neither elongation nor contraction along the mean principal axis.

Let OX and OZ be the directions of maximum elongation and maximum contraction respectively. Let A be any point of the body in its primitive condition, and $A_{,}$ the same point of the altered body, so that $OA_{,} = a.OA$.

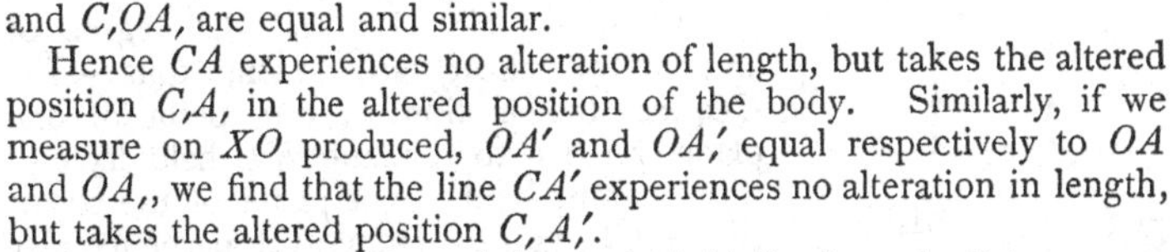

Now, if we take $OC = OA_{,}$, and if $C_{,}$ be the position of that point of the body which was in the position C initially, we shall have

$$OC_{,} = \frac{1}{a} OC, \text{ and therefore } OC_{,} = OA.$$

Hence the two triangles COA and $C_{,}OA_{,}$ are equal and similar.

Hence CA experiences no alteration of length, but takes the altered position $C_{,}A_{,}$ in the altered position of the body. Similarly, if we measure on XO produced, OA' and $OA_{,}'$ equal respectively to OA and $OA_{,}$, we find that the line CA' experiences no alteration in length, but takes the altered position $C_{,}A_{,}'$.

Consider now a plane of the body initially through CA perpendicular to the plane of the diagram, which will be altered into a plane through C_1A_1, also perpendicular to the plane of the diagram. All lines initially perpendicular to the plane of the diagram remain so, and remain unaltered in length. AC has just been proved to remain unaltered in length. Hence (§ 139) all lines in the plane we have just drawn remain unaltered in length and in mutual inclination. Similarly we see that all lines in a plane through CA', perpendicular to the plane of the diagram, altering to a plane through $C_1A'_1$, perpendicular to the plane of the diagram, remain unaltered in length and in mutual inclination.

149. The precise character of the strain we have now under consideration will be elucidated by the following:—Produce CO, and take OC' and OC'_1 respectively equal to OC and OC_1. Join $C'A$, $C'A'$, C'_1A, and $C'_1A'_1$, by plain and dotted lines as in the diagram. Then we see that the rhombus $CAC'A'$ (plain lines) of the body in its initial state becomes the rhombus $C_1A_1C'_1A'_1$ (dotted) in the altered condition. Now imagine the body thus strained to be moved as a rigid body (i. e. with its state of strain kept unchanged) till A_1 coincides with A, and C'_1 with C', keeping all the lines of the diagram still in the same plane. A'_1C_1 will take a position in CA' produced, as shown in the new diagram, and the original and the altered

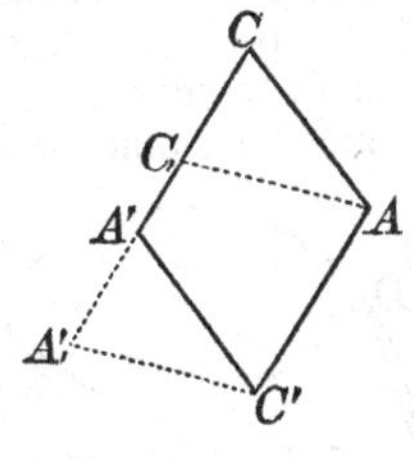

parallelogram will be on the same base AC', and between the same parallels AC' and CA'_1, and their other sides will be equally inclined on the two sides of a perpendicular to them. Hence, irrespectively of any rotation, or other absolute motion of the body not involving change of form or dimensions, the strain under consideration may be produced by holding fast and unaltered the plane of the body through AC', perpendicular to the plane of the diagram, and making every plane parallel to it slide, keeping the same distance, through a space proportional to this distance (i. e. different planes parallel to the fixed one slide through spaces proportional to their distances).

150. This kind of strain is called a *simple shear*. The plane of a shear is a plane perpendicular to the undistorted planes, and parallel to the lines of the relative motion. It has (1) the property that one set of parallel planes remain each unaltered in itself; (2) that another set of parallel planes remain each unaltered in itself. This other set is got when the first set and the degree or amount of shear are given, thus:—Let CC_1 be the motion of one point of one plane, relative to a plane KL held fixed—the diagram being in a plane of the shear. Bisect CC_1 in N. Draw NA perpendicular to it. A plane perpendicular to the plane of the diagram, initially through AC, and finally through AC_1, remains unaltered in its dimensions.

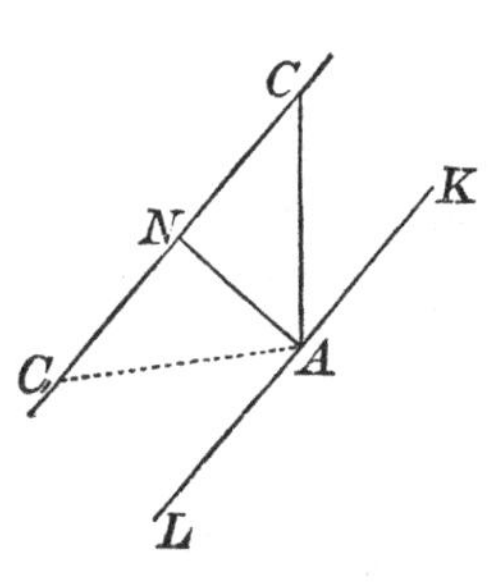

151. One set of parallel undistorted planes and the amount of their relative parallel shifting having been given, we have just seen how to find the other set. The shear may be otherwise viewed, and considered as a shifting of this second set of parallel planes, relative to any one of them. The amount of this relative shifting is of course equal to that of the first set, relatively to one of them.

152. The principal axes of a shear are the lines of maximum elongation and of maximum contraction respectively. They may be found from the preceding construction (§ 150), thus:—In the plane of the shear bisect the obtuse and acute angles between the planes destined not to become deformed. The former bisecting line is the principal axis of elongation, and the latter is the principal axis of contraction, in their initial positions. The former angle (obtuse) becomes equal to the latter, its supplement (acute), in the altered condition of the body, and the lines bisecting the altered angles are the principal axes of the strain in the altered body.

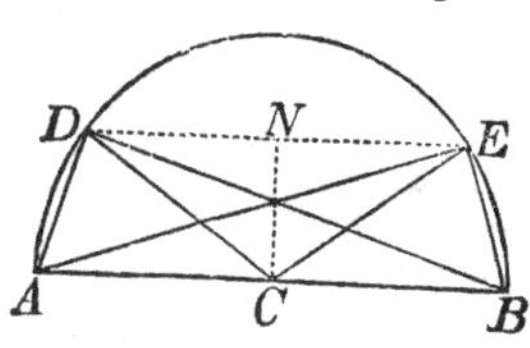

Otherwise, taking a plane of shear for the plane of the diagram, let AB be a line in which it is cut by one of either set of parallel planes of no distortion. On any portion *AB* of this as diameter, describe a semicircle. Through *C*, its middle point, draw, by the preceding construction, *CD* the initial, and *CE* the final, position of an unstretched line. Join *DA*, *DB*, *EA*, *EB*. *DA*, *DB* are the initial, and *EA*, *EB* the final, positions of the principal axes.

153. The ratio of a shear is the ratio of elongation and contraction of its principal axes. Thus if one principal axis is elongated in the ratio $1 : a$, and the other therefore (§ 148) contracted in the ratio $a : 1$, a is called the ratio of the shear. It will be convenient generally to reckon this as the ratio of elongation; that is to say, to make its numerical measure greater than unity.

In the diagram of § 152, the ratio of *DB* to *EB*, or of *EA* to *DA*, is the ratio of the shear.

154. The amount of a shear is the amount of relative motion per unit distance between planes of no distortion.

It is easily proved that this is equal to the excess of the ratio of the shear above its reciprocal.

155. The planes of no distortion in a simple shear are clearly the circular sections of the strain ellipsoid. In the ellipsoid of this case, be it remembered, the mean axis remains unaltered, and is a mean proportional between the greatest and the least axis.

156. If we now suppose all lines perpendicular to the plane of the shear to be elongated or contracted in any proportion, without altering lengths or angles in the plane of the shear, and if, lastly, we suppose every line in the body to be elongated or contracted in some other fixed ratio, we have clearly (§ 142) the most general possible kind of strain.

157. Hence any strain whatever may be viewed as compounded of a uniform dilatation in all directions, superimposed on a simple elongation in the direction of one principal axis superimposed on a simple shear in the plane of the two other principal axes.

158. It is clear that these three elementary component strains may be applied in any other order as well as that stated. Thus, if the simple elongation is made first, the body thus altered must get just the same shear in planes perpendicular to the line of elongation as the originally unaltered body gets when the order first stated is followed. Or the dilatation may be first, then the elongation, and finally the shear, and so on.

159. When the axes of the ellipsoid are lines of the body whose direction does not change, the strain is said to be *pure*, or unaccompanied by rotation. The strains we have already considered were pure strains accompanied by rotations.

160. If a body experience a succession of strains, each unaccompanied by rotation, its resulting condition will generally be producible by a strain and a rotation. From this follows the remarkable corollary that three pure strains produced one after another, in any piece of matter, each without rotation, may be so adjusted as to leave the body unstrained, but rotated through some angle about some axis. We shall have, later, most important and interesting applications to fluid motion, which will be proved to be instantaneously, or differentially, irrotational; but which may result in leaving a whole fluid mass merely turned round from its primitive position, as if it had been a rigid body. [The following elementary geometrical investigation, though not bringing out a thoroughly comprehensive view of the subject, affords a rigorous demonstration of the proposition, by proving it for a particular case.

Let us consider, as above (§ 150), a simple shearing motion. A point O being held fixed, suppose the matter of the body in a plane, cutting that of the diagram perpendicularly in CD, to move in this plane from right to left parallel to CD; and in other planes parallel to it let there be motions proportional to their distances from O. Consider first a shear from P to P_1; then from P_1 on to P_2; and let O be taken in a line through P_1, perpendicular to CD. During the shear from P to P_1 a point Q moves of course to Q_1 through a distance $QQ_1 = PP_1$. Choose Q midway between P and P_1, so that $P_1Q = QP = \frac{1}{2}P_1P$. Now, as we have seen above (§ 152), the line of the body, which is the principal axis of contraction in the shear from Q to Q_1, is OA, bisecting the angle QOE at the beginning, and OA_1, bisecting Q_1OE at the end, of the whole motion considered. The angle between these two lines is half the angle Q_1OQ, that is to say, is equal to P_1OQ. Hence, if the plane CD is rotated through an angle equal to P_1OQ, in the plane of the diagram, in the same way as the hands of a watch, during the shear from Q to Q_1, or, which is the same thing, the shear from P to P_1, this shear will be effected without final rotation of its principal axes. (Imagine the diagram turned round till OA_1 lies along OA. The actual and the newly imagined position of CD will show how this plane of the body has moved during such non-rotational shear.)

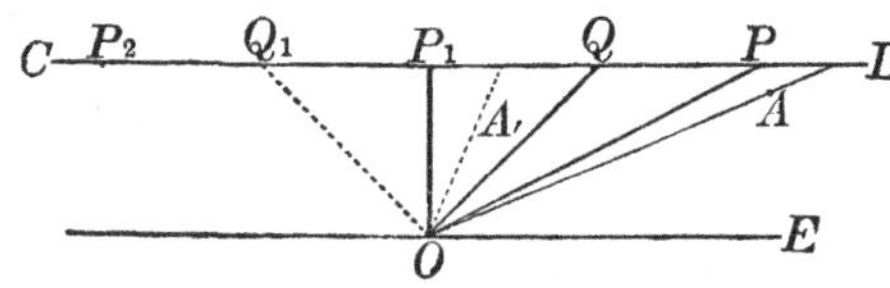

Now, let the second step, P_1 to P_2, be made so as to complete the whole shear, P to P_2, which we have proposed to consider. Such second partial shear may be made by the common shearing process parallel to the new position (imagined in the preceding parenthesis) of CD, and to make itself also non-rotational, as its predecessor has been made, we must turn further round, in the same direction, through an angle equal to Q_1OP_1. Thus in these two steps, each made non-rotational, we have turned the plane CD round through an angle equal to Q_1OQ. But now, we have a whole shear PP_2; and to make this as one non-rotational shear, we must turn CD through an angle P_1OP only, which is less than Q_1OQ by the excess of P_1OQ above QOP. Hence the resultant of the two shears, PP_1, P_1P_2, each separately deprived of rotation, is a single shear PP_2, and a rotation of its principal axes, in the direction of the hands of a watch through an angle equal to $QOP_1 - POQ$.

161. Make the two partial shears each non-rotationally. Return from their resultant in a single non-rotational shear: we conclude with the body unstrained, but turned through the angle $QOP_1 - POQ$, in the same direction as the hands of a watch.]

162. As there can be neither annihilation nor generation of matter in any natural motion or action, the whole quantity of a fluid within any space at any time must be equal to the quantity originally in that space, increased by the whole quantity that has entered it, and diminished by the whole quantity that has left it. This idea, when expressed in a perfectly comprehensive manner for every portion of a fluid in motion, constitutes what is commonly called the '*equation of continuity*.'

163. Two ways of proceeding to express this idea present themselves, each affording instructive views regarding the properties of fluids. In one we consider a definite portion of the fluid; follow it in its motions; and declare that the average density of the substance varies inversely as its volume. We thus obtain the equation of continuity in an integral form.

The form under which the equation of continuity is most commonly given, or the *differential equation of continuity*, as we may call it, expresses that the rate of diminution of the density bears to the density, at any instant, the same ratio as the rate of increase of the volume of an infinitely small portion bears to the volume of this portion at the same instant.

164. To find the differential equation of continuity, imagine a space fixed in the interior of a fluid, and consider the fluid which flows into this space, and the fluid which flows out of it, across different parts of its bounding surface, in any time. If the fluid is of the same density and incompressible, the whole quantity of matter in the space in question must remain constant at all times, and therefore the quantity flowing in must be equal to the quantity flowing out in any time. If, on the contrary, during any period of motion, more fluid enters than leaves the fixed space, there will be condensation of matter in that space; or if more fluid leaves than enters, there will be dilatation. The rate of augmentation of the average density of the fluid, per unit of time, in the fixed space in question, bears to the actual density, at any instant, the same ratio that the rate of acquisition of matter into that space bears to the whole matter in that space.

165. Several references have been made in preceding sections to the number of independent variables in a displacement, or to the degrees of *freedom* or *constraint* under which the displacement takes place. It may be well, therefore, to take a general (but cursory) view of this part of the subject itself.

166. A free point has *three* degrees of freedom, inasmuch as the most general displacement which it can take is resolvable into three, parallel respectively to any three directions, and independent of each other. It is generally convenient to choose these three directions of resolution at right angles to one another.

If the point be constrained to remain always on a given surface, *one* degree of constraint is introduced, or there are left but *two* degrees of freedom. For we may take the normal to the surface as one of three rectangular directions of resolution. No displacement can be effected parallel to it: and the other two displacements, at right angles to each other, in the tangent plane to the surface, are independent.

If the point be constrained to remain on *each* of two surfaces, it loses two degrees of freedom, and there is left but one. In fact, it is constrained to remain on the curve which is common to both surfaces, and along a curve there is at each point but one direction of displacement.

167. Taking next the case of a free rigid system, we have evidently *six* degrees of freedom to consider—*three* independent displacements or translations in rectangular directions as a point has, and three independent rotations about three mutually rectangular axes.

If it have one point fixed, the system loses *three* degrees of freedom; in fact, it has now only the rotations above mentioned.

This fixed point may be, and in general is, a point of a continuous surface of the body in contact with a continuous fixed surface. These surfaces may be supposed 'perfectly rough,' so that sliding may be impossible.

If a second point be fixed, the body loses *two* more degrees of freedom, and keeps only one freedom to rotate about the line joining the two fixed points.

If a third point, not in a line with the other two, be fixed, the body is fixed.

168. If one point of the rigid system is forced to remain on a smooth surface, *one* degree of freedom is lost; there remain *five*, two displacements in the tangent plane to the surface, and three rotations. As an additional degree of freedom is lost by each successive limitation of a point in the body to a smooth surface, *six* such conditions completely determine the position of the body. Thus if six points properly chosen on the barrel and stock of a rifle be made to rest on six convex portions of the surface of a fixed rigid body, the rifle may be replaced any number of times in precisely the same position, for the purpose of testing its accuracy.

169. If one point be constrained to remain in a curve, there remain four degrees of freedom.

If two points be constrained to remain in given curves, there are four degrees of constraint, and we have left two degrees of freedom. One of these may be regarded as being a simple rotation about the line joining the constrained parts, a motion which, it is clear, the body is free to receive. It may be shown that the other possible motion is of the most general character for one degree of freedom; that is to say, translation and rotation in any fixed proportions, as of the nut of a screw.

If one line of a rigid system be constrained to remain parallel to itself, as for instance, if the body be a three-legged stool standing on a perfectly smooth board fixed to a common window, sliding in its frame with perfect freedom, there remain *three* displacements and one rotation.

But we need not farther pursue this subject, as the number of combinations that might be considered is almost endless; and those already given suffice to show how simple is the determination of the degrees of freedom or constraint in any case that may present itself.

170. One degree of constraint of the most general character, is not producible by constraining one point of the body to a curve surface; but it consists in stopping one line of the body from longitudinal motion, except accompanied by rotation round this line, in fixed proportion to the longitudinal motion. Every other motion being left unimpeded; there remains free rotation about any axis perpendicular to that line (two degrees freedom); and translation in any direction perpendicular to the same line (two degrees freedom). These last four, with the one degree of freedom to screw, constitute the five degrees of freedom, which, with one degree of constraint, make up the six elements. This condition is realized in the following mechanical arrangement, which seems the simplest that can be imagined for the purpose:—

Let a screw be cut on one shaft of a Hooke's joint, and let the other shaft be joined to a fixed shaft by a second Hooke's joint. A nut turning on that screw-shaft has the most general kind of motion admitted when there is one degree of constraint. Or it is subjected to just one degree of constraint of the most general character. It has five degrees of freedom; for it may move, 1st, by screwing on its shaft, the two Hooke's joints being at rest; 2nd, it may rotate about either axis of the first Hooke's joint, or any axis in their plane (two more degrees of freedom: being freedom to rotate about two axes through one point); 3rd, it may, by the two Hooke's joints, each bending, have translation without rotation in any direction perpendicular to the link, or shaft between the two Hooke's joints (two more degrees of freedom). But it cannot have a motion of translation parallel to the line of the link without a definite proportion of rotation round this line; nor can it have rotation round this line without a definite proportion of translation parallel to it.

CHAPTER II.

DYNAMICAL LAWS AND PRINCIPLES.

171. In the preceding chapter we considered as a subject of pure geometry the motion of points, lines, surfaces, and volumes, whether taking place with or without change of dimensions and form; and the results we there arrived at are of course altogether independent of the idea of *matter*, and of the *forces* which matter exerts. We have heretofore assumed the existence merely of motion, distortion, etc.; we now come to the consideration, not of how we *might* consider such motion, etc., to be produced, but of the *actual* causes which in the material world *do* produce them. The axioms of the present chapter must therefore be considered to be due to actual experience, in the shape either of observation or experiment. How such experience is to be conducted will form the subject of a subsequent chapter.

172. We cannot do better, at all events in commencing, than follow Newton somewhat closely. Indeed the introduction to the *Principia* contains in a most lucid form the general foundations of dynamics. The *Definitiones* and *Axiomata, sive Leges Motûs*, there laid down, require only a few amplifications and additional illustrations, suggested by subsequent developments, to suit them to the present state of science, and to make a much better introduction to dynamics than we find in even some of the best modern treatises.

173. We cannot, of course, give a definition of *Matter* which will satisfy the metaphysician; but the naturalist may be content to know matter as *that which can be perceived by the senses*, or as *that which can be acted upon by, or can exert, force.* The latter, and indeed the former also, of these definitions involves the idea of *Force*, which, in point of fact, is a direct object of sense; probably of all our senses, and certainly of the 'muscular sense.' To our chapter on Properties of Matter we must refer for further discussion of the question, *What is matter?*

174. The *Quantity of Matter* in a body, or, as we now call it, the *Mass* of a body, is proportional, according to Newton, to the *Volume* and the *Density* conjointly. In reality, the definition gives us the meaning of density rather than of mass; for it shows us that if twice the original quantity of matter, air for example, be forced into a vessel of given capacity, the density will be doubled, and so on. But it also shows us that, of matter of uniform density, the mass or quantity is proportional to the volume or space it occupies.

Let M be the mass, ρ the density, and V the volume, of a homogeneous body. Then

$$M = V\rho;$$

if we so take our units that unit of mass is that of unit volume of a body of unit density.

If the density be not uniform, the equation

$$M = V\rho$$

gives the *Average* (§ 26) density; or, as it is usually called, the *Mean* density, of the body.

It is worthy of particular notice that, in this definition, Newton says, if there be anything which *freely* pervades the interstices of all bodies, this is *not* taken account of in estimating their Mass or Density.

175. Newton further states, that a practical measure of the mass of a body is its *Weight.* His experiments on pendulums, by which he establishes this most important remark, will be described later, in our chapter on Properties of Matter.

As will be presently explained, the unit mass most convenient for British measurements is an imperial pound of matter.

176. The *Quantity of Motion,* or the *Momentum,* of a rigid body moving without rotation is proportional to its mass and velocity conjointly. The whole motion is the sum of the motions of its several parts. Thus a doubled mass, or a doubled velocity, would correspond to a double quantity of motion; and so on.

Hence, if we take as unit of momentum the momentum of a unit of matter moving with unit velocity, the momentum of a mass M moving with velocity v is Mv.

177. *Change of Quantity of Motion,* or *Change of Momentum,* is proportional to the mass moving and the change of its velocity conjointly.

Change of velocity is to be understood in the general sense of § 31. Thus, in the figure of that section, if a velocity represented by OA be changed to another represented by OC, the change of velocity is represented in magnitude and direction by AC.

178. *Rate of Change of Momentum,* or *Acceleration of Momentum,* is proportional to the mass moving and the acceleration of its velocity conjointly. Thus (§ 44) the rate of change of momentum of a falling body is constant, and in the vertical direction. Again (§ 36) the rate of change of momentum of a mass M, describing a circle of radius R, with uniform velocity V, is $\frac{MV^2}{R}$, and is directed to the centre of the circle; that is to say, it depends upon a change of direction, not a change of speed, of the motion.

179. The *Vis Viva,* or *Kinetic Energy,* of a moving body is proportional to the mass and the square of the velocity, conjointly. If we adopt the same units of mass and velocity as before, there is particular advantage in defining kinetic energy as *half* the product of the mass and the square of its velocity.

180. *Rate of Change of Kinetic Energy* (when defined as above) is the product of the velocity into the component of acceleration of momentum in the direction of motion.

Suppose the velocity of a mass M to be changed from v to $v_{\prime}$ in any time τ; the rate at which the kinetic energy has changed is

$$\frac{1}{\tau} \cdot \tfrac{1}{2} M (v_{\prime}^2 - v^2) = \frac{1}{\tau} M (v_{\prime} - v) \cdot \tfrac{1}{2} (v_{\prime} + v).$$

Now $\frac{1}{\tau} M (v_{\prime} - v)$ is the rate of change of momentum in the direction of motion, and $\frac{1}{2} (v_{\prime} + v)$ is equal to v, if τ be infinitely small. Hence the above statement. It is often convenient to use Newton's Fluxional notation for the rate of change of any quantity per unit of time. In this notation (§ 28) $\dot{v}$ stands for $\frac{1}{\tau} (v_{\prime} - v)$; so that the rate of change of $\frac{1}{2} M v^2$, the kinetic energy, is $M\dot{v}.v$. (See also §§ 229, 241.)

181. It is to be observed that, in what precedes, with the exception of the definition of density, we have taken no account of the dimensions of the moving body. This is of no consequence so long as it does not rotate, and so long as its parts preserve the same relative positions amongst one another. In this case we may suppose the whole of the matter in it to be condensed in one point or particle. We thus speak of a *material particle*, as distinguished from a *geometrical point*. If the body rotate, or if its parts change their relative positions, then we cannot choose any one point by whose motions alone we may determine those of the other points. In such cases the momentum and change of momentum of the whole body in any direction are, the sums of the momenta, and of the changes of momentum, of its parts, in these directions; while the kinetic energy of the whole, being non-directional, is simply the sum of the kinetic energies of the several parts or particles.

182. Matter has an innate power of resisting external influences, so that every body, so far as it can, remains at rest, or moves uniformly in a straight line.

This, the *Inertia* of matter, is proportional to the quantity of matter in the body. And it follows that some *cause* is requisite to disturb a body's uniformity of motion, or to change its direction from the natural rectilinear path.

183. *Impressed Force*, or *Force* simply, is any cause which tends to alter a body's natural state of rest, or of uniform motion in a straight line.

Force is wholly expended in the *Action* it produces; and the body, after the force ceases to act, retains by its inertia the direction of motion and the velocity which were given to it. Force may be of divers kind, as pressure, or gravity, or friction, or any of the attractive or repulsive actions of electricity, magnetism, etc.

184. The three elements specifying a force, or the three elements which must be known, before a clear notion of the force under consideration can be formed, are, its place of application, its direction, and its magnitude.

(*a*) The place of application of a force. The first case to be considered is that in which the place of application is a point. It has been shown already in what sense the term 'point' is to be taken, and, therefore, in what way a force may be imagined as acting at a point. In reality, however, the place of application of a force is always either a surface or a space of three dimensions occupied by matter. The point of the finest needle, or the edge of the sharpest knife, is still a surface, and acts as such on the bodies to which it may be applied. Even the most rigid substances, when brought together, do not touch at a point merely, but mould each other so as to produce a surface of application. On the other hand, gravity is a force of which the place of application is the whole matter of the body whose weight is considered; and the smallest particle of matter that has weight occupies some finite portion of space. Thus it is to be remarked, that there are two kinds of force, distinguishable by their place of application—force whose place of application is a surface, and force whose place of application is a solid. When a heavy body rests on the ground, or on a table, force of the second character, acting downwards, is balanced by force of the first character acting upwards.

(*b*) The second element in the specification of a force is its direction. The direction of a force is the line in which it acts. If the place of application of a force be regarded as a point, a line through that point, in the direction in which the force tends to move the body, is the direction of the force. In the case of a force distributed over a surface, it is frequently possible and convenient to assume a single point and a single line, such that a certain force acting at that point in that line would produce the same effect as is really produced.

(*c*) The third element in the specification of a force is its magnitude. This involves a consideration of the method followed in dynamics for measuring forces. Before measuring anything it is necessary to have a unit of measurement, or a standard to which to refer, and a principle of numerical specification, or a mode of referring to the standard. These will be supplied presently. See also § 224, below.

185. The *Measure of a Force* is the quantity of motion which it produces in unit of time.

The reader, who has been accustomed to speak of a force of so many pounds, or so many tons, may be reasonably startled when he finds that Newton gives no countenance to such expressions. The method is not correct unless it be specified at what part of the earth's surface the pound, or other definite quantity of matter named, is to be weighed; for the weight of a given quantity of matter differs in different latitudes.

It is often, however, convenient to use instead of the absolute unit (§ 188), the gravitation unit—which is simply the weight of unit mass. It must, of course, be specified in what latitude the observation is made. Thus, let W be the mass of a body in pounds; g the velocity it would acquire in falling for a second under the influence of its weight, or the earth's attraction diminished by centrifugal force; and P its weight measured in kinetic or absolute units. We have

$$P = Wg.$$

The force of gravity on the body, *in gravitation units*, is W.

186. According to the common system followed in modern mathematical treatises on dynamics, the unit of mass is g times the mass of the standard or unit weight. This definition, giving a varying and a very unnatural unit of mass, is exceedingly inconvenient: and its clumsiness is in great contrast to the clear and simple accuracy of the absolute method as stated above, to which we shall uniformly adhere, except when we wish, in describing results, to state forces in terms of the gravitation unit, as the *vernacular* of engineers in any locality. In reality, standards of weight are *masses*, not *forces*. It is better, though less usual, to call them *standard masses* than standard weights; as weight properly means force, and ambiguity is the worst fault of language. They are employed primarily in commerce for the purpose of measuring out a definite *quantity* of matter; not an amount of matter which shall be attracted by the earth with a given force.

Whereas a merchant, with a balance and a set of standard masses, would give his customers the same quantity of matter however the earth's attraction might vary, depending as he does upon *masses* for his measurement; another, using a spring balance, would defraud his customers in high latitudes, and himself in low, if his instrument (which depends on forces and not on masses) were correctly adjusted in London.

It is a secondary application of our standards of mass to employ them for the measurement of *forces*, such as steam pressures, muscular power, etc. In all cases where great accuracy is required, the results obtained by such a method have to be reduced to what they would have been if the measurements of force had been made by means of a perfect spring-balance, graduated so as to indicate the forces of gravity on the standard masses in some conventional locality.

It is therefore very much simpler and better to take the imperial pound, or other national or international standard mass, as, for instance, the gramme (see Chapter IV.), as the unit of mass, and to derive from it, according to Newton's definition above, the unit of force.

187. The formula, deduced by Clairault from observation, and a certain theory regarding the figure and density of the earth, may be employed to calculate the most probable value of the apparent force of gravity, being the resultant of true gravitation and centrifugal force, in any locality where no pendulum observation of sufficient accuracy has been made. This formula, with the two co-efficients which it involves, corrected according to modern pendulum observations, is as follows:—

Let G be the apparent force of gravity on a unit mass at the equator, and g that in any latitude λ; then

$$g = G\,(1 + \cdot 00513 \sin^2 \lambda).$$

The value of G, in terms of the absolute unit, to be explained immediately, is

$$32\cdot 088.$$

According to this formula, therefore, polar gravity will be

$$g = 32\cdot 088 \times 1\cdot 00513 = 32\cdot 252.$$

188. As gravity does not furnish a definite standard, independent of locality, recourse must be had to something else. The principle of measurement indicated as above by Newton, but first introduced practically by Gauss in connection with national standard masses, furnishes us with what we want. According to this principle, *the standard or unit force is that force which, acting on a national standard unit of matter during the unit of time, generates the unit of velocity.*

This is known as Gauss' absolute unit; absolute, because it furnishes a standard force independent of the differing amounts of gravity at different localities.

189. The absolute unit depends on the unit of matter, the unit of time, and the unit of velocity; and as the unit of velocity depends on the unit of space and the unit of time, there is, in the definition, a single reference to mass and space, but a *double* reference to time; and this is a point that must be particularly attended to.

190. The unit of mass may be the British imperial pound or, better, the gramme; the unit of space the British standard foot or, better, the centimetre; and the unit of time the mean solar second.

We accordingly define the British absolute unit force as 'the force which, acting on one pound of matter for one second, generates a velocity of one foot per second.'

191. To render this standard intelligible, all that has to be done is to find how many absolute units will produce, in any particular locality, the same effect as the force of gravity on a given mass. The way to do this is to measure the effect of gravity in producing acceleration on a body unresisted in any way. The most accurate method is indirect, by means of the pendulum. The result of pendulum experiments made at Leith Fort, by Captain Kater, is, that the velocity acquired by a body falling unresisted for one second is at that place 32·207 feet per second. The preceding formula gives exactly 32·2, for the latitude 55° 35′, which is approximately that of Edinburgh. The variation in the force of gravity for one degree of difference of latitude about the latitude of Edinburgh is only ·0000832 of its own amount. It is nearly the same, though somewhat more, for every degree of latitude southwards, as far as the southern limits of the British Isles. On the other hand, the variation per degree would be sensibly less, as far north as the Orkney and Shetland Isles. Hence the augmentation of gravity per degree from south to north throughout the British Isles is at most about $\frac{1}{12000}$ of its whole amount in any locality. The average for the whole of Great Britain and Ireland differs certainly but little from 32·2. Our present application is, that the force of gravity at Edinburgh is 32·2 times the force which, acting on a pound for a second, would generate a velocity of one foot per second; in other words, 32·2 is the number of absolute units which measures the weight of a pound in this latitude. Thus, speaking very roughly, the British absolute unit of force is equal to the weight of about half an ounce.

192. Forces (since they involve only direction and magnitude) may be represented, as velocities are, by straight lines in their directions, and of lengths proportional to their magnitudes, respectively.

Also the laws of composition and resolution of any number of forces acting at the same point, are, as we shall show later (§ 221), the same as those which we have already proved to hold for velocities; so that with the substitution of force for velocity, §§ 30, 31 are still true.

193. The *Component* of a force in any direction, sometimes called the *Effective Component* in that direction, is therefore found by multiplying the magnitude of the force by the cosine of the angle between the directions of the force and the component. The remaining component in this case is perpendicular to the other.

It is very generally convenient to resolve forces into components parallel to three lines at right angles to each other; each such resolution being effected by multiplying by the cosine of the angle concerned.

194. [If any number of points be placed in any positions in space, another can be found, such that its distance from any plane whatever is the mean of their distances from that plane; and if one or more of the given points be in motion, the velocity of the mean point perpendicular to the plane is the mean of the velocities of the others in the same direction.

If we take two points A_1, A_2, the middle point, P_2, of the line joining them is obviously distant from any plane whatever by a quantity equal to the mean (in this case the half sum or difference as they are on the same or on opposite sides) of their distances from that plane. Hence *twice* the distance of P_2 from any plane is equal to the (algebraic) sum of the distances of A_1, A_2 from it. Introducing a third point A_3, if we join A_3P_2 and divide it in P_3 so that $A_3P_3 = 2P_3P_2$, three times the distance of P_3 from any plane is equal to the sum of the distance of A_3 and twice that of P_2 from the same plane: i. e. to the sum of the distances of A_1, A_2, and A_3 from it; or its distance is the mean of theirs. And so on for any number of points. The proof is exceedingly simple. Thus suppose P_n to be the *mean* of the first n points $A_1, A_2, \ldots A_n$; and A_{n+1} any other point. Divide $A_{n+1}P_n$ in P_{n+1} so that $A_{n+1}P_{n+1}=nP_{n+1}P_n$. Then from P_n, P_{n+1}, A_{n+1}, draw perpendiculars to any plane, meeting it in S, T, V. Draw P_nQR parallel to STV. Then $QP_{n+1}:RA_{n+1}::P_nP_{n+1}:P_nA_{n+1}::1:n+1$. Hence $\overline{n+1}QP_{n+1}=RA_{n+1}$. Add to these $\overline{n+1}QT$ and its equal nP_nS+RV, and we get

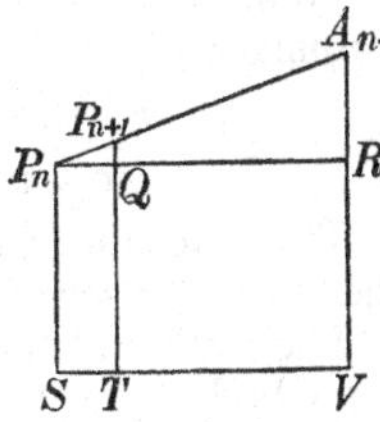

$$\overline{n+1}(QP_{n+1}+QT)=nP_nS+RV+RA_{n+1},$$

$$\text{i.e. } \overline{n+1}\,P_{n+1}T=nP_nS+A_{n+1}V.$$

In words, $\overline{n+1}$ times the distance of P_{n+1} from any plane is equal to that of A_{n+1} with n times that of P_n, i.e. equal to the sum of the distances of A_1, A_2,....A_{n+1} from the plane. Thus if the proposition be true for any number of points, it is true for one more—and so on—but it is obviously true for two, hence for three, and therefore generally. And it is obvious that the order in which the points are taken is immaterial.

As the distance of this point from any plane is the mean of the distances of the given ones, the rate of increase of that distance, i.e. the velocity perpendicular to the plane, must be the mean of the rates of increase of their distances—i.e. the mean of their velocities perpendicular to the plane.]

195. The *Centre of Inertia* or *Mass* of a system of equal material points (whether connected with one another or not) is the point whose distance is equal to their average distance from any plane whatever (§ 194).

A group of material points of unequal masses may always be imagined as composed of a greater number of equal material points, because we may imagine the given material points divided into different numbers of very small parts. In any case in which the magnitudes of the given masses are incommensurable, we may approach as near as we please to a rigorous fulfilment of the preceding statement, by making the parts into which we divide them sufficiently small.

On this understanding the preceding definition may be applied to define the centre of inertia of a system of material points, whether given equal or not. The result is equivalent to this :—

The centre of inertia of any system of material points whatever (whether rigidly connected with one another, or connected in any way, or quite detached), is a point whose distance from any plane is equal to the sum of the products of each mass into its distance from the same plane divided by the sum of the masses.

We also see, from the proposition stated above, that a point whose distance from three rectangular planes fulfils this condition, must fulfil this condition also for every other plane.

The co-ordinates of the centre of inertia, of masses w_1, w_2, etc., at points (x_1, y_1, z_1), (x_2, y_2, z_2), etc., are given by the following formulae:—

$$\bar{x} = \frac{w_1x_1 + w_2x_2 + \text{etc.}}{w_1 + w_2 + \text{etc.}} = \frac{\Sigma wx}{\Sigma w}, \quad \bar{y} = \frac{\Sigma wy}{\Sigma w}, \quad \bar{z} = \frac{\Sigma wz}{\Sigma w}.$$

These formulae are perfectly general, and can easily be put into the particular shape required for any given case.

The Centre of Inertia or Mass is thus a perfectly definite point in every body, or group of bodies. The term *Centre of Gravity* is often very inconveniently used for it. The theory of the resultant action of gravity, which will be given under Abstract Dynamics, shows that, except in a definite class of distributions of matter, there is no fixed point which can properly be called the Centre of Gravity of a rigid body. In ordinary cases of terrestrial gravitation, however, an approximate solution is available, according to which, in common parlance, the term *Centre of Gravity* may be used as equivalent to *Centre of Inertia;* but it must be carefully remembered that the fundamental ideas involved in the two definitions are essentially different.

The second proposition in § 194 may now evidently be stated thus:—The sum of the momenta of the parts of the system in any direction is equal to the momentum in the same direction of a mass equal to the sum of the masses moving with a velocity equal to the velocity of the centre of inertia.

196. The mean of the squares of the distances of the centre of inertia, I, from each of the points of a system is less than the mean of the squares of the distance of any other point, O, from them by the square of OI. Hence the centre of inertia is the point the sum of the squares of whose distances from any given points is a minimum.

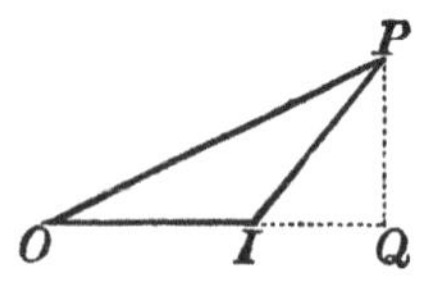

For $OP^2 = OI^2 + IP^2 + 2OI \cdot IQ$, P being any one of the points and PQ perpendicular to OI. But IQ is the distance of P from a plane through I perpendicular to OQ. Hence the mean of all distances, IQ, is zero. Hence

(mean of IP^2) = (mean of OP^2) $- OI^2$, which is the proposition.

197. Again, the mean of the squares of the distances of the points of the system from any line, exceeds the corresponding quantity for a parallel line through the centre of inertia, by the square of the distance between these lines.

For in the above figure, let the plane of the paper represent a plane through I perpendicular to these lines, O the point in which the first line meets it, P the point in which it is met by a parallel line through any one of the points of the system. Draw, as before, PQ perpendicular to OI. Then PI is the perpendicular distance, from the axis through I, of the point of the system considered, PO is its distance from the first axis, OI the distance between the two axes.

Then, as before,

$$(\text{mean of } OP^2)=OI^2+(\text{mean of } IP^2);$$

since the mean of IQ is still zero, IQ being the distance of a point of the system from the plane through I perpendicular to OI.

198. If the masses of the points be unequal, it is easy to see (as in § 195) that the first of these theorems becomes—

The sum of the squares of the distances of the parts of a system from any point, each multiplied by the mass of that part, exceeds the corresponding quantity for the centre of inertia by the product of the square of the distance of the point from the centre of inertia, by the whole mass of the system.

Also, the sum of the products of the mass of each part of a system by the square of its distance from any axis is called the *Moment of Inertia* of the system about this axis; and the second proposition above is equivalent to—

The moment of inertia of a system about any axis is equal to the moment of inertia about a parallel axis through the centre of inertia, I, together with the moment of inertia, about the first axis, of the whole mass supposed condensed at I.

199. The *Moment* of any physical agency is the numerical measure of its importance. Thus, the moment of inertia of a body round an axis (§ 198) means the importance of its inertia relatively to rotation round that axis. Again, the moment of a force round a point or round a line (§ 46), signifies the measure of its importance as regards producing or balancing rotation round that point or round that line.

It is often convenient to represent the moment of a force by a line numerically equal to it, drawn through the vertex of the triangle representing its magnitude, perpendicular to its plane, through the front of a watch held in the plane with its centre at the point, and facing so that the force tends to turn round this point in a direction opposite to the hands. The moment of a force round any axis is the moment of its component in any plane perpendicular to the axis, round the point in which the plane is cut by the axis. Here we imagine the force resolved into two components, one parallel to the axis, which is ineffective so far as rotation round the axis is concerned; the other perpendicular to the axis (that is to say, having its line in any plane perpendicular to the axis). This latter component may be called the effective component of the force, with reference to rotation round the axis. And its moment round the axis may be defined as its moment round the nearest point of the axis, which is equivalent to the preceding definition. It is clear that the moment of a force round any axis, is equal to the area of the projection on any plane perpendicular to the axis, of the figure representing its moment round any point of the axis.

200. [The projection of an area, plane or curved, on any plane, is the area included in the projection of its bounding line.

If we imagine an area divided into any number of parts, the projections of these parts on any plane make up the projection of the whole.

But in this statement it must be understood that the areas of partial projections are to be reckoned as positive if particular sides, which, for brevity, we may call the outside of the projected area and the front of the plane of projection, face the same way, and negative if they face oppositely.

Of course if the projected surface, or any part of it, be a plane area at right angles to the plane of projection, the projection vanishes. The projections of any two shells having a common edge, on any plane, are equal. The projection of a closed surface (or a shell with evanescent edge), on any plane, is nothing.

Equal areas in one plane, or in parallel planes, have equal projections on any plane, whatever may be their figures.

Hence the projection of any plane figure, or of any shell edged by a plane figure, on another plane, is equal to its area, multipled by the cosine of the angle at which its plane is inclined to the plane of projection. This angle is acute or obtuse, according as the outside of the projected area, and the front of the plane of projection, face on the whole towards the same parts, or oppositely. Hence lines representing, as above described, moments about a point in different planes, are to be compounded as forces are. See an analogous theorem in § 107.]

201. A *Couple* is a pair of equal forces acting in dissimilar directions in parallel lines. The *Moment* of a couple is the sum of the moments of its forces about any point in their plane, and is therefore equal to the product of either force into the shortest distance between their directions. This distance is called the *Arm* of the couple.

The *Axis of a Couple* is a line drawn from any chosen point of reference perpendicular to the plane of the couple, of such magnitude and in such direction as to represent the magnitude of the moment, and to indicate the direction in which the couple tends to turn. The most convenient rule for fulfilling the latter condition is this:—Hold a watch with its centre at the point of reference, and with its plane parallel to the plane of the couple. Then, according as the motion of the hands is contrary to, or along with the direction in which the couple tends to turn, draw the axis of the couple through the face or through the back of the watch. It will be found that a couple is completely represented by its axis, and that couples are to be resolved and compounded by the same geometrical constructions performed with reference to their axes as forces or velocities, with reference to the lines directly representing them.

202. By introducing in the definition of moment of velocity (§ 46) the mass of the moving body as a factor, we have an important element of dynamical science, the *Moment of Momentum*. The laws of composition and resolution are the same as those already explained.

203. [If the point of application of a force be displaced through a small space, the resolved part of the displacement in the direction of the force has been called its *Virtual Velocity*. This is positive or negative according as the virtual velocity is in the same, or in the opposite, direction to that of the force.

The product of the force, into the virtual velocity of its point of application, has been called the *Virtual Moment* of the force. These terms we have introduced since they stand in the history and developments of the science; but, as we shall show further on, they are inferior substitutes for a far more useful set of ideas clearly laid down by Newton.]

204. A force is said to *do work* if its place of application has a positive component motion in its direction; and the work done by it is measured by the product of its amount into this component motion.

Generally, unit of work is done by unit force acting through unit space. In lifting coals from a pit, the amount of work done is proportional to the weight of the coals lifted; that is, to the force overcome in raising them; and also to the height through which they are raised. The unit for the measurement of work adopted in practice by British engineers, is that required to overcome a force equal to the weight of a pound through the space of a foot; and is called a *Foot-Pound.* (See § 185.)

In purely scientific measurements, the unit of work is not the foot-pound, but the kinetic unit force (§ 190) acting through unit of space. Thus, for example, as we shall show further on, this unit is adopted in measuring the work done by an electric current, the units for electric and magnetic measurements being founded upon the kinetic unit force.

If the weight be raised obliquely, as, for instance, along a smooth inclined plane, the space through which the force has to be overcome is increased in the ratio of the length to the height of the plane; but the force to be overcome is not the whole weight, but only the resolved part of the weight parallel to the plane; and this is less than the weight in the ratio of the height of the plane to its length. By multiplying these two expressions together, we find, as we might expect, that the amount of work required is unchanged by the substitution of the oblique for the vertical path.

205. Generally, for any force, the work done during an indefinitely small displacement of the point of application is the virtual moment of the force (§ 203), or is the product of the resolved part of the force in the direction of the displacement into the displacement.

From this it appears, that if the motion of the point of application be always perpendicular to the direction in which a force acts, such a force does no work. Thus the mutual normal pressure between a fixed and moving body, the tension of the cord to which a pendulum bob is attached, or the attraction of the sun on a planet if the planet describe a circle with the sun in the centre, are all instances in which no work is done by the force.

206. The work done by a force, or by a couple, upon a body turning about an axis, is the product of the moment of either into the angle (in circular measure) through which the body acted on turns, if the moment remains the same in all positions of the body. If the moment be variable, the above assertion is only true for indefinitely small displacements, but may be made accurate by employing the proper *average* moment of the force or of the couple. The proof is obvious.

207. Work done on a body by a force is always shown by a corresponding increase of vis viva, or kinetic energy, if no other forces act on the body which can do work or have work done against them. If work be done against any forces, the increase of kinetic energy is less than in the former case by the amount of work so done. In virtue of this, however, the body possesses an equivalent in the form of *Potential Energy* (§ 239), if its physical conditions are such that these forces will act equally, and in the same directions, if the motion of the system is reversed. Thus there may be no change of kinetic energy produced, and the work done may be wholly stored up as potential energy.

Thus a weight requires work to raise it to a height, a spring requires work to bend it, air requires work to compress it, etc.; but a raised weight, a bent spring, compressed air, etc., are *stores* of energy which can be made use of at pleasure.

208. In what precedes we have given some of Newton's *Definitiones* nearly in his own words; others have been enunciated in a form more suitable to modern methods; and some terms have been introduced which were invented subsequent to the publication of the *Principia*. But the *Axiomata, sive Leges Motûs*, to which we now proceed, are given in Newton's own words. The two centuries which have nearly elapsed since he first gave them have not shown a necessity for any addition or modification. The first two, indeed, were discovered by Galileo: and the third, in some of its many forms, was known to Hooke, Huyghens, Wallis, Wren, and others, before the publication of the *Principia*. Of late there has been a tendency to divide the second law into two, called respectively the second and third, and to ignore the third entirely, though using it *directly* in every dynamical problem; but all who have done so have been forced *indirectly* to acknowledge the incompleteness of their substitute for Newton's system, by introducing as an axiom what is called D'Alembert's principle, which is really a deduction from Newton's rejected third law. Newton's own interpretation of his third law directly points out not only D'Alembert's principle, but also the modern principles of Work and Energy.

209. An Axiom is a proposition, the truth of which must be admitted as soon as the terms in which it is expressed are clearly understood. And, as we shall show in our chapter on 'Experience,' physical axioms are axiomatic to those who have sufficient knowledge of physical phenomena to enable them to understand perfectly what is asserted by them. Without further remark we shall give Newton's Three Laws; it being remembered that, as the properties of matter *might* have been such as to render a totally different set of laws axiomatic, these laws must be considered as resting on convictions drawn from observation and experiment, *not* on intuitive perception.

210. Lex I. *Corpus omne perseverare in statu suo quiescendi vel movendi uniformiter in directum, nisi quatenus illud à viribus impressis cogitur statum suum mutare.*

Every body continues in its state of rest or of uniform motion in a straight line, except in so far as it may be compelled by impressed forces to change that state.

211. The meaning of the term *Rest*, in physical science, cannot be absolutely defined, inasmuch as absolute rest nowhere exists in nature. If the universe of matter were finite, its centre of inertia might fairly be considered as absolutely at rest; or it might be imagined to be moving with any uniform velocity in any direction whatever through infinite space. But it is remarkable that the first law of motion enables us (§ 215, below) to explain what may be called *directional* rest. Also, as will be seen farther on, a perfectly smooth spherical body, made up of concentric shells, each of uniform material and density throughout, if made to revolve about an axis, will, *in spite of impressed forces*, revolve with uniform angular velocity, and will maintain its axis of revolution in an absolutely fixed direction. Or, as will soon be shown (§ 233), the plane in which the moment of momentum of the universe (if finite) round its centre of inertia is the greatest, which is clearly determinable from the actual motions at any instant, is fixed in direction in space.

212. We may logically convert the assertion of the first law of motion as to velocity into the following statements:—

The times during which any particular body, not compelled by force to alter the speed of its motion, passes through equal spaces, are equal. And, again—Every other body in the universe, not compelled by force to alter the speed of its motion, moves over equal spaces in successive intervals, during which the particular chosen body moves over equal spaces.

213. The first part merely expresses the convention universally adopted for the measurement of *Time*. The earth in its rotation about its axis, presents us with a case of motion in which the condition of not being compelled by force to alter its speed, is more nearly fulfilled than in any other which we can easily or accurately observe. And the numerical measurement of time practically rests on defining *equal intervals of time*, as *times during which the earth turns through equal angles*. This is, of course, a mere convention, and not a law of nature; and, as we now see it, is a part of Newton's first law.

214. The remainder of the law is not a convention, but a great truth of nature, which we may illustrate by referring to small and trivial cases as well as to the grandest phenomena we can conceive.

A curling-stone, projected along a horizontal surface of ice, travels equal distances, except in so far as it is retarded by friction and by the resistance of the air, in successive intervals of time during which the earth turns through equal angles. The sun moves through equal portions of interstellar space in times during which the earth turns through equal angles, except in so far as the resistance of interstellar matter, and the attraction of other bodies in the universe, alter his speed and that of the earth's rotation.

215. If two material points be projected from one position, A, at the same instant with any velocities in any directions, and each left to move uninfluenced by force, the line joining them will be always parallel to a fixed direction. For the law asserts, as we have seen, that $AP : AP' :: AQ : AQ'$, if P, Q, and again P', Q', are simultaneous positions; and therefore PQ is parallel to $P'Q'$. Hence if four material points O, P, Q, R are all projected at one instant from one position, OP, OQ, OR are fixed directions of reference ever after. But, practically, the determination of fixed directions in space (§ 233) is made to depend upon the rotation of groups of particles exerting forces on each other, and thus involves the Third Law of Motion.

216. The whole law is singularly at variance with the tenets of the ancient philosophers, who maintained that circular motion is perfect.

The last clause, '*nisi quatenus*,' etc., admirably prepares for the introduction of the second law, by conveying the idea that *it is force alone which can produce a change of motion.* How, we naturally inquire, does the change of motion produced depend on the magnitude and direction of the force which produces it? The answer is—

217. Lex II. *Mutationem motûs proportionalem esse vi motrici impressae, et fieri secundum lineam rectam quâ vis illa imprimitur.*

Change of motion is proportional to the impressed force, and takes place in the direction of the straight line in which the force acts.

218. If any force generates motion, a double force will generate double motion, and so on, whether simultaneously or successively, instantaneously or gradually, applied. And this motion, if the body was moving beforehand, is either added to the previous motion if directly conspiring with it; or is subtracted if directly opposed; or is geometrically compounded with it, according to the kinematical principles already explained, if the line of previous motion and the direction of the force are inclined to each other at any angle. (This is a paraphrase of Newton's own comments on the second law.)

219. In Chapter I. we have considered change of velocity, or acceleration, as a purely geometrical element, and have seen how it may be at once inferred from the given initial and final velocities of a body. By the definition of a quantity of motion (§ 211), we see that, if we multiply the change of velocity, thus geometrically determined, by the mass of the body, we have the change of motion referred to in Newton's law as the measure of the force which produces it.

It is to be particularly noticed, that in this statement there is nothing said about the actual motion of the body before it was acted on by the force: it is only the *change* of motion that concerns us. Thus the same force will produce precisely the same change of motion in a body, whether the body be at rest, or in motion with any velocity whatever.

220. Again, it is to be noticed that nothing is said as to the body being under the action of *one* force only; so that we may logically put a part of the second law in the following (apparently) amplified form:—

When any forces whatever act on a body, then, whether the body be originally at rest or moving with any velocity and in any direction, each force produces in the body the exact change of motion which it would have produced if it had acted singly on the body originally at rest.

221. A remarkable consequence follows immediately from this view of the second law. Since forces are measured by the changes of motion they produce, and their directions assigned by the directions in which these changes are produced; and since the changes of motion of one and the same body are in the directions of, and proportional to, the changes of velocity—a single force, measured by the resultant change of velocity, and in its direction, will be the equivalent of any number of simultaneously acting forces. Hence

The resultant of any number of forces (applied at one point) is to be found by the same geometrical process as the resultant of any number of simultaneous velocities.

222. From this follows at once (§ 31) the construction of the *Parallelogram of Forces* for finding the resultant of two forces, and the *Polygon of Forces* for the resultant of any number of forces, in lines all through one point.

The case of the equilibrium of a number of forces acting at one point, is evidently deducible at once from this; for if we introduce one other force equal and opposite to their resultant, this will produce a change of motion equal and opposite to the resultant change of motion produced by the given forces; that is to say, will produce a condition in which the point experiences no change of motion, which, as we have already seen, is the only kind of rest of which we can ever be conscious.

223. Though Newton perceived that the Parallelogram of Forces, or the fundamental principle of Statics, is essentially involved in the second law of motion, and gave a proof which is virtually the same as the preceding, subsequent writers on Statics (especially in this country) have very generally ignored the fact; and the consequence has been the introduction of various unnecessary Dynamical Axioms, more or less obvious, but in reality included in or dependent upon Newton's laws of motion. We have retained Newton's method, not only on account of its admirable simplicity, but because we believe it contains the most philosophical foundation for the static as well as for the kinetic branch of the dynamic science.

224. But the second law gives us the means of measuring force, and also of measuring the mass of a body.

For, if we consider the actions of various forces upon the same body for equal times, we evidently have changes of velocity produced which are *proportional* to the forces. The changes of velocity, then, give us in this case the means of comparing the magnitudes of different forces. Thus the velocities acquired in one second by the same mass (falling freely) at different parts of the earth's surface, give us the relative amounts of the earth's attraction at these places.

Again, if equal forces be exerted on different bodies, the changes of velocity produced in equal times must be *inversely* as the masses of the various bodies. This is approximately the case, for instance, with trains of various lengths started by the same locomotive: it is exactly realized in such cases as the action of an electrified body on a number of solid or hollow spheres of the same external diameter, and of different metals.

Again, if we find a case in which different bodies, each acted on by a force, acquire in the same time the same changes of velocity, the forces must be proportional to the masses of the bodies. This, when the resistance of the air is removed, is the case of falling bodies; and from it we conclude that the weight of a body in any given locality, or the force with which the earth attracts it, is proportional to its mass; a most important physical truth, which will be treated of more carefully in the chapter devoted to Properties of Matter.

225. It appears, lastly, from this law, that every theorem of Kinematics connected with acceleration has its counterpart in Kinetics. Thus, for instance (§ 38), we see that the force under which a particle describes any curve, may be resolved into two components, one in the tangent to the curve, the other *towards* the centre of curvature; their magnitudes being the acceleration of momentum, and the product of the momentum and the angular velocity about the centre of curvature, respectively. In the case of uniform motion, the first of these vanishes, or the whole force is perpendicular to the direction of motion. When there is no force perpendicular to the direction of motion, there is no curvature, or the path is a straight line.

226. We have, by means of the first two laws, arrived at a *definition* and a *measure* of force; and have also found how to compound, and therefore also how to resolve, forces: and also how to investigate the motion of a single particle subjected to given forces. But more is required before we can completely understand the more complex cases of motion, especially those in which we have mutual actions between or amongst two or more bodies; such as, for instance, attractions, or pressures, or transferrence of energy in any form. This is perfectly supplied by

227. Lex III. *Actioni contrariam semper et aequalem esse reactionem: sive corporum duorum actiones in se mutuò semper esse aequales et in partes contrarias dirigi.*

To every action there is always an equal and contrary reaction: or, the mutual actions of any two bodies are always equal and oppositely directed.

228. If one body presses or draws another, it is pressed or drawn by this other with an equal force in the opposite direction. If any one presses a stone with his finger, his finger is pressed with the same force in the opposite direction by the stone. A horse towing a boat on a canal is dragged backwards by a force equal to that which he impresses on the towing-rope forwards. By whatever amount, and in whatever direction, one body has its motion changed by impact upon another, this other body has its motion changed by the same amount in the opposite direction; for at each instant during the impact the force between them was equal and opposite on the two. When neither of the two bodies has any rotation, whether before or after impact, the changes of velocity which they experience are inversely as their masses.

When one body attracts another from a distance, this other attracts it with an equal and opposite force. This law holds not only for the attraction of gravitation, but also, as Newton himself remarked and verified by experiment, for magnetic attractions: also for electric forces, as tested by Otto-Guericke.

229. What precedes is founded upon Newton's own comments on the third law, and the actions and reactions contemplated are simple forces. In the scholium appended, he makes the following remarkable statement, introducing another specification of actions and reactions subject to his third law, the full meaning of which seems to have escaped the notice of commentators:—

Si aestimetur agentis actio ex ejus vi et velocitate conjunctim; et similiter resistentis reactio aestimetur conjunctim ex ejus partium singularum velocitatibus et viribus resistendi ab earum attritione, cohaesione, pondere, et acceleratione oriundis; erunt actio et reactio, in omni instrumentorum usu, sibi invicem semper aequales.

In a previous discussion Newton has shown what is to be understood by the velocity of a force or resistance; i.e. that it is the velocity of the point of application of the force *resolved in the direction of the force*, in fact proportional to the virtual velocity. Bearing this in mind, we may read the above statement as follows:—

If the action of an agent be measured by the product of its force into its velocity; and if, similarly, the reaction of the resistance be measured by the velocities of its several parts into their several forces, whether these arise from friction, cohesion, weight, or acceleration;—action and reaction, in all combinations of machines, will be equal and opposite.

Farther on we shall give a full development of the consequences of this most important remark.

230. Newton, in the passage just quoted, points out that forces of resistance against acceleration are to be reckoned as reactions equal and opposite to the actions by which the acceleration is produced. Thus, if we consider any one material point of a system, its reaction against acceleration must be equal and opposite to the resultant of the forces which that point experiences, whether by the actions of other parts of the system upon it, or by the influence of matter not belonging to the system. In other words, it must be in equilibrium with these forces. Hence Newton's view amounts to this, that all the forces of the system, with the reactions against acceleration of the material points composing it, form groups of equilibrating systems for these points considered individually. Hence, by the principle of superposition of forces in equilibrium, all the forces acting on points of the system form, with the reactions against acceleration, an equilibrating set of forces on the whole system. This is the celebrated principle first explicitly stated, and very usefully applied, by D'Alembert in 1742, and still known by his name. We have seen, however, that it is very distinctly implied in Newton's own interpretation of his third law of motion. As it is usual to investigate the general equations or conditions of equilibrium, in treatises on Analytical Dynamics, before entering in detail on the kinetic branch of the subject, this principle is found practically most useful in showing how we may write down at once the equations of motion for any system for which the equations of equilibrium have been investigated.

231. Every rigid body may be imagined to be divided into indefinitely small parts. Now, in whatever form we may eventually find a *physical* explanation of the origin of the forces which act between these parts, it is certain that each such small part may be considered to be held in its position relatively to the others by mutual forces in lines joining them.

232. From this we have, as immediate consequences of the second and third laws, and of the preceding theorems relating to centre of inertia and moment of momentum, a number of important propositions such as the following:—

(*a*) The centre of inertia of a rigid body moving in any manner, but free from external forces, moves uniformly in a straight line.

(*b*) When any forces whatever act on the body, the motion of the centre of inertia is the same as it would have been had these forces been applied with their proper magnitudes and directions at that point itself.

(*c*) Since the moment of a force acting on a particle is the same as the moment of momentum it produces in unit of time, the changes of moment of momentum in any two parts of a rigid body due to their mutual action are equal and opposite. Hence the moment of momentum of a rigid body, about any axis which is fixed in direction, and passes through a point which is either fixed in space or moves uniformly in a straight line, is unaltered by the mutual actions of the parts of the body.

(*d*) The rate of increase of moment of momentum, when the body is acted on by external forces, is the sum of the moments of these forces about the axis.

233. We shall for the present take for granted, that the mutual action between two rigid bodies may in every case be imagined as composed of pairs of equal and opposite forces in straight lines. From this it follows that the sum of the quantities of motion, parallel to any fixed direction, of two rigid bodies influencing one another in any possible way, remains unchanged by their mutual action; also that the sum of the moments of momentum of all the particles of the two bodies, round any line in a fixed direction in space, and passing through any point moving uniformly in a straight line in any direction, remains constant. From the first of these propositions we infer that the centre of inertia of any number of mutually influencing bodies, if in motion, continues moving uniformly in a straight line, unless in so far as the direction or velocity of its motion is changed by forces acting mutually between them and some other matter not belonging to them; also that the centre of inertia of any body or system of bodies moves just as all their matter, if concentrated in a point, would move under the influence of forces equal and parallel to the forces really acting on its different parts. From the second we infer that the axis of resultant rotation through the centre of inertia of any system of bodies, or through any point either at rest or moving uniformly in a straight line, remains unchanged in direction, and the sum of moments of momenta round it remains constant if the system experiences no force from without. This principle is sometimes called *Conservation of Areas*, a not very convenient designation.

234. The kinetic energy of any system is equal to the sum of the kinetic energies of a mass equal to the sum of the masses of the system, moving with a velocity equal to that of its centre of inertia, and of the motions of the separate parts relatively to the centre of inertia.

Let OI represent the velocity of the centre of inertia, IP that of any point of the system relative to O. Then the actual velocity of that point is OP, and the proof of § 196 applies at once—it being remembered that the mean of IQ, i. e. the mean of the velocities relative to the centre of inertia and parallel to OI, is zero by § 65.

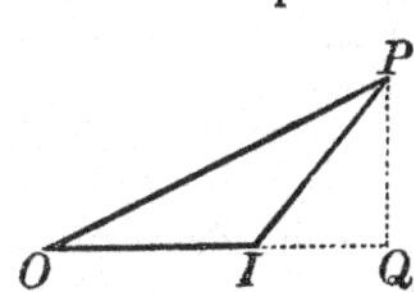

235. The kinetic energy of rotation of a rigid system about any axis is (§§ 55, 179) expressed by $\frac{1}{2}\Sigma mr^2\omega^2$, where m is the mass of any part, r its distance from the axis, and ω the angular velocity of rotation. It may evidently be written in the form $\frac{1}{2}\omega^2\Sigma mr^2$. The factor Σmr^2 is of course (§ 198) the *Moment of Inertia* of the system about the axis in question.

It is worth while to notice that the moment of momentum of any rigid system about an axis, being $\Sigma mvr = \omega\Sigma mr^2$, is the product of the angular velocity into the moment of inertia; while, as above, the half product of the moment of inertia by the square of the angular velocity is the kinetic energy.

If we take a quantity k, such that

$$k^2\Sigma m = \Sigma mr^2,$$

k is called the *Radius of Gyration* about the axis from which r is measured. The radius of gyration about any axis is therefore the distance from that axis at which, if the whole mass were placed, it would have the same moment of inertia as before. In a fly-wheel, where it is desirable to have as great a moment of inertia with as small a mass as possible, within certain limits of dimensions, the greater part of the mass is formed into a ring of the largest admissible diameter, and the radius of this ring is then approximately the radius of gyration of the whole.

236. The rate of increase of moment of momentum is thus, in Newton's notation (§ 28), $\dot{\omega}\Sigma mr^2$; and, in the case of a body free to rotate about a fixed axis, is equal to the moment of the couple about that axis. Hence a constant couple gives uniform acceleration of angular velocity; or $\dot{\omega} = \frac{\text{Couple}}{Mk^2}$. By § 178 we see that the corresponding formula for linear acceleration is $\ddot{s} = \dot{v} = \frac{\text{Force}}{M}$.

237. For every rigid body there may be described about any point as centre, an ellipsoid (called *Poinsot's Momental Ellipsoid*) which is such that the length of any radius-vector is inversely proportional to the radius of gyration of the body about that radius-vector as axis.

The axes of the ellipsoid are the *Principal Axes* of inertia of the body at the point in question.

When the moments of inertia about two of these are equal, the ellipsoid becomes a spheroid, and the radius of gyration is the same for every axis in the plane of its equator.

When all three principal moments are equal, the ellipsoid becomes a sphere, and every axis has the same radius of gyration.

238. The principal axes at any point of a rigid body are normals to the three surfaces of the second order which pass through that point, and are confocal with an ellipsoid, having its centre at the centre of inertia, and its three principal diameters coincident with the three principal axes through these points, and equal respectively to the doubles of the radii of gyration round them. This ellipsoid is called the *Central Ellipsoid.*

239. A rigid body is said to be kinetically symmetrical about its centre of inertia when its moments of inertia about three principal axes through that point are equal; and therefore necessarily the moments of inertia about *all* axes through that point equal (§ 237), and all these axes principal axes. About it uniform spheres, cubes, and in general any complete crystalline solid of the first system (see chapter on Properties of Matter) are kinetically symmetrical.

A rigid body is kinetically symmetrical about an *axis* when this axis is one of the principal axes through the centre of inertia, and the moments of inertia about the other two, and therefore about any line in their plane, are equal. A spheroid, a square or equilateral triangular prism or plate, a circular ring, disc, or cylinder, or any complete crystal of the second or fourth system, is kinetically symmetrical about its axis.

240. The foundation of the abstract theory of energy is laid by Newton in an admirably distinct and compact manner in the sentence of his scholium already quoted (§ 229), in which he points out its application to mechanics[1]. The *actio agentis*, as he defines it, which is evidently equivalent to the product of the effective component of the force, into the velocity of the point on which it acts, is simply, in modern English phraseology, the rate at which the agent works. The subject for measurement here is precisely the same as that for which Watt, a hundred years later, introduced the practical unit of a '*Horse-power*,' or the rate at which an agent works when overcoming 33,000 times the weight of a pound through the space of a foot in a minute; that is, producing 550 foot-pounds of work per second. The unit, however, which is most generally convenient is that which Newton's definition implies, namely, the rate of doing work in which the unit of energy is produced in the unit of time.

241. Looking at Newton's words (§ 229) in this light, we see that they may be logically converted into the following form:—

Work done on any system of bodies (in Newton's statement, the parts of any machine) *has its equivalent in work done against friction, molecular forces, or gravity, if there be no acceleration; but if there be acceleration, part of the work is expended in overcoming the resistance to acceleration, and the additional kinetic energy developed is equivalent to the work so spent.* This is evident from § 180.

When part of the work is done against molecular forces, as in bending a spring; or against gravity, as in raising a weight; the recoil of the spring, and the fall of the weight, are capable at any future time, of reproducing the work originally expended (§ 207). But in Newton's day, and long afterwards, it was supposed that work was *absolutely lost* by friction; and, indeed, this statement is still to be found even in recent authoritative treatises. But we must defer the examination of this point till we consider in its modern form the principle of *Conservation of Energy*.

242. If a system of bodies, given either at rest or in motion, be influenced by no forces from without, the sum of the kinetic energies of all its parts is augmented in any time by an amount equal to the whole work done in that time by the mutual forces, which we may imagine as acting between its points. When the lines in which these forces act remain all unchanged in length, the forces do no work, and the sum of the kinetic energies of the whole system remains constant. If, on the other hand, one of these lines varies in length during the motion, the mutual forces in it will do work, or will consume work, according as the distance varies with or against them.

243. A limited system of bodies is said to be *dynamically conservative* (or simply *conservative*, when force is understood to be the subject), if the mutual forces between its parts always perform, or always consume, the same amount of work during any motion

[1] The reader will remember that we use the word 'mechanics' in its true classical sense, the science of machines, the sense in which Newton himself used it, when he dismissed the further consideration of it by saying (in the scholium referred to), *Caeterum mechanicam tractare non est hujus instituti.*

whatever, by which it can pass from one particular configuration to another.

244. The whole theory of energy in physical science is founded on the following proposition:—

If the mutual forces between the parts of a material system are independent of their velocities, whether relative to one another, or relative to any external matter, the system must be dynamically conservative.

For if more work is done by the mutual forces on the different parts of the system in passing from one particular configuration to another, by one set of paths than by another set of paths, let the system be directed, by frictionless constraint, to pass from the first configuration to the second by one set of paths and return by the other, over and over again for ever. It will be a continual source of energy without any consumption of materials, which is impossible.

245. The *potential energy* of a conservative system, in the configuration which it has at any instant, is the amount of work that its mutual forces perform during the passage of the system from any one chosen configuration to the configuration at the time referred to. It is generally, but not always, convenient to fix the particular configuration chosen for the zero of reckoning of potential energy, so that the potential energy, in every other configuration practically considered, shall be positive.

246. The potential energy of a conservative system, at any instant, depends solely on its configuration at that instant, being, according to definition, the same at all times when the system is brought again and again to the same configuration. It is therefore, in mathematical language, said to be a function of the co-ordinates by which the positions of the different parts of the system are specified. If, for example, we have a conservative system consisting of two material points; or two rigid bodies, acting upon one another with force dependent only on the relative position of a point belonging to one of them, and a point belonging to the other; the potential energy of the system depends upon the co-ordinates of one of these points relatively to lines of reference in fixed directions through the other. It will therefore, in general, depend on three independent co-ordinates, which we may conveniently take as the distance between the two points, and two angles specifying the absolute direction of the line joining them. Thus, for example, let the bodies be two uniform metal globes, electrified with any given quantities of electricity, and placed in an insulating medium such as air, in a region of space under the influence of a vast distant electrified body. The mutual action between these two spheres will depend solely on the relative position of their centres. It will consist partly of gravitation, depending solely on the distance between their centres, and of electric force, which will depend on the distance between them, but also, in virtue of the inductive action of the distant body, will depend on the absolute direction of the line joining their centres. Or again, if the system consist of two balls of soft iron, in any locality of the earth's surface, their mutual action will be partly gravitation, and partly due to the magnetism induced in them by terrestrial magnetic force. The portion of the potential energy depending on the latter cause, will be a function of the distance between their centres and the inclination of this line to the direction of the terrestrial magnetic force.

247. In nature the hypothetical condition of § 243 is *apparently violated* in all circumstances of motion. A material system can never be brought through any returning cycle of motion without spending more work against the mutual forces of its parts than is gained from these forces, because no relative motion can take place without meeting with frictional or other forms of resistance; among which are included (1) mutual friction between solids sliding upon one another; (2) resistances due to the viscosity of fluids, or imperfect elasticity of solids; (3) resistances due to the induction of electric currents; (4) resistances due to varying magnetization under the influence of imperfect magnetic retentiveness. No motion in nature can take place without meeting resistance due to some, if not to all, of these influences. It is matter of everyday experience that friction and imperfect elasticity of solids impede the action of all artificial mechanisms; and that even when bodies are detached, and left to move freely in the air, as falling bodies, or as projectiles, they experience resistance owing to the viscosity of the air.

The greater masses, planets and comets, moving in a less resisting medium, show less indications of resistance[1]. Indeed it cannot be said that observation upon any one of these bodies, with the possible exception of Encke's comet, has demonstrated resistance. But the analogies of nature, and the ascertained facts of physical science, forbid us to doubt that every one of them, every star, and every body of any kind moving in any part of space, has its relative motion impeded by the air, gas, vapour, medium, or whatever we choose to call the substance occupying the space immediately round it; just as the motion of a rifle-bullet is impeded by the resistance of the air.

248. There are also indirect resistances, owing to friction impeding the tidal motions, on all bodies which, like the earth, have portions of their free surfaces covered by liquid, which, as long as these bodies move relatively to neighbouring bodies, must keep drawing off energy from their relative motions. Thus, if we consider, in the first place, the action of the moon alone, on the earth with its oceans, lakes, and rivers, we perceive that it must tend to equalize the periods of the earth's rotation about its axis, and of the revolution of the two bodies about their centre of inertia; because as long as these periods differ, the tidal action of the earth's surface must keep subtracting energy from their motions. To view the subject more in detail, and, at the same time, to avoid unnecessary complications, let us suppose the

[1] Newton, *Principia*. (Remarks on the first law of motion.) 'Majora autem Planetarum et Cometarum corpora motus suos et progressivos et circulares, in spatiis minus resistentibus factos, conservant diutius.'

moon to be a uniform spherical body. The mutual action and reaction of gravitation between her mass and the earth's, will be equivalent to a single force in some line through her centre; and must be such as to impede the earth's rotation as long as this is performed in a shorter period than the moon's motion round the earth. It must therefore lie in some such direction as the line MQ in the diagram, which represents, necessarily with enormous exaggeration, its deviation, OQ, from the earth's centre. Now the actual force on the moon in the line MQ, may be regarded as consisting of a force in the line MO towards the earth's centre, sensibly equal in amount to the whole force, and a comparatively very small force in the line MT perpendicular to MO. This latter is very nearly tangential to the moon's path, and is in the direction *with* her motion. Such a force, if suddenly commencing to act, would, in the first place, increase the moon's velocity; but after a certain time she would have moved so much farther from the earth, in virtue of this acceleration, as to have lost, by moving against the earth's attraction, as much velocity as she had gained by the tangential accelerating force. The integral effect on the moon's motion, of the particular disturbing cause now under consideration, is most easily found by using the principle of moments of momenta (§ 233). Thus we see that as much moment of momentum is gained in any time by the motions of the centres of inertia of the moon and earth relatively to their common centre of inertia, as is lost by the earth's rotation about its axis. It is found that the distance would be increased to about 347,100 miles, and the period lengthened to 48·36 days. Were there no other body in the universe but the earth and the moon, these two bodies might go on moving thus for ever, in circular orbits round their common centre of inertia, and the earth rotating about its axis in the same period, so as always to turn the same face to the moon, and therefore to have all the liquids at its surface at rest relatively to the solid. But the existence of the sun would prevent any such state of things from being permanent. There would be solar tides—twice high water and twice low water—in the period of the earth's revolution relatively to the sun (that is to say, twice in the solar day, or, which would be the same thing, the month). This could not go on without loss of energy by fluid friction. It is not easy to trace the whole course of the disturbance in the earth's and moon's motions which this cause would produce, but its ultimate effect must be to bring the earth, moon, and sun to rotate round their common centre of inertia, like parts of one rigid body. It is probable that the moon, in ancient times liquid or viscous in its outer layer if not throughout, was thus brought to turn always the same face to the earth.

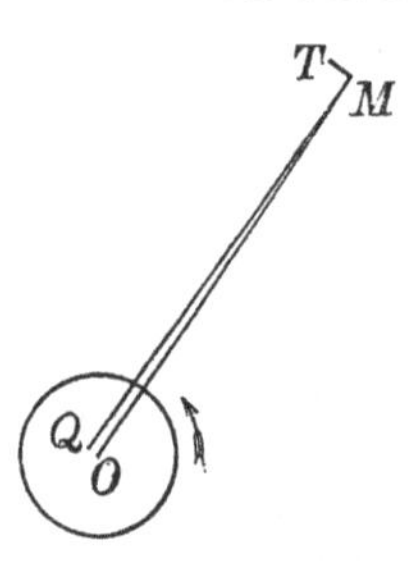

249. We have no data in the present state of science for estimating the relative importance of tidal friction, and of the resistance of the resisting medium through which the earth and moon move; but whatever it may be, there can be but one ultimate result for such a system as that of the sun and planets, if continuing long enough under existing laws, and not disturbed by meeting with other moving masses in space. That result is the falling together of all into one mass, which, although rotating for a time, must in the end come to rest relatively to the surrounding medium.

250. The theory of energy cannot be completed until we are able to examine the physical influences which accompany loss of energy in each of the classes of resistance mentioned above (§ 247). We shall then see that in every case in which energy is lost by resistance, heat is generated; and we shall learn from Joule's investigations that the quantity of heat so generated is a perfectly definite equivalent for the energy lost. Also that in no natural action is there ever a development of energy which cannot be accounted for by the disappearance of an equal amount elsewhere by means of some known physical agency. Thus we shall conclude, that if any limited portion of the material universe could be perfectly isolated, so as to be prevented from either giving energy to, or taking energy from, matter external to it, the sum of its potential and kinetic energies would be the same at all times: in other words, that every material system subject to no other forces than actions and reactions between its parts, is a dynamically conservative system, as defined above (§ 243). But it is only when the inscrutably minute motions among small parts, possibly the ultimate molecules of matter, which constitute light, heat, and magnetism; and the intermolecular forces of chemical affinity; are taken into account, along with the palpable motions and measurable forces of which we become cognizant by direct observation, that we can recognize the universally conservative character of all natural dynamic action, and perceive the bearing of the principle of reversibility on the whole class of natural actions involving resistance, which seem to violate it. In the meantime, in our studies of abstract dynamics, it will be sufficient to introduce a special reckoning for energy lost in working against, or gained from work done by, forces not belonging palpably to the conservative class.

251. The only actions and reactions between the parts of a system, not belonging palpably to the conservative class, which we shall consider in abstract dynamics, are those of friction between solids sliding on solids, except in a few instances in which we shall consider the general character and ultimate results of effects produced by viscosity of fluids, imperfect elasticity of solids, imperfect electric conduction, or imperfect magnetic retentiveness. We shall also, in abstract dynamics, consider forces as applied to parts of a limited system arbitrarily from without. These we shall call, for brevity, the applied forces.

252. The law of energy may then, in abstract dynamics, be expressed as follows:—

The whole work done in any time, on any limited material system, by applied forces, is equal to the whole effect in the forms of potential and kinetic energy produced in the system, together with the work lost in friction.

253. This principle may be regarded as comprehending the whole of abstract dynamics, because, as we now proceed to show, the conditions of equilibrium and of motion, in every possible case, may be derived from it.

254. A material system, whose relative motions are unresisted by friction, is in equilibrium in any particular configuration if, and is not in equilibrium unless, the rate at which the applied forces perform work at the instant of passing through it is equal to that at which potential energy is gained, in every possible motion through that configuration. This is the celebrated principle of virtual velocities which Lagrange made the basis of his *Mécanique Analytique.*

255. To prove it, we have first to remark that the system cannot possibly move away from any particular configuration except by work being done upon it by the forces to which it is subject: it is therefore in equilibrium if the stated condition is fulfilled. To ascertain that nothing less than this condition can secure the equilibrium, let us first consider a system having only one degree of freedom to move. Whatever forces act on the whole system, we may always hold it in equilibrium by a single force applied to any one point of the system in its line of motion, opposite to the direction in which it tends to move, and of such magnitude that, in any infinitely small motion in either direction, it shall resist, or shall do, as much work as the other forces, whether applied or internal, altogether do or resist. Now, by the principle of superposition of forces in equilibrium, we might, without altering their effect, apply to any one point of the system such a force as we have just seen would hold the system in equilibrium, and another force equal and opposite to it. All the other forces being balanced by one of these two, they and it might again, by the principle of superposition of forces in equilibrium, be removed; and therefore the whole set of given forces would produce the same effect, whether for equilibrium or for motion, as the single force which is left acting alone. This single force, since it is in a line in which the point of its application is free to move, must move the system. Hence the given forces, to which the single force has been proved equivalent, cannot possibly be in equilibrium unless their whole work for an infinitely small motion is nothing, in which case the single equivalent force is reduced to nothing. But whatever amount of freedom to move the whole system may have, we may always, by the application of frictionless constraint, limit it to one degree of freedom only;—and this may be freedom to execute any particular motion whatever, possible under the given conditions of the system. If, therefore, in any such infinitely small motion, there is variation of potential energy uncompensated by work of the applied forces, constraint limiting the freedom of the system to only this motion will bring us to the case in which we have just demonstrated there cannot be equilibrium. But the application of constraints limiting motion cannot possibly disturb equilibrium, and therefore the given system under the actual conditions cannot be in equilibrium in any particular configuration if the rate of doing work is greater than that at which potential energy is stored up in any possible motion through that configuration.

256. If a material system, under the influence of internal and applied forces, varying according to some definite law, is balanced by them in any position in which it may be placed, its equilibrium is said to be neutral. This is the case with any spherical body of uniform material resting on a horizontal plane. A right cylinder or cone, bounded by plane ends perpendicular to the axis, is also in neutral equilibrium on a horizontal plane. Practically, any mass of moderate dimensions is in neutral equilibrium when its centre of inertia only is fixed, since, when its longest dimension is small in comparison with the earth's radius, gravity is, as we shall see, approximately equivalent to a single force through this point.

But if, when displaced infinitely little in any direction from a particular position of equilibrium, and left to itself, it commences and continues vibrating, without ever experiencing more than infinitely small deviation in any of its parts, from the position of equilibrium, the equilibrium in this position is said to be stable. A weight suspended by a string, a uniform sphere in a hollow bowl, a loaded sphere resting on a horizontal plane with the loaded side lowest, an oblate body resting with one end of its shortest diameter on a horizontal plane, a plank, whose thickness is small compared with its length and breadth, floating on water, are all cases of stable equilibrium; if we neglect the motions of rotation about a vertical axis in the second, third, and fourth cases, and horizontal motion in general, in the fifth, for all of which the equilibrium is neutral.

If, on the other hand, the system can be displaced in any way from a position of equilibrium, so that when left to itself it will not vibrate within infinitely small limits about the position of equilibrium, but will move farther and farther away from it, the equilibrium in this position is said to be unstable. Thus a loaded sphere resting on a horizontal plane with its load as high as possible, an egg-shaped body standing on one end, a board floating edgewise in water, would present, if they could be realized in practice, cases of unstable equilibrium.

When, as in many cases, the nature of the equilibrium varies with the direction of displacement, if unstable for any possible displacement it is practically unstable on the whole. Thus a circular disc standing on its edge, though in neutral equilibrium for displacements in its plane, yet being in unstable equilibrium for those perpendicular to its plane, is practically unstable. A sphere resting in equilibrium on a saddle presents a case in which there is stable, neutral, or unstable equilibrium, according to the direction in which it may be displaced by rolling; but practically it is unstable.

257. The theory of energy shows a very clear and simple test for discriminating these characters, or determining whether the equilibrium is neutral, stable, or unstable, in any case. If there is just as much potential energy stored up as there is work performed by the applied and internal forces in any possible displacement, the equilibrium is neutral, but not unless. If in every possible infinitely small displacement from a position of equilibrium there is more potential energy stored up than work done, the equilibrium is thoroughly stable, and not unless. If in any or in every infinitely small displacement from a position of equilibrium there is more work done than energy stored up, the equilibrium is unstable. It follows that if the system is influenced only by internal forces, or if the applied forces follow the law of doing always the same amount of work upon the system passing from one configuration to another by all possible paths, the whole potential energy must be constant, in all positions, for neutral equilibrium; must be a minimum for positions of thoroughly stable equilibrium; must be either a maximum for all displacements, or a maximum for some displacements and a minimum for others when there is unstable equilibrium.

258. We have seen that, according to D'Alembert's principle, as explained above (§ 230), forces acting on the different points of a material system, and their reactions against the accelerations which they actually experience in any case of motion, are in equilibrium with one another. Hence in any actual case of motion, not only is the actual work done by the forces equal to the kinetic energy produced in any infinitely small time, in virtue of the actual accelerations; but so also is the work which would be done by the forces, in any infinitely small time, if the velocities of the points constituting the system were at any instant changed to any possible infinitely small velocities, and the accelerations unchanged. This statement, when put into the concise language of mathematical analysis, constitutes Lagrange's application of the 'principle of virtual velocities' to express the conditions of D'Alembert's equilibrium between the forces acting, and the resistances of the masses to acceleration. It comprehends, as we have seen, every possible condition of every case of motion. The 'equations of motion' in any particular case are, as Lagrange has shown, deduced from it with great ease.

259. When two bodies, in relative motion, come into contact, pressure begins to act between them to prevent any parts of them from jointly occupying the same space. This force commences from nothing at the first point of collision, and gradually increases per unit of area on a gradually increasing surface of contact. If, as is always the case in nature, each body possesses some degree of elasticity, and if they are not kept together after the impact by cohesion, or by some artificial appliance, the mutual pressure between them will reach a maximum, will begin to diminish, and in the end will come to nothing, by gradually diminishing in amount per unit of area on a gradually diminishing surface of contact. The whole process would occupy not greatly more or less than an hour if the bodies were of such dimensions as the earth, and such degrees of rigidity as copper, steel, or glass. It is finished, probably, within a thousandth of a second, if they are globes of any of these substances not exceeding a yard in diameter.

260. The whole amount, and the direction, of the '*Impact*' experienced by either body in any such case, are reckoned according to the 'change of momentum which it experiences. The amount of the impact is measured by the amount, and its direction by the direction of the change of momentum, which is produced. The component of an impact in a direction parallel to any fixed line is similarly reckoned according to the component change of momentum in that direction.

261. If we imagine the whole time of an impact divided into a very great number of equal intervals, each so short that the force does not vary sensibly during it, the component change of momentum in any direction during any one of these intervals will (§ 185) be equal to the force multiplied by the measure of the interval. Hence the component of the impact is equal to the sum of the forces in all the intervals, multiplied by the length of each interval.

262. Any force in a constant direction acting in any circumstances, for any time great or small, may be reckoned on the same principle; so that what we may call its whole amount during any time, or its '*time-integral*,' will measure, or be measured by, the whole momentum which it generates in the time in question. But this reckoning is not often convenient or useful except when the whole operation considered is over before the position of the body, or configuration of the system of bodies, involved, has altered to such a degree as to bring any other forces into play, or alter forces previously acting, to such an extent as to produce any sensible effect on the momentum measured. Thus if a person presses gently with his hand, during a few seconds, upon a mass suspended by a cord or chain, he produces an effect which, if we know the degree of the force at each instant, may be thoroughly calculated on elementary principles. No approximation to a full determination of the motion, or to answering such a partial question as 'how great will be the whole deflection produced?' can be founded on a knowledge of the '*time-integral*' alone. If, for instance, the force be at first very great and gradually diminish, the effect will be very different from what it would be if the force were to increase very gradually and to cease suddenly, even although the time-integral were the same in the two cases. But if the same body is 'struck a blow,' in a horizontal direction, either by the hand, or by a mallet or other somewhat hard mass, the action of the force is finished before the suspending cord has experienced any sensible deflection from the vertical. Neither gravity nor any other force sensibly alters the effect of the blow. And therefore the whole momentum at the end of the blow is sensibly equal to the 'amount of the impact,' which is, in this case, simply the time-integral.

263. Such is the case of Robins' *Ballistic Pendulum*, a massive block of wood movable about a horizontal axis at a considerable distance above it—employed to measure the velocity of a cannon or musket-shot. The shot is fired into the block in a horizontal direction perpendicular to the axis. The impulsive penetration is so nearly instantaneous, and the inertia of the block so large compared with the momentum of the shot, that the ball and pendulum are moving on as one mass *before the pendulum has been sensibly deflected from the position of equilibrium.* This is the essential peculiarity of the ballistic method; which is used also extensively in electro-magnetic researches and in practical electric testing, when the integral quantity of the electricity which has passed in a current of short duration is to be measured. The ballistic formula (§ 272) is applicable, with the proper change of notation, to all such cases.

264. Other illustrations of the cases in which the time-integral gives us the complete solution of the problem may be given without limit. They include all cases in which the direction of the force is always coincident with the direction of motion of the moving body, and those special cases in which the time of action of the force is so short that the body's motion does not, during its lapse, sensibly alter its relation to the direction of the force, or the action of any other forces to which it may be subject. Thus, in the vertical fall of a body, the time-integral gives us at once the change of momentum; and the same rule applies in most cases of forces of brief duration, as in a 'drive' in cricket or golf.

265. The simplest case which we can consider, and the one usually treated as an introduction to the subject, is that of the collision of two smooth spherical bodies whose centres before collision were moving in the same straight line. The force between them at each instant must be in this line, because of the symmetry of circumstances round it; and by the third law it must be equal in amount on the two bodies. Hence (Lex II.) they must experience changes of motion at equal rates in contrary directions; and at any instant of the impact the integral amounts of these changes of motion must be equal. Let us suppose, to fix the ideas, the two bodies to be moving both before and after impact in the same direction in one line: one of them gaining on the other before impact, and either following it at a less speed, or moving along with it, as the case may be, after the impact is completed. Cases in which the former is driven backwards by the force of the collision, or in which the two moving in opposite directions meet in collision, are easily reduced to dependence on the same formula by the ordinary algebraic convention with regard to positive and negative signs.

In the standard case, then, the quantity of motion lost, up to any instant of the impact, by one of the bodies, is equal to that gained by the other. Hence at the instant when their velocities are equalized they move as one mass with a momentum equal to the sum of the momenta of the two before impact. That is to say, if v denote the common velocity at this instant, we have

$$(M+M')v=MV+M'V',$$

or $$v=\frac{MV+M'V'}{M+M'},$$

if M, M' denote the masses of the two bodies, and V, V' their velocities before impact.

During this first period of the impact the bodies have been, on the whole, coming into closer contact with one another, through a compression or deformation experienced by each, and resulting, as remarked above, in a fitting together of the two surfaces over a finite area. No body in nature is perfectly inelastic; and hence, at the instant of closest approximation, the mutual force called into action between the two bodies continues, and tends to separate them. Unless prevented by natural surface cohesion or welding (such as is always found, as we shall see later in our chapter on Properties of Matter, however hard and well polished the surfaces may be), or by artificial appliances (such as a coating of wax, applied in one of the common illustrative experiments; or the coupling applied between two railway-carriages when run together so as to push in the springs, according to the usual practice at railway-stations), the two bodies are actually separated by this force, and move away from one another. Newton found that, *provided the impact is not so violent as to make any sensible permanent indentation in either body*, the relative velocity of separation after the impact bears a proportion to their previous relative velocity of approach, which is constant for the same two bodies. This proportion, always less than unity, approaches more and more nearly to it the harder the bodies are. Thus with balls of compressed wool he found it $\frac{5}{9}$, iron nearly the same, glass $\frac{15}{16}$. The results of more recent experiments on the same subject have confirmed Newton's law. These will be described later. In any case of the collision of two balls, let e denote this proportion, to which we give the name *Co-efficient of Restitution*[1]; and, with previous notation, let in addition U, U' denote the velocities of the two bodies after the conclusion of the impact; in the standard case each being positive, but $U' > U$. Then we have

$$U'-U=e(V-V')$$

and, as before, since one has lost as much momentum as the other has gained, $$MU+M'U'=MV+M'V'.$$

From these equations we find

$$(M+M')U=MV+M'V'-eM'(V-V'),$$

with a similar expression for U'.

Also we have, as above,

$$(M+M')v=MV+M'V'.$$

Hence, by subtraction,

$$(M+M')(v-U)=eM'(V-V')=e\{M'V-(M+M')v+MV\},$$

[1] In most modern treatises this is called a 'co-efficient of elasticity;' a misnomer, suggested, it may be, by Newton's words, but utterly at variance with modern language and modern knowledge regarding elasticity.

and therefore $v-U=e(V-v)$.
Of course we have also $U'-v=e(v-V')$.
These results may be put in words thus:—The *relative* velocity of either of the bodies with regard to the centre of inertia of the two is, after the completion of the impact, reversed in direction, and diminished in the ratio $e:1$.

266. Hence the loss of kinetic energy, being, according to §§ 233, 234, due only to change of kinetic energy relative to the centre of inertia, is to this part of the whole as $1-e^2:1$.

Thus by § 234,

$$\text{Initial kinetic energy} = \tfrac{1}{2}(M+M')v^2+\tfrac{1}{2}M(V-v)^2+\tfrac{1}{2}M'(v-V')^2.$$
$$\text{Final} \quad ,, \quad ,, \quad = \tfrac{1}{2}(M+M')v^2+\tfrac{1}{2}M(v-U)^2+\tfrac{1}{2}M'(U'-v)^2.$$
$$\text{Loss} = \tfrac{1}{2}(1-e^2)\{M(V-v)^2+M'(v-V')^2\}.$$

267. When two elastic bodies, the two balls supposed above for instance, impinge, some portion of their previous kinetic energy will always remain in them as vibrations. A *portion* of the loss of energy (miscalled the effect of imperfect elasticity alone) is necessarily due to this cause in every real case.

Later, in our chapter on the Properties of Matter, it will be shown as a result of experiment, that forces of elasticity are, to a very close degree of accuracy, simply proportional to the strains (§ 135), within the limits of elasticity, in elastic solids which, like metals, glass, etc., bear but small deformations without permanent change. Hence when two such bodies come into collision, sometimes with greater and sometimes with less mutual velocity, but with all other circumstances similar, the velocities of all particles of either body, at corresponding times of the impacts, will be always in the same proportion. Hence the velocity of separation of the centres of inertia after impact will bear a constant proportion to the previous velocity of approach; which agrees with the Newtonian law. It is therefore probable that a very sensible portion, if not the whole, of the loss of energy in the visible motions of two elastic bodies, after impact, experimented on by Newton, may have been due to vibrations; but unless some other cause also was largely operative, it is difficult to see how the loss was so much greater with iron balls than with glass.

268. In certain definite extreme cases, imaginable although not realizable, no energy will be spent in vibrations, and the two bodies will separate, each moving simply as a rigid body, and having in this simple motion the whole energy of work done on it by elastic force during the collision. For instance, let the two bodies be cylinders, or prismatic bars with flat ends, of the same kind of substance, and of equal and similar transverse sections; and let this substance have the property of compressibility with perfect elasticity, in the direction of the length of the bar, and of absolute resistance to change in every transverse dimension. Before impact, let the two bodies be placed with their lengths in one line, and their transverse sections (if not circular) similarly situated, and let one or both be set in motion in this line. Then, if the lengths of the two be equal, they will separate after impact with the same relative velocity as that with which they approached, and neither will retain any vibratory motion after the end of the collision. The result, as regards the motions of the two bodies after the collision, will be sensibly the same if they are of any real ordinary elastic solid material, provided the greatest transverse diameter of each is very small in comparison of its length.

269. If the two bars are of an unequal length, the shorter will, after the impact, be in exactly the same state as if it had struck another of its own length, and it therefore will move as a rigid body after the collision. But the other will, along with a motion of its centre of gravity, calculable from the principle that its whole momentum must (§ 233) be changed by an amount equal exactly to the momentum gained or lost by the first, have also a vibratory motion, of which the whole kinetic and potential energy will make up the deficiency of energy which we shall presently calculate in the motions of the centres of inertia. For simplicity, let the longer body be supposed to be at rest before the collision. Then the shorter on striking it will be left at rest; this being clearly the result in the case of the $e = 1$ in the preceding formulae (§ 265) applied to the impact of one body striking another of equal mass previously at rest. The longer bar will move away with the same momentum, and therefore with less velocity of its centre of inertia, and less kinetic energy of this motion, than the other body had before impact, in the ratio of the smaller to the greater mass. It will also have a very remarkable vibratory motion, which, when its length is more than double of that of the other, will consist of a wave running backwards and forwards through its length, and causing the motion of its ends, and, in fact, of every particle of it, to take place by 'fits and starts,' not continuously. The full analysis of these circumstances, though very simple, must be reserved until we are especially occupied with waves, and the kinetics of elastic solids. It is sufficient at present to remark, that the motions of the centres of inertia of the two bodies after impact, whatever they may have been previously, are given by the preceding formulae with for e the value $\frac{M'}{M}$, where M and M' are the smaller and larger mass respectively.

270. The mathematical theory of the vibrations of solid elastic spheres has not yet been worked out; and its application to the case of the vibrations produced by impact presents considerable difficulty. Experiment, however, renders it certain, that but a small part of the whole kinetic energy of the previous motions can remain in the form of vibrations after the impact of two equal spheres of glass or of ivory. This is proved, for instance, by the common observation, that one of them remains nearly motionless after striking the other previously at rest; since, the velocity of the common centre of inertia of the two being necessarily unchanged by the impact, we infer that the second ball acquires a velocity nearly equal to that which the first had before striking it. But it is to be expected that unequal balls of the same substance coming into collision will, by impact, convert a very sensible proportion of the kinetic energy of their previous motions into energy of vibrations; and generally, that the same will be the case when equal or unequal masses of different substances come into collision; although for one particular proportion of their diameters, depending on their densities and elastic qualities, this effect will be a minimum, and possibly not much more sensible than it is when the substances are the same and the diameters equal.

271. It need scarcely be said that in such cases of impact as that of the tongue of a bell, or of a clock-hammer striking its bell (or spiral spring as in the American clocks), or of pianoforte-hammers striking the strings, or of the drum struck with the proper implement, a large part of the kinetic energy of the blow is spent in generating vibrations.

272. The *Moment of an Impact* about any axis is derived from the line and amount of the impact in the same way as the moment of a velocity or force is determined from the line and amount of the velocity or force, § 46. If a body is struck, the change of its moment of momentum about any axis is equal to the moment of the impact round that axis. But, without considering the measure of the impact, we see (§ 233) that the moment of momentum round any axis, lost by one body in striking another, is, as in every case of mutual action, equal to that gained by the other.

Thus, to recur to the ballistic pendulum—the line of motion of the bullet at impact may be in any direction whatever, but the only part which is effective is the component in a plane perpendicular to the axis. We may therefore, for simplicity, consider the motion to be in a line perpendicular to the axis, though not necessarily horizontal. Let m be the mass of the bullet, v its velocity, and p the distance of its line of motion from the axis. Let M be the mass of the pendulum with the bullet lodged in it, and k its radius of gyration. Then if ω be the angular velocity of the pendulum when the impact is complete,

$$mvp = Mk^2\omega,$$

from which the solution of the question is easily determined.

For the kinetic energy after impact is changed (§ 207) into its equivalent in potential energy when the pendulum reaches its position of greatest deflection. Let this be given by the angle θ: then the height to which the centre of inertia is raised is $h(1-\cos\theta)$ if h be its distance from the axis. Thus

$$Mgh(1-\cos\theta) = \tfrac{1}{2}Mk^2\omega^2 = \tfrac{1}{2}\frac{m^2v^2p^2}{Mk^2},$$

or $$2\sin\frac{\theta}{2} = \frac{m}{M}\cdot\frac{p}{k}\cdot\frac{v}{\sqrt{gh}} = \frac{m}{M}\cdot\frac{p}{h}\cdot\frac{\pi v}{gT}, \text{ if } T = \pi\sqrt{\frac{k^2}{gh}}.$$

an expression for the chord of the angle of deflection. In practice the chord of the angle θ is measured by means of a light tape or cord attached to a point of the pendulum, and slipping with small friction through a clip fixed close to the position occupied by that point when the pendulum hangs at rest.

273. *Work done by an impact* is, in general, the product of the impact into half the sum of the initial and final velocities of the point at which it is applied, resolved in the direction of the impact. In the case of direct impact, such as that treated in § 265, the initial kinetic energy of the body is $\frac{1}{2}MV^2$, the final $\frac{1}{2}MU^2$, and therefore the gain, by the impact is

$$\tfrac{1}{2}M(U^2 - V^2),$$

or, which is the same,

$$M(U-V).\tfrac{1}{2}(U+V).$$

But $M(U-V)$ is (§ 260) equal to the amount of the impact. Hence the proposition: the extension of which to the most general circumstances is not difficult, but requires somewhat higher analysis than can be admitted here.

274. It is worthy of remark, that if any number of impacts be applied to a body, their whole effect will be the same whether they be applied together or successively (provided that the whole time occupied by them be infinitely short), although the work done by each particular impact is, in general, different according to the order in which the several impacts are applied. The whole amount of work is the sum of the products obtained by multiplying each impact by half the sum of the components of the initial and final velocities of the point to which it is applied.

275. The effect of any stated impulses, applied to a rigid body, or to a system of material points or rigid bodies connected in any way, is to be found most readily by the aid of D'Alembert's principle; according to which the given impulses, and the impulsive reaction against the generation of motion, measured in amount by the momenta generated, are in equilibrium; and are, therefore, to be dealt with mathematically by applying to them the equations of equilibrium of the system.

276. [A material system of any kind, given at rest, and subjected to an impulse in any specified direction, and of any given magnitude, moves off so as to take the greatest amount of kinetic energy which the specified impulse can give it.

277. If the system is guided to take, under the action of a given impulse, any motion different from the natural motion, it will have less kinetic energy than that of the natural motion, by a difference equal to the kinetic energy of the motion represented by the resultant (§ 67) of those two motions, one of them reversed.

Cor. If a set of material points are struck independently by impulses each given in amount, more kinetic energy is generated if the points are perfectly free to move each independently of all the others, than if they are connected in any way. And the deficiency of energy in the latter case is equal to the amount of the kinetic energy of the motion which geometrically compounded with the motion of either case would give that of the other.

278. Given any material system at rest. Let any parts of it be set in motion suddenly with any specified velocities, possible according to the conditions of the system; and let its other parts be influenced only by its connections with those. It is required to find the motion. The solution of the problem is—The motion actually taken by the system is that which has less kinetic energy than any other motion fulfilling the prescribed velocity conditions. And the excess of the energy of any other such motion, above that of the actual motion, is equal to the energy of the motion that would be generated by the action alone of the impulse which, if compounded with the impulse producing the actual motion, would produce this other supposed motion.]

279. Maupertuis' celebrated principle of *Least Action* has been, even up to the present time, regarded rather as a curious and somewhat perplexing property of motion, than as a useful guide in kinetic investigations. We are strongly impressed with the conviction that a much more profound importance will be attached to it, not only in abstract dynamics, but in the theory of the several branches of physical science now beginning to receive dynamic explanations. As an extension of it, Sir W. R. Hamilton[1] has evolved his method of *Varying Action*, which undoubtedly must become a most valuable aid in future generalizations.

What is meant by 'Action' in these expressions is, unfortunately, something very different from the *Actio Agentis* defined by Newton, and, it must be admitted, is a much less judiciously chosen word. Taking it, however, as we find it, now universally used by writers on dynamics, we define the *Action of a Moving System* as proportional to the average kinetic energy, which it has possessed during the time from any convenient epoch of reckoning, multiplied by the time. According to the unit generally adopted, the action of a system which has not varied in its kinetic energy, is twice the amount of the energy multiplied by the time from the epoch. Or if the energy has been sometimes greater and sometimes less, the action at time t is the double of what we may call the *time-integral* of the energy; that is to say, the action of a system is equal to the sum of the *average momenta for the spaces* described by the particles from any era each multiplied by the length of its path.

280. The principle of Least Action is this:—Of all the different sets of paths along which a conservative system may be guided to move from one configuration to another, with the sum of its potential and kinetic energies equal to a given constant, that one for which the action is the least is such that the system will require only to be started with the proper velocities, to move along it unguided.

281. [In any unguided motion whatever, of a conservative system, the Action from any one stated position to any other, though not necessarily a minimum, fulfils the *stationary condition*, that is to say,

[1] *Phil. Trans.*, 1834–1835.

the condition that the variation vanishes, which secures either a minimum or maximum, or maximum-minimum.]

282. From this principle of stationary action, founded, as we have seen, on a comparison between a natural motion, and any other motion, arbitrarily guided and subject only to the law of energy, the initial and final configurations of the system being the same in each case; Hamilton passes to the consideration of the variation of the action in a natural or unguided motion of the system produced by varying the initial and final configurations, and the sum of the potential and kinetic energies. The result is, that—

283. The rate of *decrease* of the action per unit of increase of any one of the free (generalized) co-ordinates specifying the initial configuration, is equal to the corresponding (generalized) component momentum of the actual motion from that configuration: the rate of *increase* of the action per unit increase of any one of the free co-ordinates specifying the final configuration, is equal to the corresponding component momentum of the actual motion towards this second configuration: and the rate of increase of the action per unit increase of the constant sum of the potential and kinetic energies, is equal to the time occupied by the motion of which the action is reckoned.

284. The determination of the motion of any conservative system from one to another of any two configurations, when the sum of its potential and kinetic energies is given, depends on the determination of a single function of the co-ordinates specifying those configurations by means of two quadratic, partial differential equations of the first order, with reference to those two sets of co-ordinates respectively, with the condition that the corresponding terms of the two differential equations become separately equal when the values of the two sets of co-ordinates agree. The function thus determined and employed to express the solution of the kinetic problem was called the *Characteristic Function*, by Sir W. R. Hamilton, to whom the method is due. It is, as we have seen, the 'action' from one of the configurations to the other; but its peculiarity in Hamilton's system is, that it is to be expressed as a function of the co-ordinates and a constant, the whole energy, as explained above. It is evidently symmetrical with respect to the two configurations, changing only in sign if their co-ordinates are interchanged.

285. The most general possible solution of the quadratic, partial differential equation of the first order, satisfied by Hamilton's Characteristic Function (either terminal configuration alone varying), when interpreted for the case of a single free particle, expresses the action up to any point from some point of a certain arbitrarily given surface, from which the particle has been projected, in the direction of the normal, and with the proper velocity to make the sum of the potential and actual energies have a given value. In other words, the physical problem solved by the most general solution of that partial differential equation, for a single free particle, is this:—

Let free particles, not mutually influencing one another, be projected normally from all points of a certain arbitrarily given surface, each with the proper velocity to make the sum of its potential and kinetic energies have a given value. To find, for that one of the particles which passes through a given point, the 'action' in its course from the surface of projection to this point. The Hamiltonian principles stated above, show that the surfaces of equal action cut the paths of the particles at right angles; and give also the following remarkable properties of the motion:—

If, from all points of an arbitrary surface, particles not mutually influencing one another be projected with the proper velocities in the directions of the normals; points which they reach with equal actions lie on a surface cutting the paths at right angles. The infinitely small thickness of the space between any two such surfaces corresponding to amounts of action differing by any infinitely small quantity, is inversely proportional to the velocity of the particle traversing it; being equal to the infinitely small difference of action divided by the whole momentum of the particle.

286. Irrespectively of methods for finding the 'characteristic function' in kinetic problems, the fact that any case of motion whatever can be represented by means of a single function in the manner explained in § 284, is most remarkable, and, when geometrically interpreted, leads to highly important and interesting properties of motion, which have valuable applications in various branches of Natural Philosophy; one of which, explained below, led Hamilton[1] to a general theory of optical instruments, comprehending the whole in one expression. Some of the most direct applications of the general principle to the motions of planets, comets, etc., considered as free points, and to the celebrated problem of perturbations, known as the Problem of Three Bodies, are worked out in considerable detail by Hamilton (*Phil. Trans.*, 1834–5), and in various memoirs by Jacobi, Liouville, Bour, Donkin, Cayley, Boole, etc.

The now abandoned, but still interesting, corpuscular theory of light furnishes the most convenient language for expressing the optical application. In this theory light is supposed to consist of material particles not mutually influencing one another; but subject to molecular forces from the particles of bodies, not sensible at sensible distances, and therefore not causing any deviation from uniform rectilinear motion in a homogeneous medium, except within an indefinitely small distance from its boundary. The laws of reflection and of single refraction follow correctly from this hypothesis, which therefore suffices for what is called geometrical optics.

We hope to return to this subject, with sufficient detail, in treating of Optics. At present we limit ourselves to state a theorem comprehending the known rule for measuring the magnifying power of a telescope or microscope (by comparing the diameter of the object-

[1] *On the Theory of Systems of Rays.* Trans. R. I. A., 1824, 1830, 1832.

glass with the diameter of pencil of parallel rays emerging from the eye-piece, when a point of light is placed at a great distance in front of the object-glass), as a particular case.

287. Let any number of attracting or repelling masses, or perfectly smooth elastic objects, be fixed in space. Let two stations, O and O', be chosen. Let a shot be fired with a stated velocity, V, from O, in such a direction as to pass through O'. There may clearly be more than one natural path by which this may be done; but, generally speaking, when one such path is chosen, no other, not sensibly diverging from it, can be found; and any infinitely small deviation in the line of fire from O, will cause the bullet to pass infinitely near to, but not through O'. Now let a circle, with infinitely small radius r, be described round O as centre, in a plane perpendicular to the line of fire from this point, and let—all with infinitely nearly the same velocity, but fulfilling the condition that the sum of the potential and kinetic energies is the same as that of the shot from O—bullets be fired from all points of this circle, all directed infinitely nearly parallel to the line of fire from O, but each precisely so as to pass through O'. Let a target be held at an infinitely small distance, a', beyond O', in a plane perpendicular to the line of the shot reaching it from O. The bullets fired from the cirumference of the circle round O, will, after passing through O', strike this target in the circumference of an exceedingly small ellipse, each with a velocity (corresponding of course to its position, under the law of energy) differing infinitely little from V', the common velocity with which they pass through O'. Let now a circle, equal to the former, be described round O', in the plane perpendicular to the central path through O', and let bullets be fired from points in its circumference, each with the proper velocity, and in such a direction infinitely nearly parallel to the central path as to make it pass through O. These bullets, if a target is held to receive them perpendicularly at a distance $a = a'\frac{V}{V'}$, beyond O, will strike it along the circumference of an ellipse equal to the former and placed in a corresponding position; and the points struck by the individual bullets will correspond in the manner explained below. Let P and P' be points of the first and second circles, and Q and Q' the points on the first and second targets which bullets from them strike; then if P' be in a plane containing the central path through O', and the position which Q would take if its ellipse were made circular by a pure strain (§ 159); Q and Q' are similarly situated on the two ellipses.

288. The most obvious optical application of this remarkable result is, that in the use of any optical apparatus whatever, if the eye and the object be interchanged without altering the position of the instrument, the magnifying power is unaltered. This is easily understood when, as in an ordinary telescope, microscope, or opera-glass (Galilean telescope), the instrument is symmetrical about an axis, and is curiously contradictory of the common idea that a telescope 'diminishes' when looked through the wrong way, which no doubt is true if the telescope is simply reversed about the middle of its length, eye and object remaining fixed. But if the telescope be removed from the eye till its eye-piece is close to the object, the part of the object seen will be seen enlarged to the same extent as when viewed with the telescope held in the usual manner. This is easily verified by looking from a distance of a few yards, in through the object-glass of an opera-glass, at the eye of another person holding it to his eye in the usual way.

The more general application may be illustrated thus:—Let the points, O, O' (the centres of the two circles described in the preceding enunciation), be the optic centres of the eyes of two persons looking at one another through any set of lenses, prisms, or transparent media arranged in any way between them. If their pupils are of equal sizes in reality, they will be seen as similar ellipses of equal apparent dimensions by the two observers. Here the imagined particles of light, projected from the circumference of the pupil of either eye, are substituted for the projectiles from the circumference of either circle, and the retina of the other eye takes the place of the target receiving them, in the general kinetic statement.

289. If instead of one free particle we have a conservative system of any number of mutually influencing free particles, the same statement may be applied with reference to the initial position of one of the particles and the final position of another, or with reference to the initial positions, or to the final positions of two of the particles. It thus serves to show how the influence of an infinitely small change in one of those positions, on the direction of the other particle passing through the other position, is related to the influence on the direction of the former particle passing through the former position produced by an infinitely small change in the latter position, and is of immense use in physical astronomy. A corresponding statement, in terms of generalized co-ordinates, may of course be adapted to a system of rigid bodies or particles connected in any way. All such statements are included in the following very general proposition:—

The rate of increase of the component momentum relative to any one of the co-ordinates, per unit of increase of any other co-ordinate, is equal to the rate of increase of the component momentum relative to the latter per unit increase or diminution of the former co-ordinate, according as the two co-ordinates chosen belong to one configuration of the system, or one of them belongs to the initial configuration and the other to the final.

290. If a conservative system is infinitely little displaced from a configuration of stable equilibrium, it will ever after vibrate about this configuration, remaining infinitely near it; each particle of the system performing a motion which is composed of simple harmonic vibrations. If there are i degrees of freedom to move, and we consider any system of generalized co-ordinates specifying its position at any time, the deviation of any one of these co-ordinates from its value for the configuration of equilibrium will vary according to a complex harmonic function (§ 88), composed in general of i simple harmonics of incommensurable periods, and therefore (§ 85) the whole motion of the system will not recur periodically through the same series of configurations. There are in general, however, i distinct determinate displacements, which we shall call *the normal displacements*, fulfilling the condition, that if any one of them be produced alone, and the system then left to itself for an instant at rest, this displacement will diminish and increase periodically according to a simple harmonic function of the time, and consequently every particle of the system will execute a simple harmonic movement in the same period. This result, we shall see later, includes cases in which there are an infinite number of degrees of freedom; as, for instance, a stretched cord; a mass of air in a closed vessel; waves in water, or oscillations in a vessel of water of limited extent, or in an elastic solid; and in these applications it gives the theory of the so-called 'fundamental vibration,' and successive 'harmonics' of the cord, and of all the different possible simple modes of vibration in the other cases.

291. If, as may be in particular cases, the periods of the vibrations for two or more of the normal displacements are equal, any displacement compounded of them will also fulfil the condition of a normal displacement. And if the system be displaced according to any one such normal displacement, and projected with velocity corresponding to another, it will execute a movement, the resultant of two simple harmonic movements in equal periods. The graphic representation of the variation of the corresponding co-ordinates of the system, laid down as two rectangular co-ordinates in a plane diagram, will consequently (§ 82) be a circle or an ellipse; which will therefore, of course, be the form of the orbit of any particle of the system which has a distinct direction of motion, for two of the displacements in question. But it must be remembered that some of the principal parts may have only one degree of freedom; or even that each part of the system may have only one degree of freedom (as, for instance, if the system is composed of a set of particles each constrained to remain on a given line, or of rigid bodies on fixed axes, mutually influencing one another by elastic cords or otherwise). In such a case as the last, no particle of the system can move otherwise than in one line; and the ellipse, circle, or other graphical representation of the composition of the harmonic motions of the system, is merely an aid to comprehension, and not a representation of any motion actually taking place in any part of the system.

292. In nature, as has been said above (§ 250), every system uninfluenced by matter external to it is conservative, when the ultimate molecular motions constituting heat, light, and magnetism, and the potential energy of chemical affinities, are taken into account along with the palpable motions and measurable forces. But (§ 247) practically we are obliged to admit forces of friction, and resistances of the other classes there enumerated, as causing losses of energy to be reckoned, in abstract dynamics, without regard to the equivalents of heat or other molecular actions which they generate. Hence when such resistances are to be taken into account, forces opposed to the motions of various parts of a system must be introduced into the equations. According to the approximate knowledge which we have from experiment, these forces are independent of the velocities when due to the friction of solids; and are simply proportional to the velocities when due to fluid viscosity directly, or to electric or magnetic influences, with corrections depending on varying temperature, and on the varying configuration of the system. In consequence of the last-mentioned cause, the resistance of a *real liquid* (which is always more or less viscous) against a body moving very rapidly through it, and leaving a great deal of irregular motion, such as 'eddies,' in its wake, seems to be nearly in proportion to the square of the velocity; although, as Stokes has shown, at the lowest speeds the resistance is probably in simple proportion to the velocity, and for all speeds may, it is probable, be approximately expressed as the sum of two terms, one simply as the velocity, and the other as the square of the velocity.

293. The effect of friction of solids rubbing one against another is simply to render impossible the *infinitely* small vibrations with which we are now particularly concerned; and to allow any system in which it is present, to rest balanced when displaced within certain finite limits, from a configuration of frictionless equilibrium. In mechanics it is easy to estimate its effects with sufficient accuracy when any practical case of finite oscillations is in question. But the other classes of dissipative agencies give rise to resistances simply as the velocities, without the corrections referred to, when the motions are infinitely small, and can never balance the system in a configuration deviating to any extent, however small, from a configuration of equilibrium without friction. In the theory of infinitely small vibrations, they are to be taken into account by adding to the expressions for the generalized components of force, terms consisting of the generalized velocities each multiplied by a constant, which gives us equations still remarkably amenable to rigorous mathematical treatment. The result of the integration for the case of a single degree of freedom is very simple; and it is of extreme importance, both for the explanation of many natural phenomena, and for use in a large variety of experimental investigations in Natural Philosophy. Partial conclusions from it, in the first place, stated in general terms, are as follows:—

294. If the resistance is less than a certain limit, in any particular case, the motion is a simple harmonic oscillation, with amplitude decreasing by equal proportions in equal successive intervals of time. But if the resistance exceeds this limit, the system, when displaced from its position of equilibrium and left to itself, returns gradually towards its position of equilibrium, never oscillating through it to the other side, and only reaching it after an infinite time.

In the unresisted motion, let n^2 be the rate of acceleration, when the displacement is unity; so that (§ 74) we have $T=\frac{2\pi}{n}$: and let the rate of retardation due to the resistance corresponding to unit velocity be k. Then the motion is of the oscillatory or non-oscillatory class according as $k < 2n$ or $k > 2n$. In the first case, the period of the oscillation is increased, by the resistance, from T to $T\frac{n}{(\frac{1}{4}k^2-n^2)^{\frac{1}{2}}}$; and the rate at which the Napierian logarithm of the amplitude diminishes per unit of time is $\frac{1}{2}k$.

295. An indirect but very simple proof of this important proposition may be obtained by means of elementary mathematics as follows:—A point describes a logarithmic spiral with uniform angular velocity about the pole—find the acceleration.

Since the angular velocity of SP and the inclination of this line to the tangent are each constant, the linear velocity of P is as SP. Take a length PT, equal to $n\,SP$, to represent it. Then the hodograph, the locus of p, where Sp is parallel and equal to PT, is evidently another logarithmic spiral similar to the former, and described with the same uniform angular velocity. Hence (§§ 35, 49) pt, the acceleration required, is equal to $n\,Sp$, and makes with Sp an angle Spt equal to SPT. Hence, if Pu be drawn parallel and equal to pt, and uv parallel to PT, the whole acceleration pt or Pu may be resolved into Pv and vu. Now Pvu is an isosceles triangle, whose base angles (v, u) are each equal to the constant angle of the spiral. Hence Pv and vu bear constant ratios to Pu, and therefore to SP and to PT respectively.

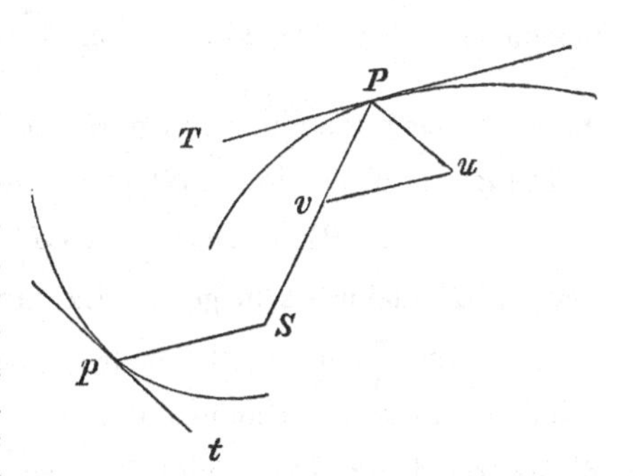

The acceleration, therefore, is composed of a central attractive part proportional to the distance, and a tangential retarding part proportional to the velocity.

And, if the resolved part of P's motion parallel to any line in the plane of the spiral be considered, it is obvious that in it also the acceleration will consist of two parts—one directed towards a point in the line (the projection of the pole of the spiral), and proportional to the distance from it; the other proportional to the velocity, but retarding the motion.

Hence a particle which, unresisted, would have a simple harmonic motion, has, when subject to resistance proportional to its velocity, a motion represented by the resolved part of the spiral motion just described.

296. If a be the constant angle of the spiral, ω the angular velocity of SP, we have evidently

$$PT \,.\, \sin a = SP \,.\, \omega. \quad \text{But } PT = n\, SP, \text{ so that } n = \frac{\omega}{\sin a}.$$

Hence

$$Pv = Pu = pt = nSp = nPT = n^2 \,.\, SP$$

$$\text{and } vu = 2Pv \,.\, \cos a = 2n \cos a\, PT = k \,.\, PT \text{ (suppose.)}$$

Thus the central force at unit distance is $n^2 = \dfrac{\omega^2}{\sin^2 a}$, and the co-efficient of resistance is $k = 2n \cos a = \dfrac{2\omega \cos a}{\sin a}$.

The time of oscillation in the resolved motion is evidently $\dfrac{2\pi}{\omega}$; but, if there had been no resistance, the properties of simple harmonic motion show that it would have been $\dfrac{2\pi}{n}$; so that it is increased by the resistance in the ratio cosec a to 1, or n to $\sqrt{n^2 - \dfrac{k^2}{4}}$.

The rate of diminution of SP is evidently

$$PT.\ \cos a = n \cos a\, SP = \frac{k}{2} SP;$$

that is, SP diminishes in geometrical progression as time increases, the rate being $\dfrac{k}{2}$ per unit of time per unit of length. By an ordinary result of arithmetic (compound interest payable every instant) the diminution of log. SP in unit of time is $\dfrac{k}{2}$.

This process of solution is only applicable to resisted harmonic vibrations when n is greater than $\dfrac{k}{2}$. When n is not greater than $\dfrac{k}{2}$ the auxiliary curve can no longer be a logarithmic spiral, for the moving particle never describes more than a finite angle about the pole. A curve, derived from an equilateral hyperbola, by a process somewhat resembling that by which the logarithmic spiral is deduced from a circle, may be introduced; but then the geometrical method ceases to be simpler than the analytical one, so that it is useless to pursue the investigation farther, at least from this point of view.

297. The general solution of the problem, to find the motion of a system having any number, i, of degrees of freedom, when infinitely little disturbed from a position of equilibrium, and left to move subject to resistances proportional to velocities, shows that the whole motion may be resolved, in general determinately, into $2i$ different motions each either simple harmonic with amplitude diminishing according to the law stated above (§ 294), or non-oscillatory, and consisting of equi-proportionate diminutions of the components of displacement in equal successive intervals of time.

298. When the forces of a system depending on configuration, and not on motion, or, as we may call them for brevity, the forces of position, violate the law of conservatism, we have seen (§ 244) that energy without limit may be drawn from it by guiding it perpetually through a returning cycle of configurations, and we have inferred that in every real system, not supplied with energy from without, the forces of position fulfil the conservative law. But it is easy to arrange a system artificially, in connexion with a source of energy, so that its forces of position shall be non-conservative; and the consideration of the kinetic effects of such an arrangement, especially of its oscillations about or motions round a configuration of equilibrium, is most instructive, by the contrasts which it presents to the phenomena of a natural system.

299. But although, when the equilibrium is stable, no possible infinitely small displacement and velocity given to the system can cause it, when left to itself, to go on moving either farther and farther away till a finite displacement is reached, or till a finite velocity is acquired; it is very remarkable that stability should be possible, considering that even in the case of stability an endless increase of velocity may, as is easily seen from § 244, be obtained merely by *constraining* the system to a particular closed course, or circuit of configurations, nowhere deviating by more than an infinitely small amount from the configuration of equilibrium, and leaving it at rest anywhere in a certain part of this circuit. This result, and the distinct peculiarities of the cases of stability and instability, is sufficiently illustrated by the simplest possible example,—that of a material particle moving in a plane.

300. There is scarcely any question in dynamics more important for Natural Philosophy than the stability or instability of motion. We therefore, before concluding this chapter, propose to give some general explanations and leading principles regarding it.

A 'conservative disturbance of motion' is a disturbance in the motion or configuration of a conservative system, not altering the sum of the potential and kinetic energies. A conservative disturbance of the motion through any particular configuration is a change in velocities, or component velocities, not altering the whole kinetic energy. Thus, for example, a conservative disturbance of the motion of a particle through any point, is a change in the direction of its motion, unaccompanied by change of speed.

301. The actual motion of a system, from any particular configuration, is said to be *stable* if every possible infinitely small conservative disturbance of its motion through that configuration may be compounded of conservative disturbances, any one of which would give rise to an alteration of motion which would bring the system again to some configuration belonging to the undisturbed path, in a finite time, and without more than an infinitely small digression. If this condition is not fulfilled, the motion is said to be *unstable*.

302. For example, if a body, A, be supported on a fixed vertical axis; if a second, B, be supported on a parallel axis belonging to the first; a third, C, similarly supported on B, and so on; and if B, C, etc., be so placed as to have each its centre of inertia as far as possible from the fixed axis, and the whole set in motion with a common angular velocity about this axis, the motion will be stable, from every configuration, as is evident from the principles regarding the resultant centrifugal force on a rigid body, to be proved later. If, for instance, each of the bodies is a flat rectangular board hinged on one edge, it is obvious that the whole system will be kept stable by centrifugal force, when all are in one plane and as far out from the axis as possible. But if A consist partly of a shaft and crank, as a common spinning-wheel, or the fly-wheel and crank of a steam-engine, and if B be supported on the crank-pin as axis, and turned inwards (towards the fixed axis, or across the fixed axis), then, even although the centres of inertia of C, D, etc., are placed as far from the fixed axis as possible, consistent with this position of B, the motion of the system will be unstable.

303. The rectilinear motion of an elongated body lengthwise, or of a flat disc edgewise, through a fluid is unstable. But the motion of either body, with its length or its broadside perpendicular to the direction of motion, is stable. Observation proves the assertion we have just made, for real fluids, air and water, and for a great variety of circumstances affecting the motion; and we shall return to the subject later, as being not only of great practical importance, but profoundly interesting, and by no means difficult in theory.

304. The motion of a single particle affords simpler and not less instructive illustrations of stability and instability. Thus if a weight, hung from a fixed point by a light inextensible cord, be set in motion so as to describe a circle about a vertical line through its position of equilibrium, its motion is stable. For, as we shall see later, if disturbed infinitely little in direction without gain or loss of energy, it will describe a sinuous path, cutting the undisturbed circle at points successively distant from one another by definite fractions of the circumference, depending upon the angle of inclination of the string to the vertical. When this angle is very small, the motion is sensibly the same as that of a particle confined to one plane and moving under the influence of an attractive force towards a fixed point, simply proportional to the distance; and the disturbed path cuts the undisturbed circle four times in a revolution. Or if a particle confined to one plane, move under the influence of a centre in this plane, attracting with a force inversely as the square of the distance, a path infinitely little disturbed from a circle will cut the circle twice in a revolution. Or if the law of central force be the nth power of the distance, and if $n+3$ be positive, the disturbed path will cut the undisturbed circular orbit at successive angular intervals, each equal to $\frac{\pi}{\sqrt{n+3}}$. But the motion will be unstable if n be negative, and $-n>3$.

305. The case of a particle moving on a smooth fixed surface under the influence of no other force than that of the constraint, and therefore always moving along a geodetic line of the surface, affords extremely simple illustrations of stability and instability. For instance, a particle placed on the inner circle of the surface of an anchor-ring, and projected in the plane of the ring, would move perpetually in that circle, but unstably, as the smallest disturbance would clearly send it away from this path, never to return until after a digression round the outer edge. (We suppose of course that the particle is held to the surface, as if it were placed in the infinitely narrow space between a solid ring and a hollow one enclosing it.) But if a particle is placed on the outermost, or greatest, circle of the ring, and projected in its plane, an infinitely small disturbance will cause it to describe a sinuous path cutting the circle at points round it successively distant by angles each equal to $\pi\sqrt{\frac{b}{a}}$, and therefore at intervals of time, each equal to $\frac{\pi}{\omega}\sqrt{\frac{b}{a}}$, where a denotes the radius of that circle, ω the angular velocity in it, and b the radius of the circular cross section of the ring. This is proved by remarking that an infinitely narrow band from the outermost part of the ring has, at each point, a and b from its principal radii of curvature, and therefore (§ 134) has for its geodetic lines the great circles of a sphere of radius $\sqrt{ab}$, upon which it may be bent.

306. In all these cases the undisturbed motion has been circular or rectilineal, and, when the motion has been stable, the effect of a disturbance has been *periodic*, or recurring with the same phases in equal successive intervals of time. An illustration of thoroughly stable motion in which the effect of a disturbance is not 'periodic,' is presented by a particle sliding down an inclined groove under the action of gravity. To take the simplest case, we may consider a particle sliding down along the lowest straight line of an inclined hollow cylinder. If slightly disturbed from this straight line, it will oscillate on each side of it perpetually in its descent, but not with a uniform periodic motion, though the durations of its excursions to each side of the straight line are all equal.

307. A very curious case of stable motion is presented by a particle constrained to remain on the surface of an anchor-ring fixed in a vertical plane, and projected along the great circle from any point of it, with any velocity. An infinitely small disturbance will give rise to a disturbed motion of which the path will cut the vertical circle over and over again for ever, at unequal intervals of time, and unequal angles of the circle; and obviously not recurring periodically in any cycle, except with definite particular values for the whole energy, some of which are less and an infinite number are greater than that which just suffices to bring the particle to the highest point of the ring. The full mathematical investigation of these circumstances would afford an excellent exercise in the theory of differential equations, but it is not necessary for our present illustrations.

308. In this case, as in all of stable motion with only two degrees of freedom, which we have just considered, there has been stability throughout the motion; and an infinitely small disturbance from any point of the motion has given a disturbed path which intersects the undisturbed path over and over again at finite intervals of time. But, for the sake of simplicity, at present confining our attention to two degrees of freedom, we have a *limited* stability in the motion of an unresisted projectile, which satisfies the criterion of stability only at points of its upward, not of its downward, path. Thus if $MOPQ$ be the path of a projectile, and if at O it be disturbed by an infinitely

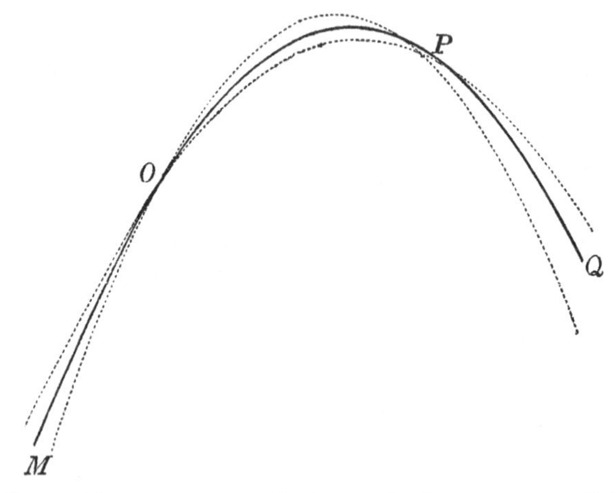

small force either way perpendicular to its instantaneous direction of motion, the disturbed path will cut the undisturbed infinitely near the point P where the direction of motion is perpendicular to that at O: as we easily see by considering that the line joining two particles projected from one point at the same instant with equal velocities in the directions of any two lines, will always remain perpendicular to the line bisecting the angle between these two.

309. The principle of varying action gives a mathematical criterion for stability or instability in every case of motion. Thus in the first place it is obvious (§§ 308, 311), that if the action is a true minimum in the motion of a system from any one configuration to the configuration reached at any other time, however much later, the motion is thoroughly unstable. For instance, in the motion of a particle constrained to remain on a smooth fixed surface, and uninfluenced by gravity, the action is simply the length of the path, multiplied by the constant velocity. Hence in the particular case of a particle uninfluenced by gravity, moving round the inner circle in the plane of an anchor-ring considered above, the action, or length of path, is clearly a minimum for any one point to the point reached at any subsequent time. (The action is not merely a minimum, but is the least possible, from any point of the circular path to any other, through less than half a circumference of the circle.) On the other hand, although the path from any point in the greatest circle of the ring to any other at a distance from it along the circle, less than $\pi\sqrt{ab}$, is clearly least possible if along the circumference; the path of absolutely least length is not along the circumference between two points at a greater circular distance than $\pi\sqrt{ab}$ from one another, nor is the path along the circumference between them a minimum at all in this latter case. On any surface whatever which is everywhere anticlastic, or along a geodetic of any surface which passes altogether through an anticlastic region, the motion is thoroughly unstable. For if it were stable from any point O, we should have the given undisturbed path, and the disturbed path from O cutting it at some point Q—two different geodetic lines joining two points; which is impossible on any anticlastic surface, inasmuch as the sum of the exterior angles of any closed figure of geodetic lines exceeds four right angles when the integral curvature of the enclosed area is negative, which is the case for every portion of surface thoroughly anticlastic. But, on the other hand, it is easily proved that if we have an endless rigid band of curved surface everywhere synclastic, with a geodetic line running through its middle, the motion of a particle projected along this line will be stable throughout, and an infinitely slight disturbance will give a disturbed path cutting the given undisturbed path again and again for ever at successive distances differing according to the different specific curvatures of the intermediate portions of the surface.

310. If, from any one configuration, two courses differing infinitely little from one another, have again a configuration in common, this second configuration will be called a kinetic focus relatively to the first: or (because of the reversibility of the motion) these two configurations will be called conjugate kinetic foci. Optic foci, if for a moment we adopt the corpuscular theory of light, are included as a particular case of kinetic foci in general. But it is not difficult to prove that there must be finite intervals of space and time between two conjugate foci in every motion of every kind of system, only provided the kinetic energy does not vanish.

311. Now it is obvious that, provided only a sufficiently short course is considered, the *action*, in any natural motion of a system, is less than for any other course between its terminal configurations. It will be proved presently (§ 318) that the first configuration up to which the action, reckoned from a given initial configuration, ceases to be a minimum, is the first kinetic focus; and conversely, that when the first kinetic focus is passed, the action, reckoned from the initial configuration, ceases to be a minimum; and therefore of course can never again be a minimum, because a course of shorter action, deviating infinitely little from it, can be found for a part, without altering the remainder of the whole, natural course.

312. In such statements as this it will frequently be convenient to indicate particular configurations of the system by single letters, as O, P, Q, R; and any particular course, in which it moves through configurations thus indicated, will be called the course $O \ldots P \ldots Q \ldots R$. The *action* in any natural course will be denoted simply by the terminal letters, taken in the order of the motion. Thus OR will denote the action from O to R; and therefore $OR = -RO$. When there are more real natural courses from O to R than one, the analytical expression for OR will have more than one real value; and it may be necessary to specify for which of these courses the action is reckoned. Thus we may have

$$OR \text{ for } O \ldots E \ldots R,$$
$$OR \text{ for } O \ldots E' \ldots R,$$
$$OR \text{ for } O \ldots E'' \ldots R,$$

three different values of one algebraic irrational expression.

313. In terms of this notation the preceding statement (§ 311) may be expressed thus:—If, for a conservative system, moving on a certain course $O \ldots P \ldots O' \ldots P'$, the first kinetic focus conjugate to O be O', the action OP, in this course, will be less than the action along any other course deviating infinitely little from it: but, on the other hand, OP' is greater than the actions in some courses from O to P' deviating infinitely little from the specified natural course $O \ldots P \ldots O' \ldots P'$.

314. It must not be supposed that the action along OP is necessarily *the least possible* from O to P. There are, in fact, cases in which the action ceases to be *least of all possible*, before a kinetic focus is reached. Thus if $OEAPO'E'A'$ be a sinuous geodetic line cutting the outer circle of an anchor-ring, or the equator of an oblate spheroid, in successive points O, A, A', it is easily seen that O', the first kinetic focus conjugate to O, must lie somewhat beyond A. But the length $OEAP$, although a *minimum* (a stable position for a stretched string), is not the shortest distance on the surface from O to P, as *this* must obviously be a line lying entirely on one side

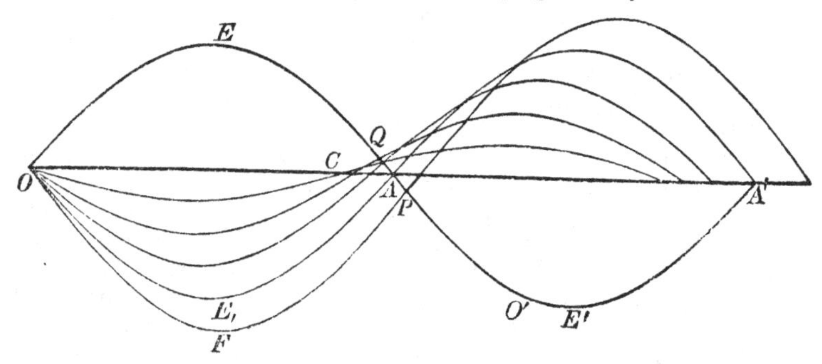

of the great circle. From O to any point, Q, short of A, the distance along the geodetic $OEQA$ is clearly the least possible: but if Q be near enough to A (that is to say, between A and the point in which the envelop of the geodetics drawn from O, cuts OEA), there will also be two other geodetics from O to Q. The length of one of these will be a minimum, and that of the other not a minimum. If Q is moved forward to A, the former becomes OE,A, equal and similar to OEA, but on the other side of the great circle: and the latter becomes the great circle from O to A. If now Q be moved on to P, beyond A, the minimum geodetic $OEAP$ ceases to be the less of the two minima, and the geodetic OFP lying altogether on the other side of the great circle becomes the least possible line from O to P. But until P is advanced beyond the point O', in which it is cut by another geodetic from O lying infinitely nearly along it, the length $OEAP$ remains a minimum according to the general proposition of § 311.

315. As it has been proved that the action from any configuration ceases to be a minimum at the first conjugate kinetic focus, we see immediately that if O' be the first kinetic focus conjugate to O, reached after passing O, no two configurations on this course from O to O' can be kinetic foci to one another. For, the action from O just ceasing to be a minimum when O' is reached, the action between any two intermediate configurations of the same course is necessarily a minimum.

316. When there are i degrees of freedom to move there are in general, on any natural course from any particular configuration, O, at least $i-1$ kinetic foci conjugate to O. Thus, for example, on the course of a ray of light emanating from a luminous point O, and passing through the centre of a convex lens held obliquely to its path, there are two kinetic foci conjugate to O, as defined above, being the points in which the line of the central ray is cut by the so-called 'focal lines'[1] of a pencil of rays diverging from O and made convergent after passing through the lens. But some or all of these kinetic foci may be on the course previous to O; as, for instance, in the case of a common projectile when its course passes obliquely downwards through O. Or some or all may be lost, as when, in the optical illustration just referred to, the lens is only strong enough to produce convergence in one of the principal planes, or too weak to produce convergence in either. Thus also in the case of the undisturbed rectilineal motion of a point, or in the motion of a point uninfluenced by force, on an anticlastic surface (§ 309), there are no real kinetic foci. In the motion of a projectile (not confined to one vertical plane) there can be only one kinetic focus on each path, conjugate to one given point; though there are three degrees of freedom. Again, there may be any number

[1] In our second volume we hope to give all necessary elementary explanations on this subject.

more than $i-1$ of foci in one course, all conjugate to one configuration, as for instance on the course of a particle, uninfluenced by force moving round the surface of an anchor-ring, along either the outer great circle, or along a sinuous geodetic such as we have considered in § 311, in which clearly there are an infinite number of foci each conjugate to any one point of the path, at equal successive distances from one another.

317. If $i-1$ distinct[1] courses from a configuration O, each differing infinitely little from a certain natural course $O \,.\,.\, E \,.\,.\, O_1 \,.\,.\, O_2 \,.\,.\,.\,.\,.\, O_{i-1} \,.\,.\, Q$, cut it in configurations $O_1, O_2, O_3, \ldots O_{i-1}$, and if, besides these, there are not on it any other kinetic foci conjugate to O, between O and Q, and no focus at all, conjugate to E, between E and Q, the action in this natural course from O to Q is the maximum for all courses $O \ldots P_{,,} P_{,} \ldots Q$; $P_{,}$ being a configuration infinitely nearly agreeing with some configuration between E and O_1 of the standard course $O \,.\,.\, E \,.\,.\, O_1 \,.\,.\, O_2 \,.\,.\,.\,.\, O_{i-1} \,.\,.\, Q$, and $O \ldots P_{,,} P_{,} \ldots Q$ denoting the natural courses between O and $P_{,,}$ and $P_{,}$ and Q, which deviate infinitely little from this standard course.

318. Considering now, for simplicity, only cases in which there are but two degrees (§ 165) of freedom to move, we see that after any infinitely small conservative disturbance of a system in passing through a certain configuration, the system will first again pass through a configuration of the undisturbed course, at the first configuration of the latter at which the action in the undisturbed motion ceases to be a minimum. For instance, in the case of a particle, confined to a surface, and subject to any conservative system of force, an infinitely small conservative disturbance of its motion through any point, O, produces a disturbed path, which cuts the undisturbed path at the first point, O', at which the action in the undisturbed path from O ceases to be a minimum. Or, if projectiles, under the influence of gravity alone, be thrown from one point, O, in all directions with equal velocities, in one vertical plane, their paths, as is easily proved, intersect one another consecutively in a parabola, of which the focus is O, and the vertex the point reached by the particle projected directly upwards. The actual course of each particle from O is the course of least possible action to any point, P, reached before the enveloping parabola, but is not a course of minimum action to any point, Q, in its path after the envelop is passed.

319. Or again, if a particle slides round along the greatest circle of the smooth inner surface of a hollow anchor-ring, the 'action,' or simply the length of path, from point to point, will be least possible for lengths (§ 305) less than $\pi\sqrt{ab}$. Thus if a string be tied round outside on the greatest circle of a perfectly smooth anchor-ring, it will slip off unless held in position by staples, or checks of some kind, at

[1] Two courses are here called not distinct if they differ from one another only in the absolute magnitude, not in the proportions, of the components of the deviations by which they differ from the standard course.

distances of not less than this amount, $\pi\sqrt{ab}$, from one another in succession round the circle. With reference to this example, see also § 314, above.

Or, if a particle slides down an inclined hollow cylinder, the action from any point will be the least possible along the straight path to any other point reached in a time less than that of the vibration one way of a simple pendulum of length equal to the radius of the cylinder, and influenced by a force equal to $g \cos i$, instead of g the whole force of gravity. But the action will not be a minimum from any point, along the straight path, to any other point reached in a longer time than this. The case in which the groove is horizontal ($i=0$) and the particle is projected along it, is particularly simple and instructive, and may be worked out in detail with great ease, without assuming any of the general theorems regarding action.

CHAPTER III.

EXPERIENCE.

320. By the term Experience, in physical science, we designate, according to a suggestion of Herschel's, our means of becoming acquainted with the material universe and the laws which regulate it. In general the actions which we see ever taking place around us are *complex*, or due to the simultaneous action of many causes. When, as in astronomy, we endeavour to ascertain these causes by simply watching their effects, we *observe;* when, as in our laboratories, we interfere arbitrarily with the causes or circumstances of a phenomenon, we are said to *experiment.*

321. For instance, supposing that we are possessed of instrumental means of measuring time and angles, we may trace out by successive observations the relative position of the sun and earth at different instants; and (the method is not susceptible of any accuracy, but is alluded to here only for the sake of illustration) from the variations in the apparent diameter of the former we may calculate the ratios of our distances from it at those instants. We have thus a set of observations involving time, angular position with reference to the sun, and ratios of distances from it; sufficient (if numerous enough) to enable us to discover the laws which connect the variations of these co-ordinates.

Similar methods may be imagined as applicable to the motion of any planet about the sun, of a satellite about its primary, or of one star about another in a binary group.

322. In general all the data of Astronomy are determined in this way, and the same may be said of such subjects as Tides and Meteorology. Isothermal Lines, Lines of Equal Dip or Intensity, Lines of No Declination, the Connexion of Solar Spots with Terrestrial Magnetism, and a host of other data and phenomena, to be explained under the proper heads in the course of the work, are thus deducible from *Observation* merely. In these cases the apparatus for the gigantic experiments is found ready arranged in Nature, and all that the philosopher has to do is to watch and measure their progress to its last details.

323. Even in the instance we have chosen above, that of the planetary motions, the observed effects are complex; because, unless possibly in the case of a double star, we have no instance of the *undisturbed* action of one heavenly body on another; but to a first approximation the motion of a planet about the sun is found to be the same as if no other bodies than these two existed; and the approximation is sufficient to indicate the probable law of mutual action, whose full confirmation is obtained when, *its* truth being assumed, the disturbing effects thus calculated are allowed for, and found to account completely for the observed deviations from the consequences of the first supposition. This may serve to give an idea of the mode of obtaining the laws of phenomena, which can only be observed in a complex form; and the method can always be directly applied when one cause is known to be pre-eminent.

324. Let us take a case of the other kind—that in which the effects are so complex that we cannot deduce the causes from the observation of combinations arranged in Nature, but must endeavour to form for ourselves other combinations which may enable us to study the effects of each cause separately, or at least with only slight modification from the interference of other causes.

A stone, when dropped, falls to the ground; a brick and a boulder, if dropped from the top of a cliff at the same moment, fall side by side, and reach the ground together. But a brick and a slate do not; and while the former falls in a nearly vertical direction, the latter describes a most complex path. A sheet of paper or a fragment of gold-leaf presents even greater irregularities than the slate. But by a slight modification of the circumstances, we gain a considerable insight into the nature of the question. The paper and gold-leaf, if rolled into balls, fall nearly in a vertical line. Here, then, there are evidently at least two causes at work, one which tends to make all bodies fall, and that vertically; and another which depends on the form and substance of the body, and tends to retard its fall and alter its vertical direction. How can we study the effects of the former on all bodies without sensible complication from the latter? The effects of Wind, etc., at once point out *what* the latter cause is, the air (whose existence we may indeed suppose to have been discovered by such effects); and to study the nature of the action of the former it is necessary to get rid of the complications arising from the presence of air. Hence the necessity for *Experiment.* By means of an apparatus to be afterwards described, we remove the greater part of the air from the interior of a vessel, and in *that* we try again our experiments on the fall of bodies; and now a general law, simple in the extreme, though most important in its consequences, is at once apparent—viz. that *all* bodies, of whatever size, shape, or material, if dropped side by side at the same instant, fall side by side in a space void of air. Before experiment had thus separated the phenomena, hasty philosophers had rushed to the conclusion that some bodies possess the quality of *heaviness*, others that of *lightness*, etc. Had this state of things remained, the law of gravitation, vigorous though its action be throughout the universe, could never have been recognized as a general principle by the human mind.

Mere observation of lightning and its effects could never have led to the discovery of their relation to the phenomena presented by rubbed amber. A modification of the course of Nature, such as the bringing down of atmospheric electricity into our laboratories, was necessary. Without experiment we could never even have learned the existence of terrestrial magnetism.

325. In all cases when a particular agent or cause is to be studied, experiments should be arranged in such a way as to lead, if possible, to results depending on it alone; or, if this cannot be done, they should be arranged so as to increase the effects due to the cause to be studied till these so far exceed the unavoidable concomitants, that the latter may be considered as only disturbing, not essentially modifying, the effects of the principal agent.

Thus, in order to find the nature of the action of a galvanic current upon a magnetized needle, we may adopt either of these methods. For instance, we may neutralize the disturbing effects of the earth's magnetism on the needle by properly placing a magnetized bar in its neighbourhood. This is an instance of the first method.

Or we may, by increasing the strength of the current, or by coiling the wire many times about the needle (as will be explained when we describe the galvanometer), multiply the effects of the current so that those of the earth's magnetism may be negligible in comparison.

326. In some cases, however, the latter mode of procedure is utterly deceptive—as, for instance, in the use of multiplying condensers for the detection of very small electro-motive forces. In this case the friction between the parts of the condenser often produces more electricity than that which is to be measured, so that the true results cannot be deduced: a feeble positive charge, for instance, may be trebled, neutralized, or even changed to a negative one, by variations of manipulation so delicate as to be undiscoverable, and therefore unavoidable.

327. We thus see that it is uncertain which of these methods may be preferable in any particular case; and indeed, in discovery, he is the most likely to succeed who, not allowing himself to be disheartened by the non-success of one form of experiment, carefully varies his methods, and thus interrogates in every conceivable manner the subject of his investigations.

328. A most important remark, due to Herschel, regards what are called *residual* phenomena. When, in an experiment, all known causes being allowed for, there remain certain unexplained effects (excessively slight it may be), these must be carefully investigated, and every conceivable variation of arrangement of apparatus, etc., tried; until, if possible, we manage so to exaggerate the residual phenomenon as to be able to detect its cause. It is here, perhaps, that in the present state of science we may most reasonably look for extensions of our knowledge; at all events we are warranted by the recent history of Natural Philosophy in so doing. Thus, to take only a very few instances, and to say nothing of the discovery of electricity and magnetism by the ancients, the peculiar smell observed in a room in which an electrical machine is kept in action, was long ago observed, but called the 'smell of electricity,' and thus left unexplained. The sagacity of Schönbein led to the discovery that this is due to the formation of Ozone, a most extraordinary body, of enormous chemical energies; whose nature is still uncertain, though the attention of chemists has for years been directed to it.

329. Slight anomalies in the motion of Uranus led Adams and Le Verrier to the discovery of a new planet; and the fact that a magnetized needle comes to rest sooner when vibrating above a copper plate than when the latter is removed, led Arago to what was once called magnetism of rotation, but has since been explained, immensely extended, and applied to most important purposes. In fact, this accidental remark about the oscillation of a needle led to facts from which, in Faraday's hands, was evolved the grand discovery of the Induction of Electrical Currents by magnets or by other currents. We need not enlarge upon this point, as in the following pages the proofs of the truth and usefulness of the principle will continually recur. Our object has been not so much to give applications as methods, and to show, if possible, how to attack a new combination, with the view of separating and studying in detail the various causes which generally conspire to produce observed phenomena, even those which are apparently the simplest.

330. If, on repetition several times, an experiment continually gives different results, it must either have been very carelessly performed, or there must be some disturbing cause not taken account of. And, on the other hand, in cases where no very great coincidence is likely on repeated trials, an unexpected degree of agreement between the results of various trials should be regarded with the utmost suspicion, as probably due to some unnoticed peculiarity of the apparatus employed. In either of these cases, however, careful observation cannot fail to detect the cause of the discrepancies or of the unexpected agreement, and may possibly lead to discoveries in a totally unthought-of quarter. Instances of this kind may be given without limit; one or two must suffice.

331. Thus, with a *very* good achromatic telescope a star appears to have a sensible disc. But, as it is observed that the discs of all stars appear to be of equal angular diameter, we of course suspect some common error. Limiting the aperture of the object-glass *increases* the appearance in question, which, on full investigation, is found to have nothing to do with discs at all. It is, in fact, a diffraction phenomenon, and will be explained in our chapters on Light.

Again, in measuring the velocity of Sound by experiments conducted at night with cannon, the results at one station were never found to agree exactly with those at the other; sometimes, indeed, the differences were very considerable. But a little consideration led to the remark, that on those nights in which the discordance was greatest a strong wind was blowing nearly from one station to the other. Allowing for the obvious effect of this, or rather eliminating it altogether, the mean velocities on different evenings were found to agree very closely.

332. It may perhaps be advisable to say a few words here about the use of hypotheses, and especially those of very different gradations of value which are promulgated in the form of Mathematical Theories of different branches of Natural Philosophy.

333. Where, as in the case of the planetary motions and disturbances, the forces concerned are thoroughly known, the mathematical theory is absolutely true, and requires only analysis to work out its remotest details. It is thus, in general, far ahead of observation, and is competent to predict effects not yet even observed—as, for instance, Lunar Inequalities due to the action of Venus upon the Earth, etc. etc., to which no amount of observation, unaided by theory, would ever have enabled us to assign the true cause. It may also, in such subjects as Geometrical Optics, be carried to developments far beyond the reach of experiment; but in this science the assumed bases of the theory are only approximate, and it fails to explain in all their peculiarities even such comparatively simple phenomena as Halos and Rainbows; though it is perfectly successful for the practical purposes of the maker of microscopes and telescopes, and has, in these cases, carried the construction of instruments to a degree of perfection which merely tentative processes never could have reached.

334. Another class of mathematical theories, based to a certain extent on experiment, is at present useful, and has even in certain cases pointed to new and important results, which experiment has subsequently verified. Such are the Dynamical Theory of Heat, the Undulatory Theory of Light, etc. etc. In the former, which is based upon the experimental fact that *heat is motion*, many formulae are at present obscure and uninterpretable, because we do not know *what* is moving or *how* it moves. Results of the theory in which these are not involved, are of course experimentally verified. The same difficulties exist in the Theory of Light. But before this obscurity can be perfectly cleared up, we must know something of the ultimate, or *molecular*, constitution of the bodies, or groups of molecules, at present known to us only in the aggregate.

335. A third class is well represented by the Mathematical Theories of Heat (Conduction), Electricity (Statical), and Magnetism (Permanent). Although we do not know *how* Heat is propagated in bodies, nor *what* Statical Electricity or Permanent Magnetism are, the laws of their forces are as certainly known as that of Gravitation, and can therefore like it be developed to their consequences, by the application of Mathematical Analysis. The works of Fourier[1],

[1] *Théorie Analytique de la Chaleur.* Paris, 1822.

Green[1], and Poisson[2], are remarkable instances of such development. Another good example is Ampère's Theory of Electro-dynamics. And this leads us to a fourth class, which, however ingenious, must be regarded as in reality pernicious rather than useful.

336. A good type of such a theory is that of Weber, which professes to supply a physical basis for Ampère's Theory of Electro-dynamics, just mentioned as one of the admirable and really useful third class. Ampère contents himself with experimental data as to the action of closed currents on each other, and from these he deduces mathematically the action which an element of one current ought to exert on an element of another—if such a case could be submitted to experiment. This cannot possibly lead to confusion. But Weber goes farther, he assumes that an electric current consists in the motion of particles of two kinds of electricity moving in opposite directions through the conducting wire; and that these particles exert forces on other such particles of electricity, when in relative motion, different from those they would exert if at relative rest. In the present state of science this is wholly unwarrantable, because it is impossible to conceive that the hypothesis of two electric fluids can be true, and besides, because the conclusions are inconsistent with the Conservation of Energy, which we have numberless experimental reasons for receiving as a general principle in nature. It only adds to the danger of such theories, when they happen to explain further phenomena, as those of induced currents are explained by that of Weber. Another of this class is the Corpuscular Theory of Light, which for a time did great mischief, and which could scarcely have been justifiable unless a luminous corpuscle had been actually seen and examined. As such speculations, though dangerous, are interesting, and often beautiful (as, for instance, that of Weber), we will refer to them again under the proper heads.

337. Mathematical theories of physical forces are, in general, of one of two species. First, those in which the fundamental assumption is far more general than is necessary. Thus the celebrated equation of Laplace's Functions contains the mathematical foundation of the theories of Gravitation, Statical Electricity, Permanent Magnetism, Permanent Flux of Heat, Motion of Incompressible Fluids, etc. etc., and has therefore to be accompanied by limiting considerations when applied to any one of these subjects.

Again, there are those which are built upon a few experiments, or simple but inexact hypotheses, only; and which require to be modified in the way of extension rather than limitation. As a notable example, we may refer to the whole subject of Abstract Dynamics, which requires extensive modifications (explained in Division III.) before it can, in general, be applied to practical purposes.

338. When the most probable result is required from a number of

[1] *Essay on the Application of Mathematical Analysis to the Theories of Electricity and Magnetism.* Nottingham, 1828. Reprinted in *Crelle's Journal.*
[2] *Mémoires sur le Magnétisme. Mém. de l'Acad. des Sciences,* 1811.

observations of the same quantity which do not exactly agree, we must appeal to the mathematical theory of probabilities to guide us to a method of combining the results of experience, so as to eliminate from them, as far as possible, the inaccuracies of observation. But it must be explained that we do not at present class as *inaccuracies of observation* any errors which may affect alike every one of a series of observations, such as the inexact determination of a zero-point or of the essential units of time and space, the personal equation of the observer, etc. The process, whatever it may be, which is to be employed in the elimination of errors, is applicable even to these, but only when *several distinct series* of observations have been made, with a change of instrument, or of observer, or of both.

339. We understand as inaccuracies of observation the whole class of errors which are as likely to lie in one direction as another in successive trials, and which we may fairly presume would, on the average of an infinite number of repetitions, exactly balance each other in excess and defect. Moreover, we consider only errors of such a kind that their probability is the less the greater they are; so that such errors as an accidental reading of a wrong number of whole degrees on a divided circle (which, by the way, can in general be probably corrected by comparison with other observations) are not to be included.

340. Mathematically considered, the subject is by no means an easy one, and many high authorities have asserted that the reasoning employed by Laplace, Gauss, and others, is not well founded; although the results of their analysis have been generally accepted. As an excellent treatise on the subject has recently been published by Airy, it is not necessary for us to do more than sketch in the most cursory manner what is called the *Method of Least Squares*.

341. Supposing the zero-point and the graduation of an instrument (micrometer, mural circle, thermometer, electrometer, galvanometer, etc.) to be *absolutely* accurate, successive readings of the value of a quantity (linear distance, altitude of a star, temperature, potential, strength of an electric current, etc.) may, and in general do, continually differ. What is most probably the true value of the observed quantity?

The most probable value, in all such cases, if the observations are all equally reliable, will evidently be the simple mean; or if they are not equally reliable, the mean found by attributing *weights* to the several observations in proportion to their presumed exactness. But if several such means have been taken, or several single observations, and if these several means or observations have been differently qualified for the determination of the sought quantity (some of them being likely to give a more exact value than others), we must assign *theoretically* the best method of combining them in practice.

342. Inaccuracies of observation are, in general, as likely to be in excess as in defect. They are also (as before observed) more likely to be small than great; and (practically) large errors are not to be expected at all, as such would come under the class of *avoidable mistakes*. It follows that in any one of a series of observations of the same quantity the probability of an error of magnitude x, must depend upon x^2, and must be expressed by some function whose value diminishes very rapidly as x increases. The probability that the error lies between x and $x+\delta x$, where δx is very small, must also be proportional to δx. The law of error thus found is

$$\frac{1}{\sqrt{\pi}} \epsilon^{-\frac{x^2}{h^2}} \frac{\delta x}{h}$$

where h is a constant, indicating the degree of coarseness or delicacy of the system of measurement employed. The co-efficient $\frac{1}{\sqrt{\pi}}$ secures that the sum of the probabilities of all possible errors shall be unity, as it ought to be.

343. The *Probable Error* of an observation is a numerical quantity such that the error of the observation is as likely to exceed as to fall short of it in magnitude.

If we assume the law of error just found, and call P the probable error in one trial, we have the approximate result

$$P = 0{\cdot}477h.$$

344. The probable error of any given multiple of the value of an observed quantity is evidently the same multiple of the probable error of the quantity itself.

The probable error of the sum or difference of two quantities, affected by *independent* errors, is the square root of the sum of the squares of their separate probable errors.

345. As above remarked, the principal use of this theory is in the deduction, from a large series of observations, of the values of the quantities sought in such a form as to be liable to the smallest probable error. As an instance—by the principles of physical astronomy, the place of a planet is calculated from assumed values of the elements of its orbit, and tabulated in the *Nautical Almanac*. The *observed* places do not exactly agree with the predicted places, for two reasons—first, the data for calculation are not exact (and in fact the main object of the observation is to correct their assumed values); second, the observation is in error to some unknown amount. Now the difference between the observed, and the calculated, places depends on the errors of assumed elements and of observation. Our methods are applied to eliminate as far as possible the second of these, and the resulting equations give the required corrections of the elements.

Thus if θ be the calculated R.A. of a planet: δa, δe, $\delta\varpi$, etc., the corrections required for the assumed elements: the true R.A. is

$$\theta + A\delta a + E\delta e + \Pi\delta\varpi + \text{etc.},$$

where A, E, Π, etc., are approximately known. Suppose the observed R.A. to be Θ, then

$$\theta + A\delta a + E\delta e + \Pi\delta\varpi + \ldots = \Theta,$$

or $$A\delta a + E\delta e + \Pi\delta\varpi + \ldots = \Theta - \theta,$$

a known quantity, subject to error of observation. Every observation made gives us an equation of the same *form* as this, and in general the number of observations greatly exceeds that of the quantities δa, δe, $\delta\varpi$, etc., to be found.

346. The theorems of § 344 lead to the following rule for combining any number of such equations which contain a smaller number of unknown quantities :—

Make the probable error of the second member the same in each equation, by the employment of a proper factor : multiply each equation by the co-efficient of x in it and add all, for one of the final equations ; and so, with reference to y, z, etc., for the others. The probable errors of the values of x, y, etc., found from these final equations will be less than those of the values derived from any other *linear* method of combining the equations.

This process has been called the method of *Least Squares,* because the values of the unknown quantities found by it are such as to render the sum of the squares of the errors of the original equations a minimum.

347. When a series of observations of the same quantity has been made at different times, or under different circumstances, the law connecting the value of the quantity with the time, or some other variable, may be derived from the results in several ways—all more or less approximate. Two of these methods, however, are so much more extensively used than the others, that we shall devote a page or two here to a preliminary notice of them, leaving detailed instances of their application till we come to Heat, Electricity, etc. They consist in (1) a *Curve,* giving a graphic representation of the relation between the ordinate and abscissa, and (2) an *Empirical Formula* connecting the variables.

348. Thus if the abscissae represent intervals of time, and the ordinates the corresponding height of the barometer, we may construct curves which show at a glance the dependence of barometric pressure upon the time of day; and so on. Such curves may be accurately drawn by photographic processes on a sheet of sensitive paper placed behind the mercurial column, and made to move past it with a uniform horizontal velocity by clockwork. A similar process is applied to the Temperature and Electricity of the atmosphere, and to the components of terrestrial magnetism.

349. When the observations are not, as in the last section, continuous, they give us only a series of points in the curve, from which, however, we may in general approximate very closely to the result of continuous observation by drawing, *liberâ manu,* a curve passing through these points. This process, however, must be employed with great caution; because, unless the observations are sufficiently close to each other, most important fluctuations in the curve may escape notice. It is applicable, with abundant accuracy, to all cases where the quantity observed changes very slowly. Thus, for instance, weekly observations of the temperature at depths of from 6 to 24 feet underground were found by Forbes sufficient for a very accurate approximation to the law of the phenomenon.

350. As an instance of the processes employed for obtaining an empirical formula, we may mention methods of *Interpolation*, to which the problem can always be reduced. Thus from sextant observations, at known intervals, of the altitude of the sun, it is a common problem of Astronomy to determine at what instant the altitude is greatest, and what is that greatest altitude. The first enables us to find the true solar time at the place, and the second, by the help of the *Nautical Almanac*, gives the latitude. The calculus of finite differences gives us formulae proper for various data; and Lagrange has shown how to obtain a very useful one by elementary algebra.

In finite differences we have

$$f(x+h)=f(x)+h\Delta f(x)+\frac{h(h-1)}{1.2}\Delta^2 f(x)+\ldots$$

This is useful, inasmuch as the successive differences, $\Delta f(x)$, $\Delta^2 f(x)$, etc., are easily calculated from the tabulated results of observation, provided these have been taken for equal successive increments of x.

If for values $x_1, x_2, \ldots x_n$, a function takes the values $y_1, y_2, y_3, \ldots y_n$, Lagrange gives for it the obvious expression

$$\left[\frac{y_1}{x-x_1}\frac{1}{(x_1-x_2)(x_1-x_3)\ldots(x_1-x_n)}+\frac{y_2}{x-x_2}\frac{1}{(x_2-x_1)(x_2-x_3)\ldots(x_2-x_n)}+\ldots\right](x-x_1)(x-x_2)\ldots(x-x_n).$$

Here it is assumed that the function required is a rational and integral one in x of the $n-1$th degree; and, in general, a similar limitation is in practice applied to the other formula above; for in order to find the complete expression for $f(x)$, it is necessary to determine the values of $\Delta f(x)$, $\Delta^2 f(x)$, ... If n of the co-efficients be required, so as to give the n chief terms of the general value of $f(x)$, we must have n observed simultaneous values of x and $f(x)$, and the expression becomes determinate and of the $n-1$th degree in h.

In practice it is usually sufficient to employ at most three terms of the first series. Thus to express the length l of a rod of metal as depending on its temperature t, we may assume

$$l=l_0+A(t-t_0)+B(t-t_0)^2,$$

l_0 being the measured length at any temperature t_0. A and B are to be found by the method of least squares from values of l observed for different given values of t.

351. These formulae are practically useful for calculating the probable values of any observed element, for values of the independent variable lying within the range for which observation has given values of the element. But except for values of the independent variable either actually within this range, or not far beyond it in either direction, these formulae express functions which, in general, will differ more and more widely from the truth the further their application is pushed beyond the range of observation.

In a large class of investigations the observed element is in its nature a periodic function of the independent variable. The harmonic analysis (§ 88) is suitable for all such. When the values of the independent variable for which the element has been observed are not equidifferent the co-efficients, determined according to the method of least squares, are found by a process which is necessarily very laborious; but when they are equidifferent, and especially when the difference is a submultiple of the period, the equation derived from the method of least squares becomes greatly simplified. Thus, if θ denote an angle increasing in proportion to t, the time, through four right angles in the period, T, of the phenomenon; so that

$$\theta = \frac{2\pi t}{T};$$

let
$$f(\theta) = A_0 + A_1 \cos\theta + A_2 \cos 2\theta + \dots$$
$$+ B_1 \sin\theta + B_2 \sin 2\theta + \dots,$$

where A_0, A_1, A_2, ... B_1, B_2, ... are unknown co-efficients, to be determined so that $f(\theta)$ may express the most probable value of the element, not merely at times between observations, but through all time as long as the phenomenon is strictly periodic. By taking as many of these co-efficients as there are of distinct data by observation, the formula is made to agree precisely with these data. But in most applications of the method, the periodically recurring part of the phenomenon is expressible by a small number of terms of the harmonic series, and the higher terms, calculated from a great number of data, express either irregularities of the phenomenon not likely to recur, or errors of observation. Thus a comparatively small number of terms may give values of the element even for the very times of observation, more probable than the values actually recorded as having been observed, if the observations are numerous but not minutely accurate.

The student may exercise himself in writing out the equations to determine five, or seven, or more of the co-efficients according to the method of least squares; and reducing them by proper formulae of analytical trigonometry to their simplest and most easily calculated forms where the values of θ for which $f(\theta)$ is given are equidifferent. He will thus see that when the difference is $\frac{2\pi}{i}$, i being any integer, and when the number of the data is i or any multiple of it, the equations contain each of them only one of the unknown quantities: so that the method of least squares affords the most probable values of the co-efficients, by the easiest and most direct elimination.

CHAPTER IV.

MEASURES AND INSTRUMENTS.

352. HAVING seen in the preceding chapter that for the investigation of the laws of nature we must carefully watch experiments, either those gigantic ones which the universe furnishes, or others devised and executed by man for special objects—and having seen that in all such observations accurate measurements of Time, Space, Force, etc., are absolutely necessary, we may now appropriately describe a few of the more useful of the instruments employed for these purposes, and the various standards or units which are employed in them.

353. Before going into detail we may give a rapid *résumé* of the principal Standards and Instruments to be described in this chapter. As most, if not all, of them depend on physical principles to be detailed in the course of this work—we shall assume in anticipation the establishment of such principles, giving references to the future division or chapter in which the experimental demonstrations are more particularly explained. This course will entail a slight, but unavoidable, confusion—slight, because Clocks, Balances, Screws, etc., are familiar even to those who know nothing of Natural Philosophy; unavoidable, because it is in the very nature of our subject that no one part can grow alone, each requiring for its full development the utmost resources of all the others. But if one of our departments thus borrows from others, it is satisfactory to find that it more than repays by the power which its improvement affords them.

354. We may divide our more important and fundamental instruments into four classes—

Those for measuring Time;
" " Space, linear or angular;
" " Force;
" " Mass.

Other instruments, adapted for special purposes such as the measurement of Temperature, Light, Electric Currents, etc., will come more naturally under the head of the particular physical emergencies to whose measurement they are applicable.

355. We shall now consider in order the more prominent instruments of each of these four classes, and some of their most important applications:—

Clock, Chronometer, Chronoscope, Applications to Observation and to self-registering Instruments.

Vernier and Screw-Micrometer, Cathetometer, Spherometer, Dividing Engine, Theodolite, Sextant or Circle.

Common Balance, Bifilar Balance, Torsion Balance, Pendulum, Dynamometer.

Among Standards we may mention—

1. *Time.*—Day, Hour, Minute, Second, sidereal and solar.
2. *Space.*—Yard and Mètre: Degree, Minute, Second.
3. *Force.*—Weight of a Pound or Kilogramme, etc., in any particular locality (gravitation unit); Kinetic Unit.
4. *Mass.*—Pound, Kilogramme, etc.

356. Although without instruments it is impossible to procure or apply any standard, yet, as without the standards no instrument could give us *absolute* measure, we may consider the standards first—referring to the instruments as if we already knew their principles and applications.

357. We need do no more than mention the standard of angular measure, the *Degree* or ninetieth part of a right angle, and its successive subdivisions into sixtieths called *Minutes, Seconds, Thirds,* etc. This system of division is extremely inconvenient, but it has been so long universally adopted by all Europe, that the far preferable form, the decimal division of the right angle, decreed by the French Republic when it successfully introduced other more sweeping changes, utterly failed. Seconds, however, are generally divided into decimal parts.

The decimal division is employed, of course, when *circular* measure is adopted, the unit of circular measure being the angle subtended at the centre of any circle by an arc equal in length to the radius. Thus two right angles have the circular measure π or $3{\cdot}14159\ldots$, so that π and 180° represent the same angle: and the unit angle, or the angle of which the arc is equal to radius, is $57^\circ{\cdot}29578\ldots$, or $57^\circ\ 17'\ 44''{\cdot}8$. (Compare § 41.)

Hence the number of degrees n in any angle θ given in circular measure, or the converse, will be found at once by the equation $\frac{\theta}{\pi}=\frac{n}{180}$, and therefore $n=\theta\times 57^\circ{\cdot}29578\ldots=\theta\times 57^\circ\ 17'\ 44''{\cdot}8\ldots$

358. The practical standard of time is the *Siderial Day,* being the period, nearly constant, of the earth's rotation about its axis (§ 237). It has been calculated from ancient observations of eclipses that this has not altered by $\frac{1}{10{,}000{,}000}$ of its length from 720 B.C.: but an error has been found in this calculation, and the corrected result renders it probable that the time of the earth's rotation is longer by $\frac{1}{2,700,000}$ now than at that date. From it is easily derived the *Mean Solar Day*, or the mean interval which elapses between successive passages of the sun across the meridian of any place. This is nct so nearly as the former, an absolute or invariable unit; secular changes in the period of the earth's revolution round the sun affect it, though very slightly. It is divided into 24 hours, and the hour, like the degree, is subdivided into successive sixtieths, called minutes and seconds. The usual subdivision of seconds is decimal.

It is well to observe that seconds and minutes of time are distinguished from those of angular measure by notation. Thus we have for time $13^h\ 43^m\ 27^s{\cdot}58$, but for angular measure $13°\ 43'\ 27''{\cdot}58$.

When long periods of time are to be measured, the mean solar year, consisting of 366·242203 siderial days, or 365·242242 mean solar days, or the century consisting of 100 such years, may be conveniently employed as the unit.

359. The ultimate standard of accurate chronometry must (if the human race live on the earth for a few million years) be founded on the physical properties of some body of more constant character than the earth: for instance, a carefully-arranged metallic spring, hermetically sealed in an exhausted glass vessel. The time of vibration of such a spring would be necessarily more constant from day to day than that of the balance-spring of the best possible chronometer, disturbed as this is by the train of mechanism with which it is connected: and it would certainly be more constant from age to age than the time of rotation of the earth, retarded as it now is by tidal resistance to an extent that becomes very sensible in 2000 years; and cooling and shrinking to an extent that must produce a very considerable effect on its time-keeping in fifty million years.

360. The British standard of length is the *Imperial Yard*, defined as the distance between two marks on a certain metallic bar, preserved in the Tower of London, when the whole has a temperature of 60° Fahrenheit. It was not directly derived from any fixed quantity in nature, although some important relations with natural elements have been measured with great accuracy. It has been carefully compared with the length of a second's pendulum vibrating at a certain station in the neighbourhood of London, so that if it should again be destroyed, as it was at the burning of the Houses of Parliament in 1834, and should all exact copies of it, of which several are preserved in various places, be also lost, it can be restored by pendulum observations. A less accurate, but still (unless in the event of earthquake disturbance) a very good, means of reproducing it exists in the measured base-lines of the Ordnance Survey, and the thence calculated distances between definite stations in the British Islands, which have been ascertained in terms of it with a degree of accuracy sometimes within an inch per mile, that is to say, within about $\frac{1}{60000}$.

361. In scientific investigations, we endeavour as much as possible to keep to one unit at a time, and the foot, which is defined to be one-third part of the yard, is, for British measurement, generally adopted. Unfortunately the inch, or one-twelfth of a foot, must sometimes be used, but it is subdivided decimally. The statute mile, or 1760 yards, is unfortunately often used when great lengths on land are considered; but the sea-mile, or average minute of latitude, is much to be preferred. Thus it appears that the British measurement of length is more inconvenient in its several denominations than the European measurement of time, or angles.

362. A far more perfect metrical system than the British, is the French, in which the decimal division is exclusively employed. Here the standard is the *Mètre*, defined originally as the ten-millionth part of the length of the quadrant of the earth's meridian from the pole to the equator; but now defined practically by the accurate standard mètres laid up in various national repositories in Europe. It is somewhat longer than the yard, as the following Table shows. Its great convenience is the decimal division. Thus in any expression the units represent mètres, the tens decamètres, etc.; the first decimal place represents decimètres, the second centimètres, the third millimètres, and so on.

Inch=25·39954 millimètres.	Millimètre=·03937079 inch.
Foot=3·047945 decimètres.	Decimètre=·3280899 foot.
British land-mile = 1609·315 mètres.	Kilomètre=·6213824 land-mile.
Sea-mile=1851·851 mètres.	Kilomètre=·54 sea-mile.

363. The unit of superficial measure is in Britain the square yard, in France the mètre carré. Of course we may use square inches, feet, or miles, as also square millimètres, kilomètres, etc., or the *Hectare*=10,000 square mètres.

Square inch	=	6·451367 square centimètres.
„ foot	=	9·28997 „ decimètres.
„ yard	=	83·60971 „ decimètres.
Acre	=	·4046711 of a hectare.
Square mile	=	258·9895 hectares.
Hectare	=	2·471143 acres.

364. Similar remarks apply to the cubic measure in the two countries, and we have the following Table:—

Cubic inch	=	16·38618 cubic centimètres.
„ foot	=	28·315312 „ decimètres, or *litres*.
Gallon	=	4·54346 litres.
„	=	277·274 cubic inches.
Litre	=	0·035317 cubic feet.

365. The British unit of mass is the Pound (defined by standards only); the French is the *Kilogramme*, defined originally as a litre of water at its temperature of maximum density; but now practically defined by existing standards.

Grain = 64·79896 milligrammes.	Gramme =15·43235 grains.
Pound = 453·5927 grammes.	Kilogram. =2·20362125 lbs.

Professor W. H. Miller finds (*Phil. Trans.*, 1857) that the '*kilo-*

gramme des Archives' is equal in mass to 15432·349 grains; and the '*kilogramme type laiton,*' deposited in the Ministère de l'Intérieure in Paris, as standard for French commerce, is 15432·344 grains.

366. The measurement of force, whether in terms of the weight of a stated mass in a stated locality, or in terms of the *absolute* or *kinetic* unit, has been explained in Chapter II. (See §§ 221–227.) From the measures of force and length we derive at once the measure of work or mechanical effect. That practically employed by engineers is founded on the gravitation measure of force. Neglecting the difference of gravity at London and Paris, we see from the above Tables that the following relations exist between the London and the Parisian reckoning of work:—

Foot-pound = 0·13825 kilogramme-mètre.
Kilogramme-mètre = 7·2331 foot-pounds.

367. A *Clock* is primarily an instrument which, by means of a train of wheels, records the number of vibrations executed by a pendulum; a *Chronometer* or *Watch* performs the same duty for the oscillations of a flat spiral spring—just as the train of wheel-work in a gas-meter counts the number of revolutions of the main shaft caused by the passage of the gas through the machine. As, however, it is impossible to avoid friction, resistance of air, etc., a pendulum or spring, left to itself, would not long continue its oscillations, and, while its motion continued, would perform each oscillation in less and less time as the arc of vibration diminished: a continuous supply of energy is furnished by the descent of a weight, or the uncoiling of a powerful spring. This is so applied, through the train of wheels, to the pendulum or balance-wheel by means of a mechanical contrivance called an *Escapement*, that the oscillations are maintained of nearly uniform extent, and therefore of nearly uniform duration. The construction of escapements, as well as of trains of clock-wheels, is a matter of *Mechanics*, with the details of which we are not concerned, although it may easily be made the subject of mathematical investigation. The means of avoiding errors introduced by changes of temperature, which have been carried out in *Compensation* pendulums and balances, will be more properly described in our chapters on Heat. It is to be observed that there is little inconvenience if a clock lose or gain *regularly;* that can be easily and accurately allowed for: irregular rate is fatal.

368. By means of a recent application of electricity, to be afterwards described, one good clock, carefully regulated from time to time to agree with astronomical observations, may be made (without injury to its own peformance) to control any number of other less-perfectly constructed clocks, so as to compel their pendulums to vibrate, beat for beat, with its own.

369. In astronomical observations, time is estimated to tenths of a second by a practised observer, who, while watching the phenomena, counts the beats of the clock. But for the *very* accurate measurement of short intervals, many instruments have been devised.

Thus if a small orifice be opened in a large and deep vessel full of mercury, and if we know by trial the weight of metal that escapes say in five minutes, a simple proportion gives the interval which elapses during the escape of any given weight. It is easy to contrive an adjustment by which a vessel may be placed under, and withdrawn from, the issuing stream at the time of occurrence of any two successive phenomena.

370. Other contrivances are sometimes employed, called Stop-watches, Chronoscopes, etc., which can be read off at rest, started on the occurrence of any phenomenon, and stopped at the occurrence of a second, then again read off; or which allow of the making (by pressing a stud) a slight ink-mark, on a dial revolving at a given rate, at the instant of the occurrence of each phenomenon to be noted. But, of late, these have almost entirely given place to the Electric Chronoscope, an instrument which will be fully described later, when we shall have occasion to refer to experiments in which it has been usefully employed.

371. We now come to the measurement of space, and of angles, and for these purposes the most important instruments are the *Vernier* and the *Screw*.

372. Elementary geometry, indeed, gives us the means of dividing any straight line into any assignable number of equal parts; but in practice this is by no means an accurate or reliable method. It was formerly used in the so-called Diagonal Scale, of which the construction is evident from the diagram. The reading is effected by a sliding piece whose edge is perpendicular to the length of the scale. Suppose that it is PQ whose position on the scale is required. This can evidently cut only *one* of the transverse lines. *Its* number gives the number of tenths of an inch (4 in the figure), and the horizontal line next above the point of intersection gives evidently the number of hundredths (in the present case 4). Hence the reading is 7·44. As an idea of the comparative uselessness of this method, we may mention that a quadrant of 3 feet radius, which belonged to Napier of Merchiston, and is divided on the limb by this method, reads to minutes of a degree; no higher accuracy than is now attainable by the pocket sextants made by Troughton and Simms, the radius of whose arc is virtually little more than an inch. The latter instrument is read by the help of a Vernier.

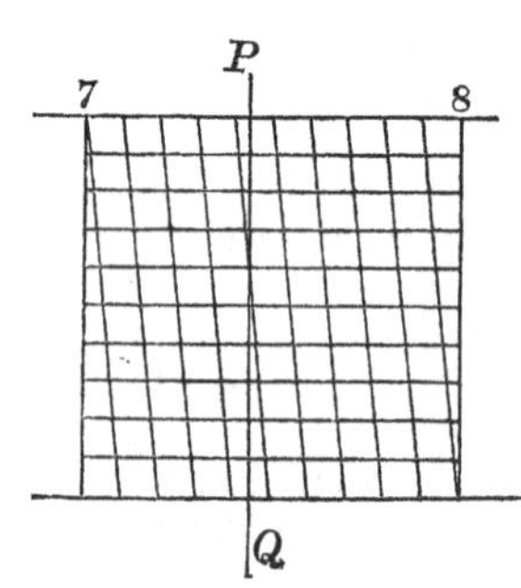

373. The *Vernier* is commonly employed for such instruments as the Barometer, Sextant, and Cathetometer, while the Screw is applied to the more delicate instruments, such as Astronomical Circles, Micrometers, and the Spherometer.

374. The vernier consists of a slip of metal which slides along a divided scale, the edges of the two being coincident. Hence, when it is applied to a divided circle, its edge is circular, and it moves about an axis passing through the centre of the divided limb.

In the sketch let 0, 1, 2, ... 10 denote the divisions on the vernier, 0, 1, 2, etc., any set of consecutive divisions on the limb or scale along whose edge it slides. If, when 0 and 0 coincide, 10 and 11 coincide also, then 10 divisions of the vernier are equal in length to 11 on the limb; and therefore each division of the vernier is $\frac{11}{10}$ths, or $1\frac{1}{10}$ of a division on the limb. If, then, the vernier be moved till 1 coincides with 1, 0 will be $\frac{1}{10}$th of a division of the limb beyond 0; if 2 coincide with 2, 0 will be $\frac{2}{10}$ths beyond 0; and so on. Hence to read the vernier in any position, note first the division next to 0, and behind it on the limb. This is the *integral* number of divisions to be read. For the fractional part, see which division of the vernier is in a line with one on the limb; if it be the 4th (as in the figure), that indicates an addition to the reading of $\frac{4}{10}$ths of a division of the limb; and so on. Thus, if the figure represent a barometer scale divided into inches and tenths, the reading is $\overset{\text{in}}{30 \cdot 34}$, the zero line of the vernier being adjusted to the level of the mercury.

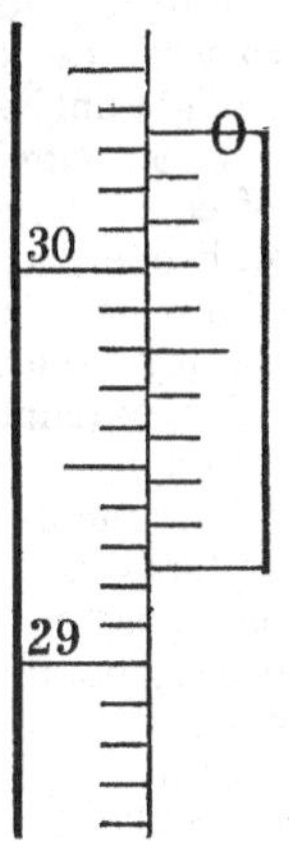

375. If the limb of a sextant be divided, as it usually is, to third-parts of a degree, and the vernier be formed by dividing twenty-one of these into twenty equal parts, the instrument can be read to twentieths of divisions on the limb, that is, to minutes of arc.

If no line on the vernier coincide with one on the limb, then since the divisions of the former are the longer there will be one of the latter included between the two lines of the vernier, and it is usual in practice to take the mean of the readings which would be given by a coincidence of either pair of bounding lines.

376. In the above sketch and description, the numbers on the scale and vernier have been supposed to run *opposite* ways. This is generally the case with British instruments. In some foreign ones the divisions run in the same direction on vernier and limb, and in that case it is easy to see that to read to tenths of a scale division we must have ten divisions of the vernier equal to *nine* of the scale.

In general to read to the nth part of a scale division, n divisions of the vernier must equal $n+1$ or $n-1$ divisions on the limb, according as these run in opposite or similar directions.

377. The principle of the *Screw* has been already noticed (§ 114). It may be used in either of two ways, i.e. the nut may be fixed, and the screw advance through it, or the screw may be prevented from moving longitudinally by a fixed collar, in which case the nut, if prevented by fixed guides from rotating, will move in the direction of the common axis. The advance in either case is evidently proportional to the angle through which the screw has turned about its axis, and this may be measured by means of a divided head fixed perpendicularly to the screw at one end, the divisions being read off by a pointer or vernier attached to the frame of the instrument. The nut carries with it either a tracing-point (as in the dividing engine) or a wire, thread, or half the object-glass of a telescope (as in micrometers), the thread or wire, or the play of the tracing-point, being at right angles to the axis of the screw.

378. Suppose it be required to divide a line into any number of equal parts. The line is placed parallel to the axis of the screw with one end exactly under the tracing-point, or under the fixed wire of a microscope carried by the nut, and the screw-head is read off. By turning the head, the tracing-point or microscope wire is brought to the other extremity of the line; and the number of turns and fractions of a turn required for the whole line is thus ascertained. Dividing this by the number of equal parts required, we find at once the number of turns and fractional parts corresponding to *one* of the required divisions, and by giving that amount of rotation to the screw over and over again, drawing a line after each rotation, the required division is effected.

379. In the *Micrometer*, the movable wire carried by the nut is parallel to a fixed wire. By bringing them into optical contact the zero reading of the head is known; hence when another reading has been obtained, we have by subtraction the number of turns corresponding to the length of the object to be measured. The *absolute* value of a turn of the screw is determined by calculation from the number of threads in an inch, or by actually applying the micrometer to an object of known dimensions.

380. For the measurement of the thickness of a plate, or the curvature of a lens, the *Spherometer* is used. It consists of a cylindrical stem through the axis of which a good screw works. The stem is supported by three feet, equidistant from each other, and having their extremities in a plane perpendicular to the axis. The lower extremity of the screw, when worked down into this plane, is equidistant from each of the feet—and the extremities of all are delicately pointed. The number of turns, whole or fractional, of the screw, is read off by a divided head and a pointer fixed to the stem. Suppose it be required to measure the thickness of a plate of glass. The three feet of the instrument are placed upon a truly flat surface, and the screw is gradually turned until its point just touches the surface. This is determinable with the utmost accuracy, by the whole system commencing to *rock*, if slightly touched, the instant that the screw-point passes below the plane of the three feet. The reason of this is, of course, that it is geometrically impossible to make a perfectly rigid body stand on four feet, without infinitely perfect fitting. At the instant at which this rocking (which is exceedingly distinct to the touch, and even to the ear) commences, the point of the screw is *just below* the plane of the feet of the instrument. The screw-head is now read off, and the screw turned backwards until room is left for the insertion, beneath its point, of the plate whose thickness is to be measured. The screw is now turned until the rocking just recommences, in which case it is evident that if the screw-point were depressed through a space equal to the thickness of the plate, it would be again *just below* the plane of the feet. From the difference of the readings of the head, we therefore easily calculate the thickness of the plate, the value of one turn of the screw having been, once for all, ascertained.

381. If the curvature of a lens is to be measured, the instrument is first placed, as before, on a plane surface, and the reading for the commencement of rocking is taken. The same operation is repeated on the spherical surface. The difference of the screw readings is evidently the greatest thickness of the glass which would be cut off by a plane passing through the three feet. This is sufficient, with the distance between each pair of feet, to enable us to calculate the radius of the spherical surface.

In fact if a be the distance between each pair of feet, l the length of screw corresponding to the difference of the two readings, R the radius of the spherical surface; we have at once $2R = \frac{a^2}{3l} + l$, or, as l is generally very small compared with a, the diameter is, very approximately, $\frac{a^2}{3l}$.

382. The *Cathetometer* is used for the accurate determination of differences of level—for instance, in measuring the height to which a fluid rises in a capillary tube above the exterior free surface. It consists of a divided metallic stem, which can (by means of levelling-screws in its three feet) be placed very nearly vertical. Upon this slides a metallic piece, bearing a telescope whose axis is rendered horizontal by means of a level. This is, of course, perpendicular to the stem; and when the latter is made to revolve in its supports, describes a horizontal plane. The adjustments are somewhat tedious, but present no other difficulty. In using the instrument the telescope is directed first to one of the objects whose difference of level is to be found, then (with its bearing-piece) it is moved by a delicate screw up or down the stem, until a horizontal wire in the focus of its eye-piece coincides with the image of the object. The vernier attached to the telescope is then read off—and, the process being repeated for the second object, a simple subtraction gives at once the required difference of level.

383. The principle of the *Balance* is known to everybody. We may note here a few of the precautions adopted in the best balances to guard against the various defects to which the instrument is liable;

and the chief points to be attended to in its construction to secure delicacy, and rapidity of weighing.

The balance-beam should be as stiff as possible, and yet not very heavy. For this purpose it is generally formed either of tubes, or of a sort of lattice-framework. To avoid friction, the axle consists of a knife-edge, as it is called; that is, a wedge of hard steel, which, when the balance is in use, rests on horizontal plates of polished agate. A similar contrivance is applied in very delicate balances at the points of the beam from which the scale-pans are suspended. When not in use, and just before use, the beam with its knife-edge is lifted by a lever arrangement from the agate plates. While thus secured it is loaded with weights as nearly as possible equal (this can be attained by previous trial with a coarser instrument), and the accurate determination is then readily effected. The last fraction of the required weight is determined by a rider, a very small weight, generally formed of wire, which can be worked (by a lever) from the outside of the glass case in which the balance is enclosed, and which may be placed in different positions upon one arm of the beam. This arm is graduated to tenths, etc., and thus shows at once the value of the rider in any case as depending on its moment or leverage, § 233.

384. The most important qualities of a good balance are—

1. *Sensibility.*—The beam should be sensibly deflected from a horizontal position by the smallest difference between the weights in the scale-pans. The definite measure of the sensibility is the angle through which the beam is deflected by a stated percentage of difference between the loads in the pans.

2. *Stability.*—This means rapidity of oscillation, and consequently speed in the performance of a weighing. It depends mainly upon the depth of the centre of gravity of the whole below the knife-edge, and the length of the beam.

3. *Constancy.*—Successive weighings of the same body must give the same result—all necessary corrections (to be explained later) depending on temperature, height of barometer, etc., being allowed for.

In our chapter on Statics we shall give the investigation of the amounts of these qualities for any given form and dimensions of the instrument.

A fine balance should turn with about a 500,000th of the greatest load which can safely be placed in either pan. In fact few measurements of any kind are correct to more than *six* significant figures.

The process of *Double Weighing*, which consists in counterpoising a mass by shot, or sand, or pieces of fine wire, and then substituting weights for it in the same pan till equilibrium is attained, is more laborious, but more accurate, than single weighing; as it eliminates all errors arising from unequal length of the arms, etc.

385. In the *Torsion-balance* invented, and used with great effect, by Coulomb, a force is measured by the torsion of a fibre of silk, a glass thread, or a metallic wire. The fibre or wire is fixed at its upper end, or at both ends, according to circumstances. In general it carries a very light horizontal rod or needle, to the extremities of which are attached the body on which is exerted the force to be measured, and a counterpoise. The upper extremity of the torsion fibre is fixed to an index passing through the centre of a divided disc, so that the angle through which that extremity moves is directly measured. If, at the same time, the angle through which the needle has turned be measured, or, more simply, if the index be always turned till the needle assumes a different position determined by marks or sights attached to the case of the instrument—we have the amount of torsion of the fibre, and it becomes a simple statical problem to determine from the latter the force to be measured; its direction, and point of application, and the dimensions of the apparatus, being known. The force of torsion as depending on the angle of torsion was found by Coulomb to follow the law of simple proportion up to the limits of perfect elasticity—as might have been expected from Hooke's Law (see *Properties of Matter*), and it only remains that we determine the amount for a particular angle in absolute measure. This determination is, in general, simple enough in theory; but in practice requires considerable care and nicety. The torsion-balance, however, being chiefly used for comparative, not absolute, measure, this determination is often unnecessary. More will be said about it when we come to its application.

386. The ordinary spiral spring-balances used for roughly comparing either small or large weights or forces, are, properly speaking, only a modified form of torsion-balance[1], as they act almost entirely by the torsion of the wire, and not by longitudinal extension or by flexure. Spring-balances we believe to be capable, if carefully constructed, of rivalling the ordinary balance in accuracy, while, for some applications, they far surpass it in sensibility and convenience. They measure directly *force*, not *mass*; and therefore if used for determining masses in different parts of the earth, a correction must be applied for the varying force of gravity. The correction for temperature must not be overlooked. These corrections may be avoided by the method of double weighing.

387. Perhaps the most delicate of all instruments for the measurement of force is the *Pendulum*. It is proved in Kinetics (see Div. II.) that for any pendulum, whether oscillating about a mean vertical position under the action of gravity, or in a horizontal plane, under the action of magnetic force, or force of torsion, the square of the number of *small* oscillations in a given time is proportional to the magnitude of the force under which these oscillations take place.

For the estimation of the relative amounts of gravity at different places, this is by far the most perfect instrument. The method of coincidences by which this process has been rendered so excessively delicate will be described later.

In fact, the kinetic measurement of force, as it is the first and

[1] J. Thomson. *Cambridge and Dublin Math. Journal*, 1848.

most truly elementary, is also far the most easy as well as perfect method in many practical cases. It admits of an easy reduction to gravitation measure.

388. Weber and Gauss, in constructing apparatus for observations of terrestrial magnetism, endeavoured so to modify them as to admit of their being read from some distance. For this purpose each bar, made at that time too ponderous, carried a plane mirror. By means of a scale, seen after reflection in the mirror and carefully read with a telescope, it was of course easy to compute the deviations which the mirror had experienced. But, for many reasons, it was deemed necessary that the deflections, even under considerable force, should be very small. With this view the *Bifilar* suspension was introduced. The bar-magnet is suspended horizontally by two vertical wires or fibres of equal length so adjusted as to share its weight equally between them. When the bar turns, the suspension-fibres become inclined to the vertical, and therefore the bar must rise. Hence, if we neglect the torsion of the fibres, the bifilar actually measures a force by comparing it with the weight of the suspended magnet.

Let a be the half length of the bar between the points of attachment of the wires, θ the angle through which the bar has been turned (in a horizontal plane) from its position of equilibrium, l the length of one of the wires.

Then if Q be the couple tending to turn the bar, and W its weight, we have

$$Q = \frac{Wa^2}{l} \frac{\sin\theta}{\sqrt{1 - \frac{4a^2}{l^2}\sin^2\frac{\theta}{2}}},$$

which gives the couple in terms of the deflection θ.

If the torsion of the fibres be taken into account, it will be sensibly equal to θ (since the greatest inclination to the vertical is small), and therefore the couple resulting from it will be $E\theta$, where E is some constant. This must be added to the value of Q just found in order to get the whole deflecting couple.

389. Dynamometers are instruments for measuring energy. *White's friction brake* measures the amount of work actually performed in any time by an engine or other 'prime mover,' by allowing it during the time of trial to waste all its work on friction. *Morin's dynamometer* measures work without wasting any of it, in the course of its transmission from the prime mover to machines in which it is usefully employed. It consists of a simple arrangement of springs, measuring at every instant the *couple* with which the prime mover turns the shaft that transmits its work, and an integrating machine from which the work done by this couple during any time can be read off.

390. White's friction brake consists of a lever clamped to the shaft, but not allowed to turn with it. The moment of the force required to prevent the lever from going round with the shaft, multiplied by the whole angle through which the shaft turns, measures the whole work done against the friction of the clamp. The same result is much more easily obtained by wrapping a rope or chain several times round the shaft, or round a cylinder or drum carried round by the shaft, and applying measured forces to its two ends in proper directions to keep it nearly steady while the shaft turns round without it. The difference of the moments of these two forces round the axis, multiplied by the angle through which the shaft turns, measures the whole work spent on friction against the rope. If we remove all other resistance to the shaft, and apply the proper amount of force at each end of the dynamometric rope or chain (which is very easily done in practice), the prime mover is kept running at the proper speed for the test, and having its whole work thus wasted for the time and measured.

DIVISION II.

ABSTRACT DYNAMICS.

CHAPTER V.—INTRODUCTORY.

391. Until we know thoroughly the nature of matter and the forces which produce its motions, it will be utterly impossible to submit to mathematical reasoning the *exact* conditions of any physical question. It has been long understood, however, that an approximate solution of almost any problem in the ordinary branches of Natural Philosophy may be easily obtained by a species of *abstraction*, or rather *limitation of the data*, such as enables us easily to solve the modified form of the question, while we are well assured that the circumstances (so modified) affect the result only in a superficial manner.

392. Take, for instance, the very simple case of a crowbar employed to move a heavy mass. The accurate mathematical investigation of the action would involve the simultaneous treatment of the motions of every part of bar, fulcrum, and mass raised; and from our almost complete ignorance of the nature of matter and molecular forces, it is clear that such a treatment of the problem is impossible.

It is a result of observation that the particles of the bar, fulcrum, and mass, separately, retain throughout the process nearly the same relative positions. Hence the idea of solving, instead of the above impossible question, another, in reality quite different, but, while infinitely simpler, obviously leading to *nearly* the same results as the former.

393. The new form is given at once by the experimental result of the trial. Imagine the masses involved to be *perfectly rigid* (i.e. incapable of changing their forms or dimensions), and the infinite multiplicity of the forces, really acting, may be left out of consideration; so that the mathematical investigation deals with a finite (and generally small) number of forces instead of a practically infinite number. Our warrant for such a substitution is established thus.

394. The only effects of the intermolecular forces would be exhibited in molecular alterations of the form or volume of the masses involved. But as these (practically) remain almost unchanged, the forces which produce, or tend to produce, changes in them may be left out of consideration. Thus we are enabled to investigate the action of machinery by supposing it to consist of separate portions whose forms and dimensions are unalterable.

395. If we go a little farther into the question, we find that the lever *bends*, some parts of it are extended and others compressed. This would lead us into a very serious and difficult inquiry if we had to take account of the whole circumstances. But (by experience) we find that a sufficiently accurate solution of this more formidable case of the problem may be obtained by supposing (what can *never* be realized in practice) the mass to be homogeneous, and the forces consequent on a dilatation, compression, or distortion, to be proportional in magnitude, and opposed in direction, to these deformations respectively. By this farther assumption, close approximations may be made to the vibrations of rods, plates, etc., as well as to the statical effects of springs, etc.

396. We may pursue the process farther. Compression, in general, develops heat, and extension, cold. These *alter* sensibly the elasticity of a body. By introducing such considerations, we reach, without great difficulty, what may be called a *third* approximation to the solution of the physical problem considered.

397. We might next introduce the conduction of the heat, so produced, from point to point of the solid, with its accompanying modifications of elasticity, and so on; and we might then consider the production of thermo-electric currents, which (as we shall see) are always developed by unequal heating in a mass if it be not perfectly homogeneous. Enough, however, has been said to show, *first*, our utter ignorance as to the true and complete solution of any physical question by the only perfect method, that of the consideration of the circumstances which affect the motion of every portion, separately, of each body concerned; and, *second*, the practically sufficient manner in which practical questions may be attacked by limiting their generality, *the limitations introduced being themselves deduced from experience*, and being therefore Nature's own solution (to a less or greater degree of accuracy) of the infinite additional number of equations by which we should otherwise have been encumbered.

398. To take another case: in the consideration of the propagation of waves on the surface of a fluid, it is impossible, not only on account of mathematical difficulties, but on account of our ignorance of *what* matter is, and what forces its particles exert on each other, to form the equations which would give us the separate motion of each. Our first approximation to a solution, and one sufficient for most practical purposes, is derived from the consideration of the motion of a homogeneous, incompressible, and perfectly plastic mass; a hypothetical substance which, of course, nowhere exists in nature.

399. Looking a little more closely, we find that the actual motion differs considerably from that given by the analytical solution of the restricted problem, and we introduce farther considerations, such as the *compressibility* of fluids, their *internal friction*, the heat generated by the latter, and its effects in dilating the mass, etc. etc. By such successive corrections we attain, at length, to a mathematical result which (at all events in the present state of experimental science) agrees, within the limits of experimental error, with observation.

400. It would be easy to give many more instances substantiating what has just been advanced, but it seems scarcely necessary to do so. We may therefore at once say that there is no question in physical science which can be *completely and accurately* investigated by mathematical reasoning (in which, be it carefully remembered, it is *not* necessary that *symbols* should be introduced), but that there are different degrees of approximation, involving assumptions more and more nearly coincident with observation, which may be arrived at in the solution of any particular question.

401. *The object of the present division of this work is to deal with the first and second of these approximations.* In it we shall suppose all solids either RIGID, i.e. unchangeable in form and volume, or ELASTIC; but in the latter case, we shall assume the law, connecting a compression or a distortion with the force which causes it, to have a particular form deduced from experiment. And we shall also leave out of consideration the thermal or electric effects which compression or distortion generally produce. We shall also suppose fluids, whether liquids or gases, to be either INCOMPRESSIBLE or compressible according to certain known laws; and we shall omit considerations of fluid friction, although we admit the consideration of friction between solids. Fluids will therefore be supposed *perfect*, i.e. such that any particle may be moved amongst the others by the slightest force.

402. When we come to Properties of Matter and the Physical Forces, we shall give in detail, as far as they are yet known, the modifications which farther approximations have introduced into the previous results.

403. The laws of friction between solids were very ably investigated by Coulomb; and, as we shall require them in the succeeding chapters, we give a brief summary of them here; reserving the more careful scrutiny of experimental results to our chapter on Properties of Matter.

404. To produce sliding of one solid body on another, the surfaces in contact being plane, requires a tangential force which depends,—(1) upon the nature of the bodies; (2) upon their polish, or the species and quantity of lubricant which may have been applied; (3) upon the normal pressure between them, to which it is in general directly proportional; (4) upon the length of time during which they have been suffered to remain in contact.

It does not (except in extreme cases where scratching or abrasion takes place) depend sensibly upon the area of the surfaces in contact. This, which is called Statical Friction, is thus capable of opposing a tangential resistance to motion which may be of any requisite amount up to μR; where R is the whole normal pressure between the bodies; and μ (which depends mainly upon the nature of the surfaces in contact) is the *co-efficient of Statical Friction.* This co-efficient varies greatly with the circumstances, being in some cases as low as 0.03, in others as high as 0.80. Later we shall give a table of its values. Where the applied forces are insufficient to produce motion, the whole amount of statical friction is not called into play; its amount then just reaches what is sufficient to equilibrate the other forces, and its direction is the opposite of that in which their resultant tends to produce motion. When the statical friction has been overcome, and sliding is produced, experiment shows that a force of friction continues to act, opposing the motion, sensibly proportional to the normal pressure, and independent of the velocity. But for the same two bodies the *co-efficient of Kinetic Friction* is less than that of Statical Friction, and is approximately the same whatever be the rate of motion.

405. When among the forces acting in any case of equilibrium, there are frictions of solids on solids, the circumstances would not be altered by doing away with all friction, and replacing its forces by forces of mutual action supposed to remain unchanged by any infinitely small relative motions of the parts between which they act. By this artifice all such cases may be brought under the general principle of Lagrange (§ 254).

406. In the following chapters on Abstract Dynamics we will confine ourselves strictly to such portions of this extensive subject as are likely to be useful to us in the rest of the work, or are of sufficient importance of themselves to warrant their introduction—except in special cases where results, more curious than useful, are given to show the nature of former applications of the methods, or to exhibit special methods of investigation adapted to the difficulties of peculiar problems. For a general view of the subject as a purely analytical problem, the reader is referred to special mathematical treatises, such as those of Poisson, Delaunay, Duhamel, Todhunter, Tait and Steele, Griffin, etc. From these little is to be learned save dexterity in the solution of problems which are in general of no great physical interest—the objects of these treatises being professedly the mathematical analysis of the subject; while in the present work we are engaged specially with those questions which best illustrate physical principles.

CHAPTER VI.

STATICS OF A PARTICLE.—ATTRACTION.

407. WE naturally divide Statics into two parts—the equilibrium of a *Particle*, and that of a rigid or elastic *Body* or *System of Particles* whether solid or fluid. The second law of motion suffices for one part—for the other, the third, and its consequences pointed out by Newton, are necessary. In the succeeding sections we shall dispose of the first of these parts, and the rest of this chapter will be devoted to a digression on the important subject of Attraction.

408. By § 221, forces acting at the same point, or on the same material particle, are to be compounded by the same laws as velocities. Therefore the sum of their resolved parts in any direction must vanish if there is equilibrium; whence the necessary and sufficient conditions.

They follow also directly from Newton's statement with regard to work, if we suppose the particle to have any velocity, constant in direction and magnitude (and § 211, this is the most general supposition we can make, since absolute rest has for us no meaning). For the work done in any time is the product of the displacement during that time into the algebraic sum of the effective components of the applied forces, and there is no change of kinetic energy. Hence this sum must vanish for *every* direction. Practically, as any displacement may be resolved into three, in any three directions not coplanar, the vanishing of the work for any one such set of three suffices for the criterion. But, in general, it is convenient to assume them in directions at right angles to each other.

Hence, for the equilibrium of a material particle, it is *necessary*, and *sufficient*, that the (algebraic) sums of the applied forces, resolved in any one set of three rectangular directions, should vanish.

409. We proceed to give a detailed exposition of the results which follow from the first clause of § 408. For three forces only we have the following statement.

The resultant of two forces, acting on a material point, is represented in direction and magnitude by the diagonal, through that point, of the parallelogram described upon lines representing the forces.

410. *Parallelogram of forces stated symmetrically as to the three forces concerned*, usually called the *Triangle of Forces*. If the lines representing three forces acting on a material point be equal and parallel to the sides of a triangle, and in directions similar to those of the three sides when taken in order round the triangle, the three forces are in equilibrium.

Let GEF be a triangle, and let MA, MB, MC, be respectively equal and parallel to the three sides EF, FG, GE of this triangle, and in directions similar to the consecutive directions of these sides in order. The point M is in equilibrium.

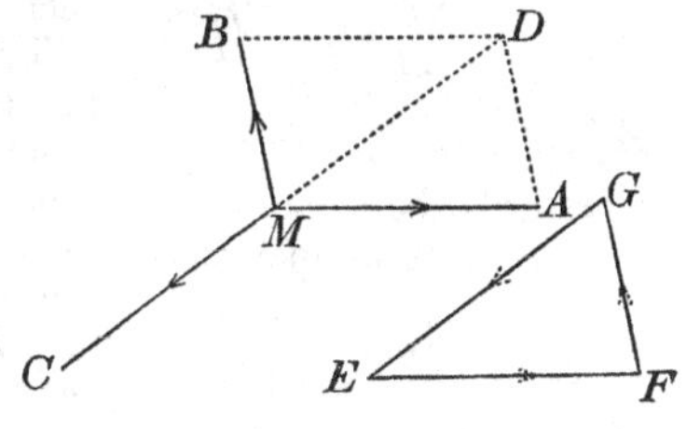

411. [*True Triangle of Forces.* Let three forces act in consecutive directions round a triangle, DEF, and be represented respectively by its sides: they are not in equilibrium, but are equivalent to a couple. To prove this, through D draw DH, equal and parallel to EF, and in it introduce a pair of balancing forces, each equal to EF. Of the five forces, three, DE, DH and FD, are in equilibrium, and may be removed; and there are then left two forces, EF and HD, equal, parallel, and in dissimilar directions, which constitute a couple.]

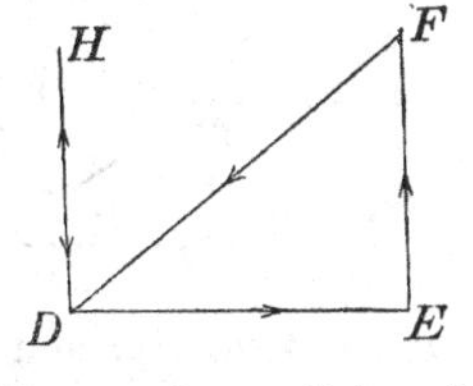

412. To find the resultant of any number of forces in lines through one point, not necessarily in one plane—

Let MA_1, MA_2, MA_3, MA_4 represent four forces acting on M, in one plane; required their resultant.

Find by the parallelogram of forces, the resultant of two of the forces, MA_1 and MA_2. It will be represented by MD'. Then similarly, find MD'', the resultant of MD' (the first subsidiary resultant), and MA_3, the third force. Lastly, find MD''', the resultant of MD'' and MA_4. MD''' represents the resultant of the given forces.

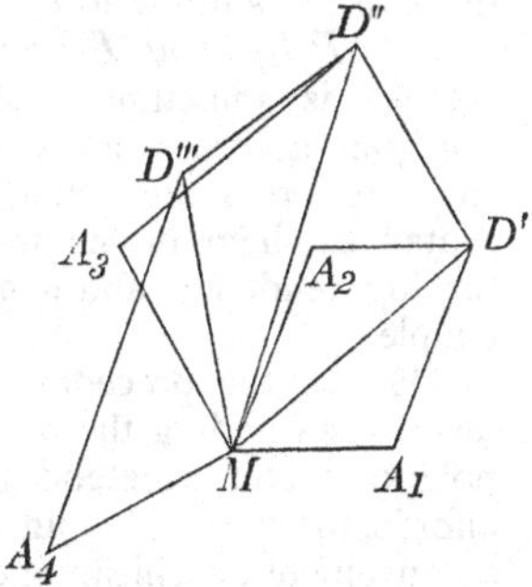

Thus, by successive applications of the fundamental proposition, the resultant of any number of forces in lines through one point can be found.

413. In executing this construction, it is not necessary to describe the successive parallelograms, or even to draw their diagonals. It is enough to draw through the given point a line equal and parallel to the representative of any one of the forces; through the point thus arrived at, to draw a line equal and parallel to the representative of another of the forces, and so on till all the forces have been taken into account. In this way we get such a diagram as the annexed.

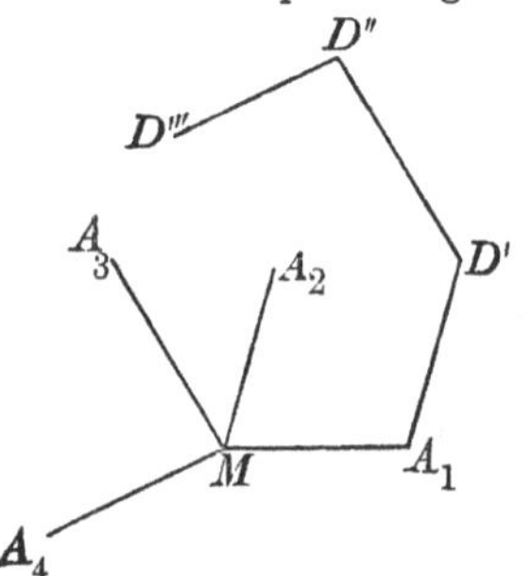

The several given forces may be taken in any order, in the construction just described. The resultant arrived at is necessarily the same, whatever be the order in which we choose to take them, as we may easily verify by elementary geometry.

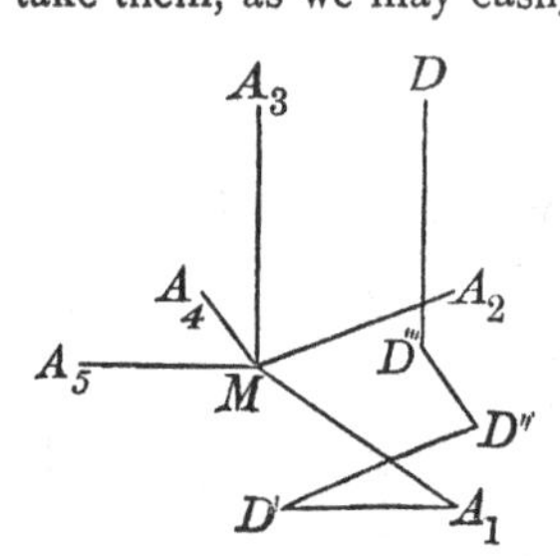

In the fig. the order is MA_1, MA_5, MA_2, MA_4, MA_3.

414. If, by drawing lines equal and parallel to the representatives of the forces, a closed figure is got, that is, if the line last drawn leads us back to the point from which we started, the forces are in equilibrium. If, on the other hand, the figure is not closed (§ 413), the resultant is obtained by drawing a line from the starting-point to the point finally reached; (from M to D): and a force represented by DM will equilibrate the system.

415. Hence, in general, a set of forces represented by lines equal and parallel to the sides of a complete polygon, are in equilibrium, provided they act in lines through one point, in directions similar to the directions followed in going round the polygon in one way.

416. *Polygon of Forces.* The construction we have just considered, is sometimes called the polygon of forces; but the true polygon of forces, as we shall call it, is something quite different. In it the forces are actually along the sides of a polygon, and represented by them in magnitude. Such a system must clearly have a turning tendency, and it may be demonstrated to be reducible to one couple.

417. In the preceding sections we have explained the principle involved in finding the resultant of any number of forces. We have now to exhibit a method, more easy than the parallelogram of forces affords, for working it out in actual cases, and especially for obtaining a convenient specification of the resultant. The instrument employed for this purpose is Trigonometry.

418. A distinction may first be pointed out between two classes of problems, direct and inverse. Direct problems are those in which the resultant of forces is to be found; inverse, those in which components of a force are to be found. The former class is fixed and determinate; the latter is quite indefinite, without limitations to be stated for each problem. A system of forces can produce only one effect; but an infinite number of systems can be obtained, which shall produce the same effect as one force. The problem, therefore, of finding components must be, in some way or other, limited. This may be done by giving the lines along which the components are to act. To find the components of a given force, in any three given directions, is, in general, as we shall see, a perfectly determinate problem.

Finding resultants is called Composition of Forces.

Finding components is called Resolution of Forces.

419. *Composition of Forces.*

Required in position and magnitude the resultant of two given forces acting in given lines on a material point.

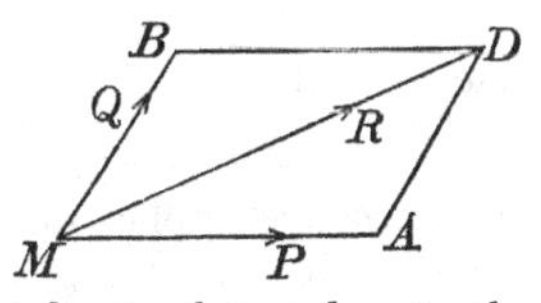

Let MA, MB represent two forces, P and Q, acting on a material point M. Let the angle BMA be denoted by ι. Required the magnitude of the resultant, and its inclination to the line of either force.

Let R denote the magnitude of the resultant; let α denote the angle DMA, at which its line MD is inclined to MA, the line of the first force P; and let β denote the angle DMB, at which it is inclined to MB, the direction of the force Q.

Given P, Q, and ι: required R, and α or β.

We have

$$MD^2 = MA^2 + MB^2 - 2\,MA.MB \times \cos MAD.$$

Hence, according to our present notation,

$$R^2 = P^2 + Q^2 - 2PQ\cos(180° - \iota),$$

or

$$R^2 = P^2 + Q^2 + 2PQ\cos\iota.$$

Hence

$$R = (P^2 + Q^2 + 2PQ\cos\iota)^{\frac{1}{2}}. \qquad (1)$$

To determine α and β after the resultant has been found; we have

$$\sin DMA = \frac{MB}{MD}\sin MAD,$$

or

$$\sin\alpha = \frac{Q}{R}\sin\iota, \qquad (2)$$

and similarly,

$$\sin\beta = \frac{P}{R}\sin\iota. \qquad (3)$$

420. These formulae are useful for many applications; but they have the inconvenience that there may be ambiguity as to the angle, whether it is to be acute or obtuse, which is to be taken when either $\sin\alpha$ or $\sin\beta$ has been calculated. If ι is acute, both α and β are acute, and there is no ambiguity. If ι is obtuse, one of the two angles, α, β, might be either acute or obtuse; but as they cannot be both obtuse, the smaller of the two must, necessarily, be acute. If, therefore, we take the formula for $\sin\alpha$, or for $\sin\beta$, according as the force P, or the force Q, is the greater, we do away with all ambiguity, and have merely to take the value of the angle shown in the table of sines. And by subtracting the value thus found, from the given value of ι, we find the value, whether acute or obtuse, of the other of the two angles, α, β.

421. To determine α and β otherwise. After the magnitude of the resultant has been found, we know the three sides, MA, AD, MD, of the triangle DMA, then we have

$$\cos DMA = \frac{MD^2+MA^2-AD^2}{2MD.MA},$$

or

$$\cos\alpha = \frac{R^2+P^2-Q^2}{2RP}, \tag{4}$$

and similarly,

$$\cos\beta = \frac{R^2+Q^2-P^2}{2RQ}, \tag{5}$$

by successive applications of the elementary trigonometrical formula used above for finding MD. Again, using this last-mentioned formula for MD^2 or R^2 in the numerators of (4) and (5), and reducing, we have

$$\cos\alpha = \frac{P+Q\cos\iota}{R}, \tag{6}$$

$$\cos\beta = \frac{Q+P\cos\iota}{R}; \tag{7}$$

formulae which are convenient in many cases. There is no ambiguity in the determination of either α or β by any of the four equations (4), (5), (6), (7).

Remark.—Either sign ($+$ or $-$) might be given to the radical in (1), and the true line of action and the direction of the force in it would be determined without ambiguity by substituting in (2) and (3) the value of R with either sign prefixed. Since, however, there can be no doubt as to the direction of the force indicated, it will be generally convenient to give the positive sign to the value of R. But in special cases, the negative sign, which with the proper interpretation of the formulae will lead to the same result as the positive, will be employed.

422. Another method of treating the general problem, which is useful in many cases, is this:
Let

$$\tfrac{1}{2}(P+Q) = F,$$
$$\tfrac{1}{2}(P-Q) = G,$$

which implies that

$$P = F+G,$$
$$Q = F-G.$$

F and G will be both positive if $P > Q$. Hence, instead of the two given forces, P and Q, we may suppose that we have on the point M four forces;—two, each equal to F, acting in the same directions, MK, ML, as the given forces, and two others, each equal to G, of which one acts in the same direction, MK, as P, and the other in ML, the direction opposite to Q. Now the resultant of the two equal forces, F, bisects the angle between them, KML; and by the investigation of § 423 below, its magnitude is found to be $2F\cos\frac{1}{2}\iota$. Again, the resultant of the two equal forces, G, is similarly seen to bisect the angle, KML', between the line of the given force, P, and the continuation through M of the line of the given force, Q; and to be equal to $2G\sin\frac{1}{2}\iota$, since the angle KLM' is the supplement of ι. Thus, instead of the two given forces in lines inclined to one another at the angle ι, which may be either an acute, an obtuse, or a right angle, we have two forces, $2F\cos\frac{1}{2}\iota$ and $2G\sin\frac{1}{2}\iota$, acting in lines, MS, MT, which bisect the angles LMK and KML', and therefore are at right angles to one another. Now, according to § 429 below, we find the resultant[1] of these two forces by means of the following formulae:—

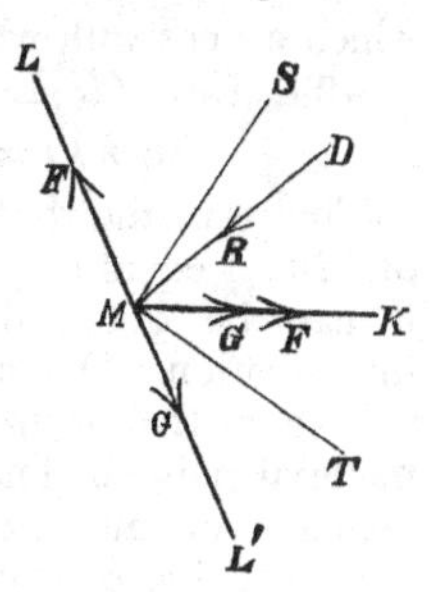

$$\tan SMD = \frac{2G\sin\frac{1}{2}\iota}{2F\cos\frac{1}{2}\iota},$$

and
$$R = 2F\cos\tfrac{1}{2}\iota\sec SMD,$$

or
$$\tan\left(\tfrac{1}{2}\iota - \alpha\right) = \frac{P-Q}{P+Q}\tan\tfrac{1}{2}\iota, \tag{8}$$

and
$$\begin{aligned} R &= (P+Q)\cos\tfrac{1}{2}\iota\sec\left(\tfrac{1}{2}\iota-\alpha\right) \\ &= (P+Q)\cos\tfrac{1}{2}(\alpha+\beta)\cos\tfrac{1}{2}(\alpha-\beta). \end{aligned} \tag{9}$$

These formulae might have been derived from the standard formulae for the solution of a plane triangle when two sides (P and Q), and the contained angle ($\pi-\iota$) are given.

423. We shall now investigate some cases of the general formulae.

Case I. Let the forces be equal, that is, let $Q=P$ in the preceding formulae.

Then, by (1), $R^2 = 2P^2 + 2P^2\cos\iota = 2P^2(1+\cos\iota)$
$$= 4P^2\cos^2\tfrac{1}{2}\iota.$$

Hence
$$R = 2P\cos\tfrac{1}{2}\iota,$$

an important result which might, of course, have been obtained directly from the proper geometrical construction in this case. Also by (2),

[1] In the diagram the direction of the balancing force is shown by the arrow-head in the line DM.

$$\sin \alpha = \frac{Q \sin \iota}{R} = \frac{Q \sin \iota}{2 Q \cos \frac{1}{2} \iota} = \sin \tfrac{1}{2} \iota,$$

which agrees with what we see intuitively, that $\alpha = \beta = \frac{1}{2} \iota$.

424. *Case II.* Let $P = Q$; and let $\iota = 120°$. Then

$$\cos \tfrac{1}{2} \iota = \cos 60° = 1, \text{ and } (\S\ 423)\ R = P.$$

The resultant, therefore, of two equal forces inclined at an angle of 120° is equal to each of them. This result is interesting, because it can be obtained very simply, and quite independently of this investigation. A consideration of the *symmetry* of the circumstances will show that if three equal forces in one plane be applied to a material point in lines dividing the space around it into three equal angles, they must be in equilibrium; which is perfectly equivalent to the preceding conclusion.

425. *Case III.* Let $\iota = 0°$; $\cos \iota = 1$; then

$$R = (P^2 + Q^2 + 2PQ)^{\frac{1}{2}},$$
$$R = P + Q.$$

426. *Case IV.* Let $\iota = 180°$; $\cos \iota = -1$; then

$$R = (P^2 + Q^2 - 2PQ)^{\frac{1}{2}},$$
$$R = P - Q.$$

This is also one of the cases in which it is convenient to give sometimes the negative sign, sometimes the positive to the expression for the resultant force: for if Q be greater than P, the preceding expression will be negative, and the interpretation will be found by considering that the force which vanishes when $P = Q$, is in the direction of P when P is the greater, and in the contrary direction, or in that of Q, when P is the less of the two forces.

427. *Case V.* Forces nearly conspiring. Let the angle ι be very small, then $\sin \iota \approx \iota$;[1] $\cos \iota \approx 1$.

The general expressions (§ 419) therefore become,

$$R \approx P + Q,$$

$$\sin \alpha \approx \frac{Q\iota}{P+Q},$$

$$\sin \beta \approx \frac{P\iota}{P+Q}.$$

To the same degree of approximation

$$\alpha \approx \frac{Q\iota}{P+Q}, \qquad (10)$$

$$\beta \approx \frac{P\iota}{P+Q}. \qquad (11)$$

Hence
$$\alpha + \beta \approx \frac{Q\iota + P\iota}{P+Q} \approx \frac{(Q+P)\,\iota}{(Q+P)} = \iota.$$

[1] The sign $\approx$ is used to denote approximate equality.

This shows that the errors in the values of α and β obtained approximately by this method compensate; one being as much above, as the other is below, the true value.

We therefore conclude that the resultant of two forces very nearly conspiring is approximately equal to their sum, and approximately divides the angle between them into parts inversely as the forces.

When the angle between the forces is infinitely small, they may either conspire in acting on one point in one line; or they may act on different points in parallel lines. In either case the resultant is precisely equal to their sum. Actually conspiring forces we have already considered; parallel forces we shall consider more particularly when we treat of the equilibrium of a rigid body. We may briefly examine the case here however. Suppose the actual points of application of the forces to be A and B, but let their lines meet in a point M; join AB, and let MAB be an isosceles triangle. Let this point M be removed gradually to an infinite distance in the direction of a perpendicular, OM, bisecting the line AB. The resultant will still divide the angle inversely as the forces: and as the circular measure of the angle is any arc described from M as centre divided by the radius, every such arc will be divided in the same proportion. Now, if M be infinitely distant, that is if the lines of the forces be parallel, the arc will become a straight line, and will be divided into parts inversely as the forces.

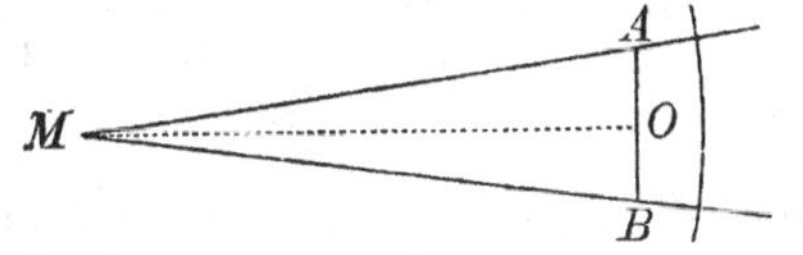

In actual cases of forces acting on a point, and very nearly conspiring, the following approximate equations show how nearly the resultant approaches the sum of the forces:—

$$\sin\theta \approx \theta; \quad \cos\theta \approx 1 - \tfrac{1}{2}\theta^2.$$

$$R^2 \approx P^2 + Q^2 + 2PQ - PQ\iota^2$$

$$R \approx \sqrt{\left\{(P+Q)^2\left[1 - \frac{PQ\iota^2}{(P+Q)^2}\right]\right\}}$$

$$R \approx (P+Q)\left[1 - \tfrac{1}{2}\frac{PQ\iota^2}{(P+Q)^2}\right]$$

$$R \approx (P+Q) - \tfrac{1}{2}\frac{PQ}{P+Q}\iota^2, \qquad (12$$

that is, the resultant of two forces very nearly conspiring falls short o their sum by the square of the angle between them multiplied into a quarter of their harmonic mean[1].

428. *Case VI.* Forces nearly opposed.

1°. Let the angle ι be very obtuse, and the two forces exactly equal

[1] The Harmonic Mean of two numbers is the reciprocal of the mean of their reciprocals. Thus the harmonic mean of P and Q is $\frac{2PQ}{P+Q}$.

Let $\iota = \pi - \theta$, where θ is very small,

then
$$\tfrac{1}{2}\iota = \tfrac{1}{2}\pi - \tfrac{1}{2}\theta,$$
$$\cos\tfrac{1}{2}\iota = \sin\tfrac{1}{2}\theta,$$
$$R = 2P\sin\tfrac{1}{2}\theta,$$
and since the sine of a very small angle is equal to the angle, in circular measure $R \approx P\theta.$

Hence the resultant of two equal very nearly opposed forces is proportional to the defalcation from direct opposition: being approximately equal to either of the forces multiplied into the supplement of the angle between them.

2°. If the forces are neither equal nor nearly equal, the resultant will be approximately equal to their difference.

We have as before,
$$\cos\iota \approx -1,$$
$$R^2 \approx P^2 + Q^2 - 2PQ.$$
Therefore
$$R \approx P - Q,$$
$$\sin\alpha \approx \frac{Q(\pi-\iota)}{P-Q}$$
$$\sin\beta \approx \frac{P(\pi-\iota)}{P-Q}.$$

The ambiguity as to whether the acute angle, shown in the table, or its supplement, is to be chosen in either case, may be removed by considering which of the two forces is the greater.

Thus, as we suppose P to be greater than Q, α is acute, and therefore
$$\sin\alpha \approx \alpha \approx \frac{Q(\pi-\iota)}{P-Q}$$
and β is obtuse.

Therefore
$$\beta \approx \pi - \frac{P(\pi-\iota)}{P-Q}$$
or
$$\beta \approx \frac{P\iota - Q\pi}{P-Q}.$$
We find, by addition,
$$\alpha + \beta = \frac{P-Q}{P-Q}\iota = \iota,$$
and conclude, as in the former case, that the errors in the approximate values of α and β compensate, one being as much above, as the other is below, the true value.

It is only when R is comparable in magnitude with P and Q, that the foregoing solution is applicable.

But if P exceeds Q, or if Q exceeds P, by any difference which is considerable in comparison with either, the formulae hold.

Let us suppose now that, while P remains of any constant magnitude, Q is made to increase from nothing, gradually, until it becomes first equal to, and then greater than P, the angle ι remaining constant. The angle a will increase very slowly, according to the approximate formula (10), until Q becomes nearly equal to P. Then as the value of Q is increased until it becomes greater than P, the value of a will increase very rapidly through nearly two right angles, until it falls but little short of ι, when its supplement will be approximately expressed by the formula (10).

In this transition, from $Q<P$ to $Q>P$, the direction and magnitude of the resultant are most conveniently found by means of (§ 422) the last of the three general methods given above for determining the resultant of two forces.

Thus, instead of the two given forces we may substitute two forces in lines bisecting respectively the obtuse angle LMK, or ι, and the acute angle KML' and of magnitudes which approximate to $\frac{1}{2}(P+Q)(\pi-\iota)$, and $P-Q$ respectively, when ι is nearly two right angles.

We infer, finally, that, however nearly P and Q are equal to one another, the approximate formulae of § 428, 2° hold, provided only $\frac{1}{2}(P+Q)(\pi-\iota)$ is a small fraction of $P-Q$.

429. *Case VII.* Let $\iota=90°$; $\cos\iota=0$, $\sin\iota=1$;

then
$$R=(P^2+Q^2)^{\frac{1}{2}}, \tag{13}$$

and
$$\left.\begin{aligned}\sin a&=\frac{Q}{R}\\ \sin\beta&=\frac{P}{R}\end{aligned}\right\}. \tag{14}$$

In this case, β being the complement of a, $\sin\beta=\cos a$.

Hence
$$\cos a=\frac{P}{R}.$$

Lastly, since
$$\tan a=\frac{\sin a}{\cos a},$$

we deduce
$$\tan a=\frac{Q}{P}, \tag{15}$$

and
$$R=P\sec a. \tag{16}$$

Remark.—These formulae have thus been derived from the general expressions (§ 419); but they can also be very readily got from a special geometrical construction, corresponding to the case in which the lines of the forces are at right angles to one another, the principles to be used being (1) the parallelogram of forces; (2) Euclid I., XLVII.; and (3) the trigonometrical definitions of sine, cosine, and tangent.

430. This case is of importance, for it affords us the formulae for rectangular resolution; by the aid of which we shall, a little later, proceed to calculate the resultant of any number of forces in one plane. We might calculate the resultant by applying the elementary formulae (§§ 419, 420, 421) to repetitions of the parallelogram of forces. But this process would be very complicated and tedious, if the forces were numerous, and their magnitudes and angles given in numbers; and we shall see that it may be avoided by resolving all the forces along two lines at right angles to one another, and thus obtaining as equivalent to them, two forces along these lines.

We shall first consider the general inverse problem (§ 418), or the resolution of forces.

431. If a force acting on a material point, and two lines in one plane with the line of that force, be given, it is possible to find determinately two forces along those lines, of which the given force is the resultant.

The two forces thus determined are called the components of the given force along the given lines, and if we substitute these two forces for the given force, we are said to resolve the given force into two forces along the given lines; or, to resolve the force along the given lines.

432. *Geometrical Solution.* Let M be the given point; R, the given force acting on it in the line, MK; and MF and MG the given lines.

It is required to find the components along MF and MG of R in MK.

Take any convenient length MD to represent the magnitude of the given force, R. Through D draw DA parallel to GM, and let it cut MF in A; and also through D draw DB parallel to FM, and let it cut MG in B; MA and MB represent the required magnitudes of the components.

433. *Trigonometrical Solution.* If the angle KMF be given $= \alpha$, and $KMG = \beta$, and if the required component of the given force R along MF be denoted by P, and the component along MG by Q, we deduce from equations (2) and (3) (§ 420), the following:—

$$P = \frac{R \sin \beta}{\sin (\alpha + \beta)}, \tag{17}$$

$$Q = \frac{R \sin \alpha}{\sin (\alpha + \beta)}. \tag{18}$$

434. When the given lines of resolution are at right angles to one another, these expressions are modified in the manner shown above (§ 429, Case VII), or we may find them at once from the geometrical construction proper for the case, thus:—

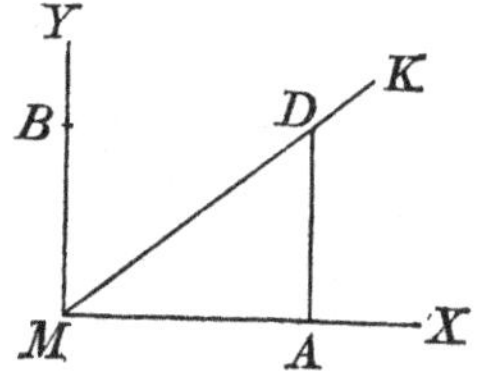

Let MX, MY be the given lines; $XMY = 90°$, and $MD = R$. Also, as before, $DMA = \alpha$, and $DMB = \beta$. Draw DA parallel to YM, or perpendicular to MX, and make $MB = AD$. Then in the right-angled triangle MAD, $MA = MD \cos DMA$, and $AD = MD \sin DMA$.

Hence, since MA represents the component along MX, and MB the component along MY,

$$P = R \cos \alpha, \tag{19}$$

$$Q = R \sin \alpha, \text{ or } Q = R \cos \beta. \tag{20}$$

Hence, in rectangular resolution, the component, along any line, of a given force, is equal to the product of the number expressing the given force, into the cosine of the angle at which its direction is inclined to that line.

435. *Application of the Resolution of Forces.* Let there be a number of forces P_1, P_2, P_3, P_4, P_5, acting respectively in lines ML_1, ML_2, ML_3, ML_4, ML_5, on a material point M; required their resultant.

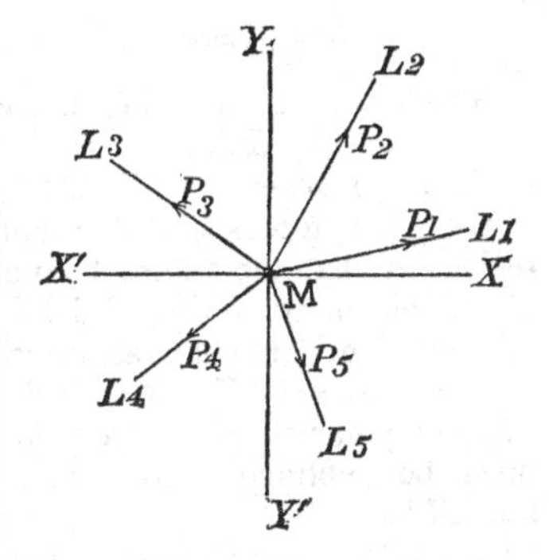

Through M, draw at right angles to each other, and in the same plane as the given forces, two lines, XX' and YY', which may be called lines or axes of resolution. Let the angle which the resultant forms with the line of resolution MX, be denoted by θ, and let the angles, which the lines of the forces make respectively with the lines of resolution, be denoted by α_1, β_1; α_2, β_2; α_3, β_3; &c.; that is, $L_1MX = \alpha_1$, $L_1MY = \beta_1$, and so on.

The angles β_1 β_2, &c., are merely the complements of α_1 α_2, &c., and, except for the sake of symmetry, they need not have been introduced into our notation.

Resolve (§ 434) the first force P_1, into two components, one along MX, and one along MY. These are

$P_1 \cos \alpha_1$ along MX, which force may be denoted by X_1,

and $P_1 \sin \alpha_1$ along MY, which force may be denoted by Y_1.

Treat all the other forces in like manner, thus reducing them to components along MX and MY; and add together the components along each of the lines of resolution. Then if X denote the sum of the components along MX, and Y the sum of the components along MY, we have

$$X = P_1 \cos \alpha_1 + P_2 \cos \alpha_2 + P_3 \cos \alpha_3 + P_4 \cos \alpha_4 + P_5 \cos \alpha_5,$$

$$Y = P_1 \sin \alpha_1 + P_2 \sin \alpha_2 + P_3 \sin \alpha_3 + P_4 \sin \alpha_4 + P_5 \sin \alpha_5.$$

Lastly, to find the resultant of X and Y.

(§ 429). $$R = \sqrt{(X^2 + Y^2)}, \tag{21}$$

and $$\cos \theta = \frac{X}{R}, \tag{22}$$

or, as is in general better for calculation,

$$\tan\theta = \frac{Y}{X}, \tag{23}$$

whence we derive the magnitude of the resultant,

$$R = X \sec\theta. \tag{24}$$

The calculation will in general be facilitated by the use of logarithms; for which purpose equations (23) and (24) are to be modified in the following manner:—

$$\text{tab. log. } \tan\theta = \text{log. } Y - \text{log. } X + 10, \tag{25}$$

$$\text{log. } R = \text{log. } X + \text{tab. log. } \sec\theta - 10. \tag{26}$$

Remark 1.—It is to be observed that the sums X of the different components X_1, X_2, &c., and Y of Y_1, Y_2, &c., are got by an algebraic addition, whatever may be the algebraic signs of the several terms. If the given forces act all round the point M, it will happen in the resolution that the different components do not all act in the same directions along XX' and YY'. It will be necessary, therefore, to fix upon one direction as positive. Thus, if MX and MY be positive directions, MX', MY' will be negative; and absolute values of the components, which act from M to X', and from M to Y', must be subtracted from, instead of added to, those along MX and MY.

Remark 2.—In choosing the axes of resolution, it simplifies the problem to fix on one of the lines which represent the forces, as one of the axes, and a line perpendicular to it, as the other.

Let ML_1, the line of the first force P_1, be the axis MX, and MY, a line perpendicular to it, the other,

a_1 in this case is nothing; and the angle $P_2MP_1 = a_2$.

Hence, if $a_1 = 0$, the resolution of the first force is

$$P_1 \begin{cases} X_1 = P_1 \cos a_1 = P_1 \\ Y_1 = P_1 \sin a_1 = 0, \end{cases}$$

that is, P_1 requires no resolution.

If two of the forces happen to be at right angles, it will be convenient to choose the axes along them, and then neither requires resolution.

Actual cases may often be simplified by observing if any two of the forces are opposite, in which case, one force, equal to the excess of the greater above the less, and acting in the direction of the greater, may be taken instead of them.

Remark 3.—When the direction of the resultant is known, and its magnitude is required, it is most convenient to make it one of the axes of resolution.

Let MK be the direction of the resultant of P_1, P_2, P_3, P_4, the different forces. Resolve each force into two, one along MK, and one in a line perpendicular to it. Add the components along MK. The sum must be the magnitude of the resultant; and the components along the other line must balance one another. Hence,

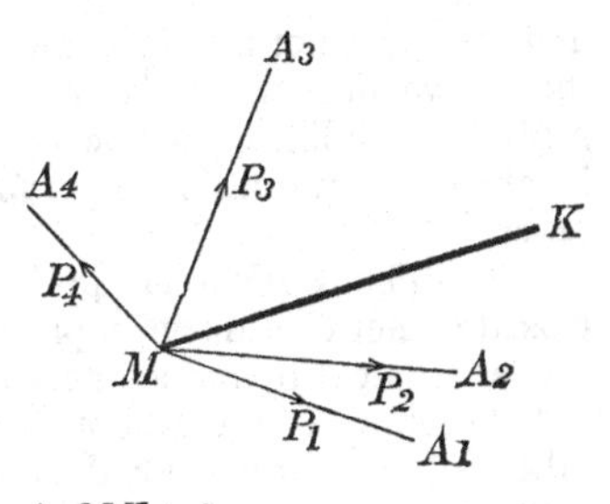

$$X = R = P_1 \cos A_1 MK + P_2 \cos A_2 MK + \&c.,$$

and $$Y = P_1 \sin A_1 MK + P_2 \sin A_2 MK + \&c. = 0.$$

Remark 4.—Equations (23) and (24) may be employed with advantage in all cases where the numbers of significant figures in the values to be used for X and Y are large.

By equations (23) and (24) the direction of the resultant is first determined, and then its magnitude, not as in equations (21) and (22), the magnitude first, and then the direction.

436. For the better understanding of what follows a slight digression (§§ 437, 464) upon projections and geometrical co-ordinates is now inserted.

437. The projection of a point on a straight line, is the point in which the latter is cut by a perpendicular to it from the former.

438. Any line, joining two points, is called an arc. It is not necessary to confine this expression to its most usual signification of a continuous curve line. It may be applied to a straight line joining two points, as an extreme case; or it may be applied to a zigzag or angular path from one point to the other; or to a self-cutting path, whether curved or polygonal; in short, to any track whatever, from one point to the other.

439. The projection of an arc on a straight line, is the portion of the latter intercepted between the projections of the extremities of the former.

440. If we imagine an arc divided into any number of parts, the projections of these parts, taken consecutively on any straight line, make up consecutively the projection of the whole. Hence, the sum of the projections of the parts is equal to the projection of the whole. But in this statement, it must be understood that, of such partial projections laid down in order, those which are drawn in one direction, or *forwards*, being reckoned as positive, those which are drawn in the other direction, or *backwards*, must be reckoned as negative.

441. The projection of an arc on any straight line, is equal to the length of the straight line joining the extremities of the former, multiplied by the cosine of the angle[1] at which it is inclined to the latter.

[1] The angle at which one line is inclined to another, is the angle between two lines

This angle, if not a right angle, will be acute or obtuse, according to the convention which is understood as to the direction reckoned *positive* in the line of projection; and the extremity of the arc which is taken *first* in drawing a *positive* line from one extremity of it to the other.

442. The orthogonal projection of a line, straight or curved, closed or not closed, on a plane, is the locus of the points in which the latter is cut by perpendiculars to it from all points of the former.

Other kinds of projections are also used in geometry; but when no other designation is applied or understood, the simple term *projection* will always mean orthogonal projection.

443. A circuit is a line returning into itself, or a line without ends in a finite space. It is (if a continuous curve) often called a *closed curve*; or if made up altogether of rectilinear parts, a *closed polygon*. A circuit in one plane may be either *simple* or self-cutting. The latter variety has been called by De Morgan, *autotomic*. But whether simple or autotomic, there is just one definite course to go round a circuit; and at double or multiple points, this course must be distinctly indicated[1] (arrow-heads being generally used for the purpose on a diagram, like the finger-posts where two or more roads cross). A circuit not confined to one plane need never be considered to be autotomic, unless as an extreme case. Thus, if we take any thread or wire, however fine, and bend it into any curve or broken line, or tie it into the most complicated knot or succession of knots, but attach its ends together; any geometrical line drawn altogether within it, from any one point of it, round through its length back to the same point, constitutes essentially a simple or not self-cutting circuit.

444. 'The area enclosed by,' or 'the area of' a simple plane circuit, is an expression which requires no explanation. But, as has been shown by De Morgan[2], a peculiar rule of interpretation is necessary to apply the same expression to an autotomic plane circuit, and it has no application, hitherto defined, to a circuit not confined to one plane.

445. The area of an autotomic plane circuit, is the sum of the areas of all its parts each multiplied by zero with unity as many times added as the circuit is crossed[3] from right to left, and unity as many

drawn parallel to them from any point, in directions similar to the directions in the given lines which are reckoned positive.

[1] 'A curve which has double or multiple points, may be in many different ways a *circuit*, or mode of proceeding from one point to the same again. Thus the figure of 8 may be traced as a *self-cutting* circuit, in the way in which it is natural if the curve be a *continuous lemniscate*, or it may be traced as a circuit presenting two coincident salient points. A determinate area requires a determinate mode of making the circuit.' De Morgan, *Cambridge and Dublin Mathematical Journal*, May, 1850.

[2] 'Extension of the word area,' *Cambridge and Dublin Mathematical Journal*, May, 1850.

[3] A moving point is said to cross a plane circuit from right to left, if it crosses from the right side to the left side as regarded by a person looking from any point of the circuit in the direction reckoned positive.

times subtracted as the circuit is crossed from left to right, when a point is carried in the plane from the outside to any position within the enclosed area in question. The diagram, which is that given by De Morgan, will show more clearly what is meant by this use of the word *area*. The reader, with this as a model, may exercise himself by drawing autotomic circuits and numbering the different portions of the enclosed area according to the rule, which he will then find no difficulty in understanding.

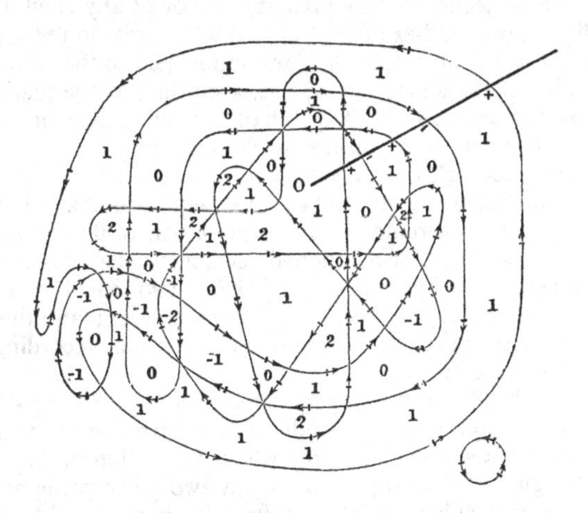

446. Any portion of surface, edged or bounded by a circuit, is called a *shell*.

A plane area may be regarded as an extreme case, but generally the surface of a shell will be supposed to be curved.

A simple shell is a shell of which the surface is single throughout. One side of the shell must always be distinguished from the other, whatever may be the convolutions of its surface. Thus we shall have a marked and unmarked side, or an outside and an inside, to distinguish from one another.

447. The projection of a shell on any plane, is the area included in the projection of its bounding line.

448. If we imagine a shell divided into any number of parts, the projections of these parts on any plane make up the projection of the whole. But in this statement it must be understood that the areas of partial projections are to be reckoned as positive only if the marked side, or, as we shall call it, the outside, of the projected area, and a marked side, which we shall call the front, of the plane of projection, face the same way.

If the outside of any portion of the projected area faces on the whole backwards, relatively to the front of the plane of projection, the projection of this portion is to be reckoned as negative in the sum.

Of course if the projected surface, or any part of it, be a plane area at right angles to the plane of projection, the projection vanishes.

Cor. The projections of any two shells having a common edge, on any plane, are equal. The projection of a closed surface (or a shell with evanescent edge), on any plane, is nothing.

449. Equal areas in one plane or in parallel planes, have equal projections on any plane, whatever may be their figures. [The proof is easily found.]

Hence the projection of any plane figure, or of any shell, edged by a plane figure, on another plane, is equal to its area, multiplied by the cosine of the angle at which its plane is inclined to the plane of projection. This angle is acute or obtuse, according as the marked sides of the projected area, and of the plane of projection face, on the whole, towards the same parts, or on the whole oppositely.

450. Two rectangles, with a common edge, but not in one plane, have their projection on any other plane, equal to that of one rectangle having their two remote sides for one pair of its opposite sides. For, the sides of this last-mentioned rectangle constitute the edge of a *shell*, which we may make by applying two equal and parallel triangular areas to the sides of the given rectangles; and the sum of the projections of these two triangles on any plane, according to the rule of § 448, is nothing.

Hence (as is shown by a very simple geometrical proof, which is left as an exercise to the student), we have the following construction to find a single plane area whose projection on any plane is equal to the sum of the projections of any two given plane areas.

From any convenient point of reference draw straight lines perpendicular to the two given plane areas *forward*, relatively to their marked sides considered as fronts. Make these lines numerically equal to the two areas respectively. On these describe a parallelogram, and draw the diagonal of this parallelogram through the point of reference. Place an area with one side marked as front, in any position perpendicular to this diagonal, facing forwards, and relatively to the direction in which it is drawn from the point of reference. Make this area equal numerically to the diagonal. Its projection on any plane will be equal to the sum of the projections of the two given areas, on the same plane.

The same construction may be continued; just as, in § 413, the geometrical construction to find the resultant of any number of forces; and thus we find a single plane area whose projection on any plane is equal to the sum of the projections on the same plane of any given plane areas. And as any shell may (if it be not composed of a finite) be regarded as composed of an infinite number of plane areas, the same construction is applicable to a shell. Hence the projection of a shell on any plane is equal to the projection on the same plane, of a certain plane area, determined by the preceding construction.

From this it appears that the projection of a shell is nothing on any plane perpendicular to the one plane on which its projection is greater than on any other; and that the projection on any intermediate plane is equal to the greatest projection multiplied by the cosine of the inclination of the plane of the supposed projection to the plane of greatest projection.

451. To specify a point is to state precisely its position. As we have no conception of position, except in so far as it is relative, the specification of a point requires definite objects of reference, that is, objects to which it may be referred. The means employed for this purpose are certain elements called co-ordinates, from the system of specification which Descartes first introduced into mathematics. This system seems to have originated in the following method, for describing a curve by a table of numbers, or by an equation.

452. Given a plane curve, a fixed line in its plane, and a fixed point in this line, choose as many points in the curve as are required to indicate sufficiently its form: draw perpendiculars from them to the fixed line, and measure the distances along it, cut off by these lines, reckoning from the fixed point. In this way any number of points in the curve were specified. The parts thus cut off along the fixed line, were termed *lineae abscissae*, and the perpendiculars, *lineae ordinatim applicatae*. The system was afterwards improved by drawing through the point of reference a line at right angles to the first, and measuring off along it, the *ordinates* of the curve. The two lines at right angles to one another are called the axes of reference, or the lines of reference. The ordinate and abscissa of any point are termed its *co-ordinates;* and an equation between them, by which either may be calculated when the other is given, expresses the curve in a perfectly full and precise manner.

453. It is not necessary that the lines of reference be chosen at right angles to each other. But when they are chosen, inclined at any other angle than a right angle, the co-ordinates of the point specified are not its perpendicular distances from them, but its distances from either, measured parallel to the other. Such oblique co-ordinates are sometimes convenient, but rectangular co-ordinates are, in general, the most useful; these we shall now consider.

454. If the points to be specified are all in one plane, the objects of reference are two lines at right angles to one another in that plane. Thus, let P be a point in a plane XOY; and let OX, OY, be two lines in the plane, cutting each other at right angles in the point O. Then will the position of the point P be known, if the perpendicular distance of the point P from the line OX, namely, the length of the line PA, and the perpendicular distance from OY, namely, the length of the line PB, be known.

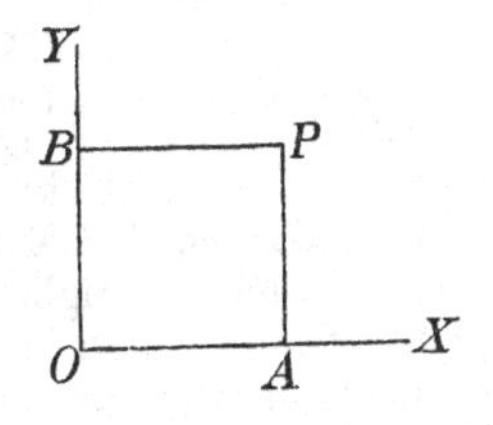

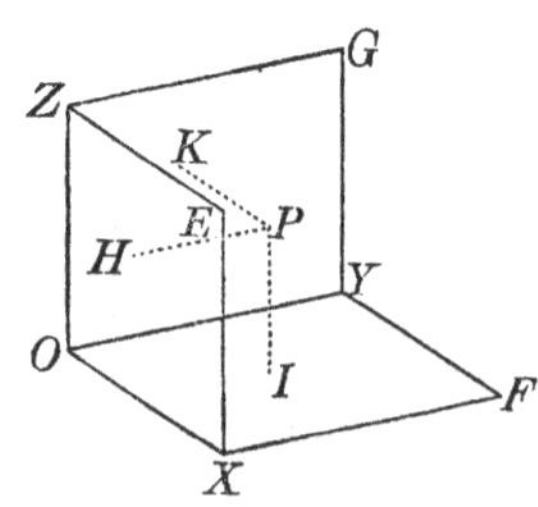

455. Again, let points, not all in one plane, but in any positions through space be considered. To specify each point now, three co-ordinates are required, and the objects of reference chosen may be three planes at right angles to one another; thus, the point P is specified by the lines PK, PH, PI, drawn perpendicular to the planes YZ, ZX, XY, respectively.

In our standard diagrams the positive directions OX, OY, OZ, are so taken that if a watch is held in the plane XOY, with its face towards OZ, an angular motion *against the hands* would carry a line from OX to OY, through the right angle XOY.

456. When the objects to be specified are lines all passing through one point, the specifying elements employed, are angles standing in definite relation to them, and to the objects of reference. There are two chief modes in which this kind of specification is carried out: the polar and the symmetrical.

457. *Polar Method.* 1°. Lines all in one plane. In this case the object of reference is any fixed line through their common point of intersection, and lying in their plane.

Let O be the common point of intersection, OX the fixed line, and OP the line to be specified. Then the position of OP will be known, if the angle XOP, which the line OP makes with OX, be known.

2°. Lines in space, all passing through one point, may be specified by reference to a plane and a line in it, both passing through their common point of intersection.

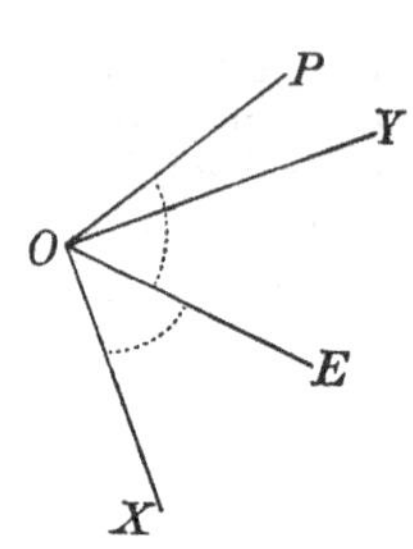

Let OP be one of a number of lines, all passing through O, to be specified with reference to the plane XOY, and the line OX in it. Through OP let a plane be drawn, cutting the plane XOY at right angles in OE. Then the line OP will be specified, if the angles XOE, EOP are given.

Corollary. Similarly, if the line OP be the locus of a series of points, any one of these points will be specified, if its distance from O and the two angles specifying the line OP, are known.

458. *Symmetrical Method.* In this method the objects of reference are three lines at right angles to each other, through the common point of intersection of the lines to be specified, and the specifying elements are the three angles which each line makes with these three lines of reference.

Thus, if O be the common point of intersection, OK one of the lines to be specified, and OX, OY, OZ, the lines of reference; then the angles XOK, YOK, ZOK, are the specifying elements.

459. From what has now been said, it will be seen that the projections of a given line on other three at right angles to each other are immediately expressible, if its direction is specified by either of the two methods.

1°. *Polar Method.* Let OK be the given line, and OX, OY, OZ, the lines along which it is to be projected. Through OZ and OK let a plane pass, cutting the plane XOY in OE. Through K draw another plane, KEA, cutting OX perpendicularly in A and KEB cutting OY perpendicularly in B. Then KE, being the intersection of two planes each perpendicular to XOY, is perpendicular to every line in this plane. Hence, OEK is a right angle.

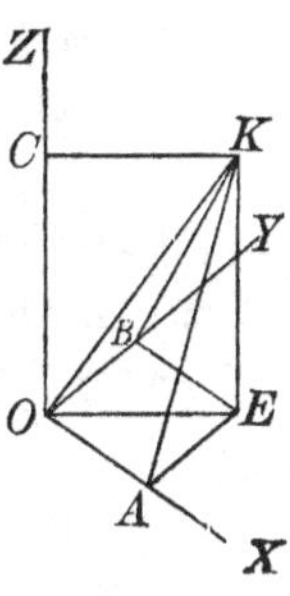

Hence,

$$OE = OK \cos KOE.$$

Again, since the plane KAE was drawn perpendicular to OX, OAE is a right angle.

Hence, $\quad OA = OE \cos EOX = OK \cos KOE \cos EOX,$

and similarly, $OB = OE \cos EOY = OK \cos KOE \cos EOY,$

or if we put

$$OK = r,\ XOE = \phi,\ EOK = i,\ KOZ = \theta = \tfrac{1}{2}\pi - i,$$

and let the required lines be denoted by x, y, z, then

$$\left.\begin{aligned} x &= r \sin\theta \cos\phi, \\ y &= r \sin\theta \sin\phi, \\ z &= r \cos\theta. \end{aligned}\right\} \tag{1}$$

2°. *Symmetrical Method.* Let the line be referred to rectangular axes by the three angles, $\alpha = KOX$, $\beta = KOY$, $\gamma = KOZ$.

Then the required projections are

$$x = r\cos\alpha, \quad y = r\cos\beta, \quad z = r\cos\gamma.$$

460. Referring again to the diagram, we have

$$OE^2 = OA^2 + OB^2,$$

and $\quad OK^2 = OE^2 + OC^2,$

therefore, $\quad OK^2 = OA^2 + OB^2 + OC^2,$

or

$$r^2 = x^2 + y^2 + z^2. \tag{2}$$

Substituting here for x, y, z, their values, in terms of r, α, β, γ, found above, and dividing both members of the resulting equation by r^2, we have

$$1 = \cos^2\alpha + \cos^2\beta + \cos^2\gamma. \tag{3}$$

461. In the symmetrical method, three angles are used; but, as we have seen, only two are necessary to fix the position of the line. We now see that, if two of the three angles, α, β, γ, are given, the third can be found. Suppose α and β given, then by § 460,

$$\cos^2\gamma = 1 - \cos^2\alpha - \cos^2\beta.$$

For logarithmic calculation, the following modification of the preceding formula is useful,

$$\cos^2\gamma = \sin^2\alpha - \cos^2\beta = -\cos(\alpha+\beta)\times\cos(\alpha-\beta),$$

whence $\cos\gamma = \sqrt{\{-\cos(\alpha+\beta)\times\cos(\alpha-\beta)\}}$
$= \sqrt{\{\cos(\pi-\alpha-\beta)\times\cos(\alpha-\beta)\}}$

$$\text{Tab. Log.} \cos\gamma = \tfrac{1}{2}\,\{\text{T. L.} \cos(\pi-\alpha-\beta) + \text{T. L.} \cos(\alpha-\beta)\}. \quad (4)$$

462. The following comparison will show in what way the two systems are related, and how it is possible to derive the specifying elements of either from those of the other. In the polar method, the fixed line in the equatorial plane, corresponds with one of the three lines of reference in the symmetrical. A line in the equatorial plane, drawn at right angles to the fixed line of the polar system, constitutes a second line of reference in the symmetrical system. The third line in the symmetrical system, is the axis of the polar system, from which the polar distance (θ) is measured. A comparison of preceding formulae shows that

$$\left.\begin{aligned}\cos\alpha &= \sin\theta\cos\phi,\\ \cos\beta &= \sin\theta\sin\phi,\\ \cos\gamma &= \cos\theta.\end{aligned}\right\} \quad (5)$$

463. The cosines of the three angles, α, β, γ, of the symmetrical system, are commonly called the *direction cosines* of the line specified. If we denote them by l, m, n, we have as above,

$$l^2 + m^2 + n^2 = 1. \quad (6)$$

A line thus specified is for brevity called *the line* (l, m, n).

If l, m, n, are the direction cosines of a certain line; it is clear that $-l$, $-m$, $-n$, are the direction cosines of the line in the opposite direction from O. Thus it appears that the direction cosines of the line, specify not only the straight line in which it lies, but the direction in it which is reckoned as positive.

464. We conclude this digression with some applications of the principles explained in it, which are useful in many dynamical investigations.

(*a*) To find the mutual inclination, θ, of two lines, (l, m, n), (l', m', n'). Measure off any length $OK = r$, along the first line (see fig. of § 459). We have, as above,

$$OA = lr, \quad AE = mr, \quad EK = nr.$$

Now (§ 441), the projection of OK on the second line, is equal to the sum of the projections of OA, AE, EK, on the same. But the cosines of the angles at which these several lengths are inclined to the line of projection, are respectively $\cos\theta$, l', m', n'. Hence

$$OK\cos\theta = OA\,l' + AE\,m' + EK\,n'.$$

If we substitute in this, for OK, OA, AE, EK, their values shown above, and divide both members by r, it becomes

$$\cos\theta = ll' + mm' + nn', \tag{7}$$

a most important and useful formula.

Sometimes it is useful to have the sine instead of the cosine of θ. To find it we have of course,

$$\sin^2\theta = 1 - (ll' + mm' + nn')^2.$$

This expression may be modified thus:—instead of 1, take what is equal to it, $(l'^2 + m'^2 + n'^2)\,(l^2 + m^2 + n^2)$,

and the second member of the preceding becomes

$$(l^2 + m^2 + n^2)\,(l'^2 + m'^2 + n'^2) - (ll' + mm' + nn')^2$$
$$= (mn')^2 + (nm')^2 - 2mm'nn' + \&c. = (mn' - nm')^2 + \&c.$$

Hence, $\sin\theta = \{(mn' - nm')^2 + (nl' - ln')^2 + (lm' - ml')^2\}^{\frac{1}{2}}$. (8)

(*b*) To find the direction cosines, λ, μ, ν, of the common perpendicular to two lines, (l, m, n), (l', m', n').

The cosine of the inclination of (λ, μ, ν) to (l, m, n) is, according to (7) above, $l\lambda + m\mu + n\nu$; and therefore

$$\left.\begin{aligned} l\lambda + m\mu + n\nu &= 0, \\ \text{similarly}\quad l'\lambda + m'\mu + n'\nu &= 0, \\ \text{also (§ 463)}\quad \lambda^2 + \mu^2 + \nu^2 &= 1. \end{aligned}\right\} \tag{9}$$

These three equations suffice to determine the three unknown quantities, λ, μ, ν. Thus, from the first two of them, we have

$$\frac{\lambda}{mn' - nm'} = \frac{\mu}{nl' - ln'} = \frac{\nu}{lm' - l'm}. \tag{10}$$

From these and the third of (9), we conclude

$$\lambda = \frac{mn' - nm'}{\{(mn' - nm')^2 + (nl' - ln')^2 + (lm' - ml')^2\}^{\frac{1}{2}}}, \quad \&c.,$$

or if we denote, as above, by θ, the mutual inclination of (l, m, n) (l', m', n');

$$\lambda = \frac{(mn' - nm')}{\sin\theta}, \quad \mu = \frac{(nl' - ln')}{\sin\theta}, \quad \nu = \frac{(lm' - ml')}{\sin\theta}. \tag{11}$$

The sign of each of these three expressions may be changed, in as much as either sign may be given to the numerical value found for $\sin\theta$ by (8). But as they stand, if $\sin\theta$ is taken positive, they express the direction cosines of the perpendicular drawn from O through the face of a watch, held in the plane $(l\ m\ n)$, $(l'\ m'\ n')$, and so facing that angular motion, *against or with the hands*, would carry a line from the direction, (l, m, n), through an angle less than 180° to the direction, (l', m', n'), according as angular motion, through a right angle from OX to OY is *against or with the hands* of a watch, held in the plane XOY, and facing towards OZ. This rule is proved by supposing, as a particular case, the lines (l, m, n), (l', m', n'), to coincide with OX and OY respectively; and then supposing them altered in their mutual inclination to any other angle between 0 and π, and their plane turned to any position whatever.

If we measure off any lengths, $OK=r$, and $OK'=r'$, along the two lines, (l, m, n) and (l', m', n'), and describe a parallelogram upon them, its area is equal to $rr' \sin\theta$, since $r' \sin\theta$ is the length of the perpendicular from K' to OK. Hence, using the preceding expression (8) for $\sin\theta$, and taking

$$lr=x, \quad mr=y, \quad nr=z,$$
$$l'r'=x', \quad m'r'=y', \quad n'r'=z',$$

we conclude the following propositions.

(*c*) The area of a parallelogram described upon lines from the origin of co-ordinates to points (x, y, z), (x', y', z') is equal to

$$\{(yz'-y'z)^2+(zx'-z'x)^2+(xy'-x'y)^2\}^{\frac{1}{2}}. \tag{12}$$

And, as λ, μ, ν, are the cosines of the angles at which the plane of this area is inclined to the planes of YZ, ZX, XY, respectively, its projections on these planes are

$$yz'-y'z, \quad zx'-z'x, \quad xy'-x'y. \tag{13}$$

The figures of these projections are parallelograms in the three planes of reference; that in the plane YZ, for instance, being described on lines drawn from the origin to the points (y, z) and (y', z'). It is easy to prove this (and, of course, the corresponding expressions for the two other planes of reference,) by elementary geometry. Thus, it is easy to obtain a simple geometrical demonstration of the equations (8) and (11). It is sufficient here to suggest this investigation as an exercise to the student. It essentially and obviously includes the *rule of signs*, stated above (§ 464 (*a*)).

(*d*) The volume of a parallelepiped described on OK, OK', OK'', three lines drawn from O to three points

(x, y, z), (x', y', z'), (x'', y'', z''), is equal to

$$x''(yz'-y'z)+y''(zx'-z'x)+z''(xy'-x'y), \tag{14}$$

an expression which is essentially positive, if OK, OK', OK'', are arranged in order similarly to OX, OY, OZ (see § 455 above). The proof is left as an exercise for the student.

In modern algebra, this expression is called a *determinant*, and is written thus :—

$$\begin{vmatrix} x, & y, & z, \\ x', & y', & z', \\ x'', & y'', & z''. \end{vmatrix} \tag{15}$$

465. To find the resultant of three forces acting on a material point in lines at right angles to one another.

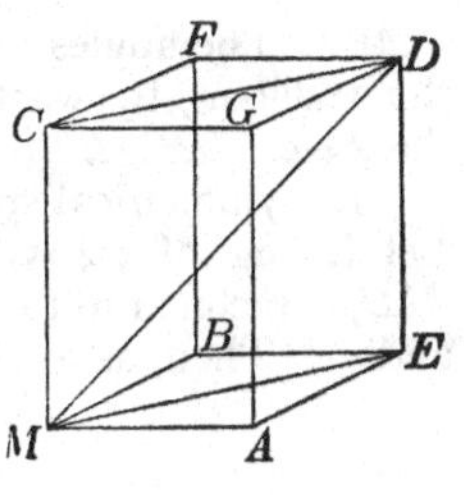

1°. To find the magnitude of the resultant. Let the forces be given numerically, X, Y, Z, and let them be represented respectively by the lines MA, MB, MC at right angles to one another.

First determine the resultant of X and Y in magnitude. If we denote it by R', we have (§ 429)

$$R' = \sqrt{(X^2+Y^2)}. \qquad (1)$$

This resultant, represented by ME, lies in the plane BMA; and since the lines of the forces X and Y are perpendicular to the line MC, the line ME must also be perpendicular to it; for, *if a line be perpendicular to two other lines, it is perpendicular to every other line in their plane;* hence R' acts perpendicularly to Z.

Next, find the resultant of R' and Z, the third force. If we denote it by R, we have

$$R = \sqrt{(R'^2+Z^2)},$$

and substituting for R'^2 its value, we have

$$R = \sqrt{(X^2+Y^2+Z^2)}. \qquad (2)$$

2°. To find the direction of the resultant. Determine first the inclination of the subsidiary resultant R' to MA or MB. Let the angle EMA be denoted by ϕ; then we have

$$\cos\phi = \frac{X}{R'}.$$

Next, let γ denote the angle at which the line MD is inclined to MC; that is, the angle CMD; we have

$$\cos\gamma = \frac{Z}{R}. \qquad (3)$$

Thus, by means of the two angles γ and ϕ, the position of the line MD, and, consequently, that of the resultant is found.

466. In the numerical solution of actual cases, it will generally be found most convenient to calculate the three elements in the following order: 1°, ϕ, 2°, γ, 3°, R.

1°. To calculate ϕ, the formula already given, may be taken

$$\tan\phi = \frac{Y}{X}. \qquad (4)$$

2°. To calculate γ. We have

$$\tan\gamma = \frac{R'}{Z}. \qquad (5)$$

But

$$R' = X\sec\phi.$$

Hence

$$\tan\gamma = \frac{X\sec\phi}{Z}. \qquad (6)$$

3°. To calculate R.

$$R = Z\sec\gamma. \qquad (7)$$

467. The angles determined by these equations specify the line of the resultant, by what was called in previous sections (§§ 457, 459) the *Polar Method.*

The symmetrical specification of the resultant is to be found thus: Let (in fig. of § 465) the angles at which the line of the resultant, *MD*, is inclined to those of the forces be respectively denoted by α, β, and γ. Then, as above (equation (3)),

$$\cos\gamma=\frac{Z}{R}. \tag{8}$$

By the same method we shall find

$$\cos\alpha=\frac{X}{R}, \tag{9}$$

and

$$\cos\beta=\frac{Y}{R}. \tag{10}$$

If, therefore, there are three forces at right angles to one another, the cosine of the inclination of their resultant to any one of them is equal to this force divided by the resultant.

This method requires that the magnitude of the resultant be known before its position is determined. For the latter purpose, any two of the angles, as was shown in Chapter V, are sufficient.

468. We shall now consider the resolution of forces along three specified lines. The most important case of all is that in which the lines are at right angles to one another.

Let the force R, given to be resolved, be represented by *MD*, and let the angles which it forms with the lines of resolution be given, either α, β, γ, or γ, ϕ. Required the components X, Y, Z.

1°. Suppose α, β, γ are given, then we deduce

from equation (9) $\quad X=R\cos\alpha$;
from equation (10) $\quad Y=R\cos\beta$;
and from equation (8) $\quad Z=R\cos\gamma$.

2°. Suppose the data are R, γ, ϕ, that is, the magnitude of the resultant, its inclination to one of the axes of resolution, and the inclination of the plane of the resultant and that axis to either of the other axes.

To find the components X and Y: resolve the force R in the vertical plane *CMED* into two rectangular components along *MC* and *ME*. Let the angle *CMD* be denoted by γ. Then we have for the component along *MC*,

$$Z=R\cos\gamma, \tag{11}$$

and for the component along *ME*,

$$ME=R\sin\gamma.$$

Next, resolve the component along *ME* in the horizontal plane *BMAE*, into two, one along *MA*, and the other along *MB*. Let the angle EMA be denoted by ϕ. Then we have for the component along MA,

$$X = ME \cos\phi = R \sin\gamma \cos\phi, \tag{12}$$

and for the component along MB,

$$Y = ME \sin\phi = R \sin\gamma \sin\phi. \tag{13}$$

469. We are now prepared to solve the general problem:—Given, any number of forces acting on one point, in lines which lie in different planes, required their resultant in position and magnitude.

Through the point acted on, draw three lines or axes of resolution at right angles to one another. Resolve each force, by § 468, 1°, or by § 468, 2°, into three components, acting respectively along the three lines. When all the forces have been thus treated, add severally the sets of components: by this means, all the forces are reduced to three at right angles to one another. Find, by equation (2), their resultant: the single force thus obtained is the resultant of the given forces, which was to be found.

Remark.—All the remarks made with reference to the resolution and composition of forces along two axes (§ 435) apply, with the necessary extension, to that of forces along three.

470. We are now prepared to answer the question which forms the first general head of Statics; *What are the conditions of Equilibrium of a material point?* The answer may be put in one or other of two forms.

1°. If a set of forces acting on a material point be in equilibrium, any one of them must be equal and opposite to the resultant of the others: or,

2°. If a set of forces acting on a material point be in equilibrium, the resultant of the whole set must be equal to nothing.

471. Let us consider the first of these statements.

Given, a set of forces, P_1, P_2, P_3, &c., in equilibrium: the force P_1, for example, is equal and opposite to the resultant of P_2, P_3, &c.; or, the resultant of P_2, P_3, &c., is $-P_1$. Omitting P_1, find the resultant of the remaining forces by the general method; the components of this resultant will be

$$\begin{aligned} &P_2 \cos\alpha_2 + P_3 \cos\alpha_3 + \text{\&c. along } MX. \\ &P_2 \cos\beta_2 + P_3 \cos\beta_3 + \text{\&c. along } MY. \\ &P_2 \cos\gamma_2 + P_3 \cos\gamma_3 + \text{\&c. along } MZ. \end{aligned}$$

Now, if $-P_1$ be the resultant, the components of $-P_1$ will be equivalent respectively to the components of this resultant, therefore

$$\begin{aligned} -P_1 \cos\alpha_1 &= P_2 \cos\alpha_2 + P_3 \cos\alpha_3 + \text{\&c.} \\ -P_1 \cos\beta_1 &= P_2 \cos\beta_2 + P_3 \cos\beta_3 + \text{\&c.} \\ -P_1 \cos\gamma_1 &= P_2 \cos\gamma_2 + P_3 \cos\gamma_3 + \text{\&c.} \end{aligned}$$

Which equations, in the following more general form, express the required conditions:

$$P_1 \cos \alpha_1 + P_2 \cos \alpha_2 + P_3 \cos \alpha_3 + \&c. = 0.$$
$$P_1 \cos \beta_1 + P_2 \cos \beta_2 + P_3 \cos \beta_3 + \&c. = 0.$$
$$P_1 \cos \gamma_1 + P_2 \cos \gamma_2 + P_3 \cos \gamma_3 + \&c. = 0.$$

472. The second form of the answer may be illustrated either *a*, dynamically, or *b*, algebraically.

(*a*) Suppose all the forces reduced to three, X, Y, Z, acting at right angles to each other. Under what circumstances will three forces give a vanishing resultant? Substitute for X and Y their resultant R', and consider R' and Z at right angles to one another. If they give a vanishing resultant, that is, if Z and R' balance, they must either be equal and directly opposed, or else they must each be equal to nothing. But they are not directly opposed, therefore each is equal to nothing. Now, since $R'=0$, X and Y, which are equivalent to R', must also each be equal to nothing: in order, therefore, that the resultant of forces acting along three lines at right angles to one another may vanish, we have

$$P_1 \cos \alpha_1 + P_2 \cos \alpha_2 + \&c. = 0.$$
$$P_1 \cos \beta_1 + P_2 \cos \beta_2 + \&c. = 0.$$
$$P_1 \cos \gamma_1 + P_2 \cos \gamma_2 + \&c. = 0.$$

(*b*) The general expression for the resultant is

$$R^2 = X^2 + Y^2 + Z^2.$$

Now, for equilibrium, $R=0$,

and therefore, $X^2 + Y^2 + Z^2 = 0.$

But the sum of three positive quantities can be equal to nothing, only when each of them is nothing: hence

$$X=0,$$
$$Y=0,$$
$$Z=0.$$

473. We may take one or two particular cases as examples of the general results above. Thus,

1. If the particle rest on a smooth curve, the resolved force along the curve must vanish.

2. If the curve be rough, the resultant force along it must be balanced by the friction.

3. If the particle rest on a smooth surface, the resultant of the applied forces must evidently be perpendicular to the surface.

4. If it rest on a rough surface, friction will be called into play, resisting motion along the surface; and there will be equilibrium at any point within a certain boundary, determined by the condition that at *it* the friction is μ times the normal pressure on the surface, while within it the friction bears a less ratio to the normal pressure. When the only applied force is gravity, we have a very simple result, which is often practically useful. Let θ be the angle between the

normal to the surface and the vertical at any point; the normal pressure on the surface is evidently $W \cos\theta$, where W is the weight of the particle; and the resolved part of the weight parallel to the surface, which must of course be balanced by the friction, is $W \sin\theta$. In the limiting position, when sliding is just about to commence, the greatest possible amount of statical friction is called into play, and we have

$$W \sin\theta = \mu W \cos\theta,$$

or

$$\tan\theta = \mu.$$

The value of θ thus found is called the *Angle of Repose*, and may be seen in nature in the case of sand-heaps, and slopes formed by débris from a disintegrating cliff (especially of a flat or laminated character), on which the lines of greatest slope are inclined to the horizon at an angle determined by this consideration.

474. A most important case of the composition of forces acting at one point is furnished by the consideration of the attraction of a body of any form upon a material particle anywhere situated. Experiment has shown that the attraction exerted by any portion of matter upon another is not modified by the neighbourhood, or even by the interposition, of other matter; and thus the attraction of a body on a particle is the resultant of the several attractions exerted by its parts. To treatises on applied mathematics we must refer for the examination of the consequences, often very curious, of various laws of attraction; but, dealing with Natural Philosophy, we confine ourselves to the law of gravitation, which, indeed, furnishes us with an ample supply of most interesting as well as useful results.

475. This law, which (as a property of matter) will be carefully considered in the next Division of this Treatise, may be thus enunciated.

Every particle of matter in the universe attracts every other particle with a force, whose direction is that of the line joining the two, and whose magnitude is directly as the product of their masses, and inversely as the square of their distance from each other.

Experiment shows (as will be seen further on) that the same law holds for electric and magnetic attractions; and it is probable that it is the fundamental law of all natural action, at least when the acting bodies are not in actual contact.

476. For the special applications of Statical principles to which we proceed, it will be convenient to use a special unit of mass, or quantity of matter, and corresponding units for the measurement of electricity and magnetism.

Thus if, in accordance with the physical law enunciated in § 475, we take as the expression for the forces exerted on each other by masses M and m, at distance D, the quantity

$$\frac{Mm}{D^2};$$

it is obvious that our *unit* force is the mutual attraction of two units of mass placed at unit of distance from each other.

477. It is convenient for many applications to speak of the *density* of a distribution of matter, electricity, etc., along a line, over a surface, or through a volume.

Here density of line is the quantity of matter per unit of length.
" " surface " " " " area.
" " volume " " " " volume.

478. In applying the succeeding investigations to electricity or magnetism, it is only necessary to premise that M and m stand for *quantities* of free electricity or magnetism, whatever these may be, and that here the idea of *mass* as depending on *inertia* is not necessarily involved. The formula $\frac{Mm}{D^2}$ will still represent the mutual action, if we take as unit of imaginary electric or magnetic matter, such a quantity as exerts unit force on an equal quantity at unit distance. Here, however, one or both of M, m may be negative; and, as in these applications like kinds *repel* each other, the mutual action will be attraction or repulsion, according as its sign is negative or positive. With these provisos, the following theory is applicable to any of the above-mentioned classes of forces. We commence with a few simple cases which can be completely treated by means of elementary geometry.

479. *If the different points of a spherical surface attract equally with forces varying inversely as the squares of the distances, a particle placed within the surface is not attracted in any direction.*

Let $HIKL$ be the spherical surface, and P the particle within it. Let two lines HK, IL, intercepting very small arcs HI, KL, be drawn through P; then, on account of the similar triangles HPI, KPL, those arcs will be proportional to the distances HP, LP; and any small elements of the spherical surface at HI and KL, each bounded all round by straight lines passing through P [and very nearly coinciding with HK], will be in the duplicate ratio of those lines. Hence the forces exercised by the matter of these elements on the particle P are equal; for they are as the quantities of matter directly, and the squares of the distances, inversely; and these two ratios compounded give that of equality. The attractions therefore, being equal and opposite, destroy one another: and a similar proof shows that all the attractions due to the whole spherical surface are destroyed by contrary attractions. Hence the particle P is not urged in any direction by these attractions.

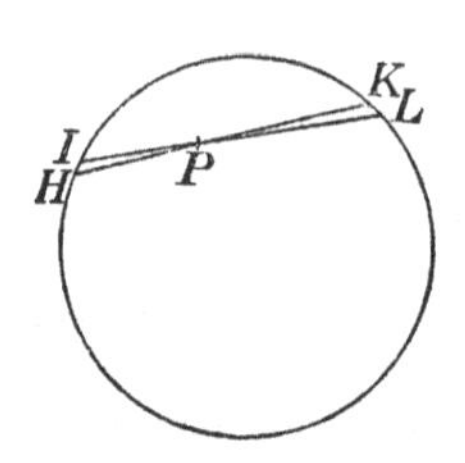

480. The division of a spherical surface into infinitely small elements, will frequently occur in the investigations which follow: and Newton's method, described in the preceding demonstration, in which the division is effected in such a manner that all the parts may be taken together in *pairs of opposite elements with reference to an internal*

point; besides other methods deduced from it, suitable to the special problems to be examined; will be repeatedly employed. The following digression (§§ 481, 486), in which some definitions and elementary geometrical propositions regarding this subject are laid down, will simplify the subsequent demonstrations, both by enabling us, through the use of convenient terms, to avoid circumlocution, and by affording us convenient means of reference for elementary principles, regarding which repeated explanations might otherwise be necessary.

481. If a straight line which constantly passes through a fixed point be moved in any manner, it is said to describe, or generate, a *conical surface* of which the fixed point is the vertex.

If the generating line be carried from a given position continuously through any series of positions, no two of which coincide, till it is brought back to the first, the entire line on the two sides of the fixed point will generate a complete conical surface, consisting of two sheets, which are called *vertical or opposite cones.* Thus the elements HI and KL, described in Newton's demonstration given above, may be considered as being cut from the spherical surface by two *opposite cones* having P for their common vertex.

482. If any number of spheres be described from the vertex of a cone as centre, the segments cut from the concentric spherical surfaces will be similar, and their areas will be as the squares of the radii. The quotient obtained by dividing the area of one of these segments by the square of the radius of the spherical surface from which it is cut, is taken as the measure of the *solid angle of the cone.* The segments of the same spherical surfaces made by the opposite cone, are respectively equal and similar to the former. Hence the solid angles of two vertical or opposite cones are equal: either may be taken as the solid angle of the complete conical surface, of which the opposite cones are the two sheets.

483. Since the area of a spherical surface is equal to the square of its radius multiplied by 4π, it follows that the sum of the solid angles of all the distinct cones which can be described with a given point as vertex, is equal to 4π.

484. The solid angles of vertical or opposite cones being equal, we may infer from what precedes that the sum of the solid angles of all the complete conical surfaces which can be described without mutual intersection, with a given point as vertex, is equal to 2π.

485. The solid angle subtended at a point by a superficial area of any kind, is the solid angle of the cone generated by a straight line passing through the point, and carried entirely round the boundary of the area.

486. A very small cone, that is, a cone such that any two positions of the generating line contain but a very small angle, is said to be cut at right angles, or orthogonally, by a spherical surface described from its vertex as centre, or by any surface, whether plane or curved, which touches the spherical surface at the part where the cone is cut by it.

A very small cone is said to be cut obliquely, when the section is inclined at any finite angle to an orthogonal section; and this angle of inclination is called the *obliquity of the section.*

The area of an orthogonal section of a very small cone is equal to the area of an oblique section in the same position, multiplied by the cosine of the obliquity.

Hence the area of an oblique section of a small cone is equal to the quotient obtained by dividing the product of the square of its distance from the vertex, into the solid angle, by the cosine of the obliquity.

487. Let E denote the area of a very small element of a spherical surface at the point E (that is to say, an element every part of which is very near the point E), let ω denote the solid angle subtended by E at any point P, and let PE, produced if necessary, meet the surface again in E'; then a denoting the radius of the spherical surface, we have

$$E=\frac{2a.\omega.PE^2}{EE'}.$$

For, the obliquity of the element E, considered as a section of the cone of which P is the vertex and the element E a section (being the angle between the given spherical surface and another described from P as centre, with PE as radius), is equal to the angle between the radii EP and EC, of the two spheres. Hence, by considering the isosceles triangle ECE', we find that the cosine of the obliquity is equal to

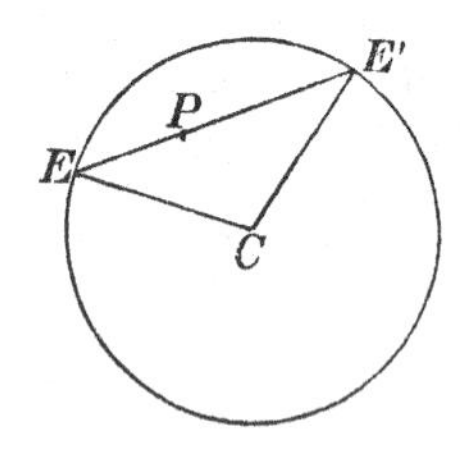

$$\frac{\frac{1}{2}EE'}{EC} \text{ or to } \frac{EE'}{2a},$$

and we arrive at the preceding expression for E.

488. *The attraction of a uniform spherical surface on an external point is the same as if the whole mass were collected at the centre*[1].

Let P be the external point, C the centre of the sphere, and CAP a straight line cutting the spherical surface in A. Take I in CP, so that CP, CA, CI may be continual proportionals, and let the

[1] This theorem, which is more comprehensive than that of Newton in his first proposition regarding attraction on an external point (Prop. LXXI.), is fully established as a corollary to a subsequent proposition (LXXIII. cor. 2). If we had considered the proportion of the forces exerted upon two external points at different instances, instead of, as in the text, investigating the absolute force on one point, and if besides we had taken together all the pairs of elements which would constitute two narrow annular portions of the surface, in planes perpendicular to PC, the theorem and its demonstration would have coincided precisely with Prop. LXXI. of the *Principia*.

whole spherical surface be divided into *pairs of opposite elements with reference to the point I.*

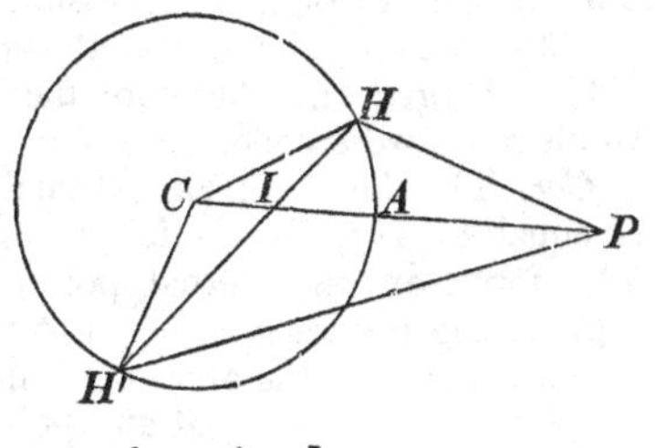

Let H and H' denote the magnitudes of a pair of such elements, situated respectively at the extremities of a chord HH'; and let ω denote the magnitude of the solid angle subtended by either of these elements at the point I.

We have (§ 486),

$$H=\frac{\omega.IH^2}{\cos CHI}, \text{ and } H'=\frac{\omega.IH'^2}{\cos CH'I}.$$

Hence, if ρ denote the density of the surface, the attractions of the two elements H and H' on P are respectively

$$\rho\frac{\omega}{\cos CHI}.\frac{IH^2}{PH^2}, \text{ and } \rho\frac{\omega}{\cos CH'I}.\frac{IH'^2}{PH'^2}.$$

Now the two triangles PCH, HCI have a common angle at C, and, since $PC : CH :: CH : CI$, the sides about this angle are proportional. Hence the triangles are similar; so that the angles CPH and CHI are equal, and

$$\frac{IH}{HP}=\frac{CH}{CP}=\frac{a}{CP}.$$

In the same way it may be proved, by considering the triangles PCH', $H'CI$, that the angles CPH' and $CH'I$ are equal, and that

$$\frac{IH'}{H'P}=\frac{CH'}{CP}=\frac{a}{CP}.$$

Hence the expressions for the attractions of the elements H and H' on P become

$$\rho\frac{\omega}{\cos CHI}.\frac{a^2}{CP^2}, \text{ and } \rho\frac{\omega}{\cos CH'I}.\frac{a^2}{CP^2},$$

which are equal, since the triangle HCH' is isosceles; and, for the same reason, the angles CPH, CPH', which have been proved to be respectively equal to the angles CHI, $CH'I$, are equal. We infer that the resultant of the forces due to the two elements is in the direction PC, and is equal to

$$2\omega.\rho.\frac{a^2}{CP^2}.$$

To find the total force on P, we must take the sum of all the forces along PC due to the pairs of opposite elements; and, since the multiplier of ω is the same for each pair, we must add all the values of ω, and we therefore obtain (§ 483), for the required resultant,

$$\frac{4\pi\rho a^2}{CP^2}.$$

The numerator of this expression (being the product of the density into the area of the spherical surface) is equal to the mass of the entire charge; and therefore the force on P is the same as if the whole mass were collected at C.

Cor. The force on an external point, infinitely near the surface, is equal to $4\pi\rho$, and is in the direction of a normal at the point. The force on an internal point, however near the surface, is, by a preceding proposition, equal to nothing.

489. Let σ be the area of an infinitely small element of the surface at any point P, and at any other point H of the surface let a small element subtending a solid angle ω, at P, be taken. The area of this element will be equal to

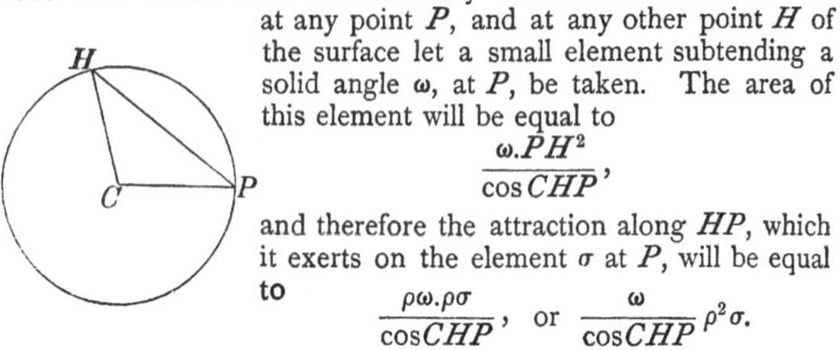

$$\frac{\omega . PH^2}{\cos CHP},$$

and therefore the attraction along HP, which it exerts on the element σ at P, will be equal to

$$\frac{\rho\omega . \rho\sigma}{\cos CHP}, \text{ or } \frac{\omega}{\cos CHP}\rho^2\sigma.$$

Now the total attraction on the element at P is in the direction CP; the component in this direction of the attraction due to the element H, is

$$\omega . \rho^2\sigma;$$

and, since all the cones corresponding to the different elements of the spherical surface lie on the same side of the tangent plane at P, we deduce, for the resultant attraction on the element σ,

$$2\pi\rho^2\sigma.$$

From the corollary to the preceding proposition, it follows that this attraction is half the force which would be exerted on an external point, possessing the same quantity of matter as the element σ, and placed infinitely near the surface.

490. In some of the most important elementary problems of the theory of electricity, spherical surfaces with densities varying inversely as the cubes of distances from excentric points occur: and it is of fundamental importance to find the attraction of such a shell on an internal or external point. This may be done synthetically as follows; the investigation being, as we shall see below, virtually the same as that of § 479, or § 488.

491. Let us first consider the case in which the given point S and the attracted point P are separated by the spherical surface. The two figures represent the varieties of this case in which, the point S being without the sphere, P is within; and, S being within, the attracted point is external. The same demonstration is applicable literally with reference to the two figures; but, for avoiding the consideration of negative quantities, some of the expressions may be conveniently modified to suit the second figure. In such instances the two expressions are given in a double line, the upper being that which is most convenient for the first figure, and the lower for the second.

Let the radius of the sphere be denoted by a, and let f be the distance of S from C, the centre of the sphere (not represented in the figures).

Join SP and take T in this line (or its continuation) so that

$$\text{(fig. 1)} \quad SP.ST = f^2 - a^2,$$
$$\text{(fig. 2)} \quad SP.TS = a^2 - f^2.$$

Through T draw any line cutting the spherical surface at K, K'. Join SK, SK', and let the lines so drawn cut the spherical surface again in E, E'.

Let the whole spherical surface be divided into pairs of opposite elements with reference to the point T. Let K and K' be a pair of such elements situated at the extremities of the chord KK', and subtending the solid angle ω at the point T; and let elements E and E' be taken subtending at S the same solid angles respectively as the elements K and K'. By this means we may divide the whole spherical surface into pairs of conjugate elements, E, E', since it is easily seen that when we have taken every pair of elements, K, K',

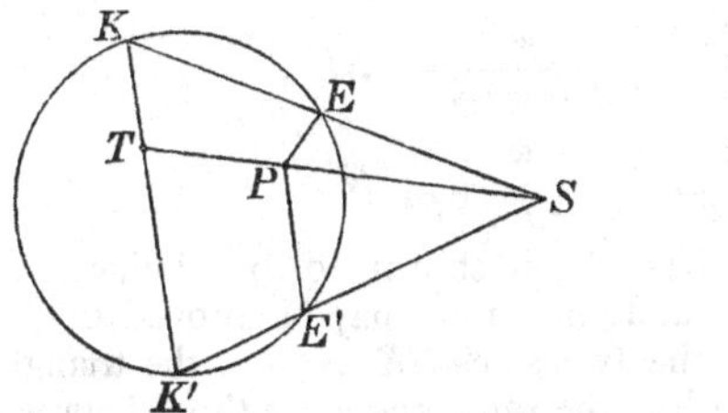

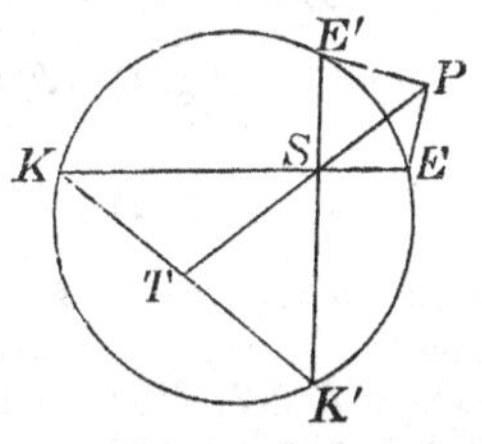

the whole surface will have been exhausted, without repetition, by the deduced elements, E, E'. Hence the attraction on P will be the final resultant of the attractions of all the pairs of elements, E, E'.

Now if ρ be the surface density at E, and if F denote the attraction of the element E on P, we have

$$F = \frac{\rho.E}{EP^2}.$$

According to the given law of density we shall have

$$\rho = \frac{\lambda}{SE^3},$$

where λ is a constant. Again, since SEK is equally inclined to the spherical surface at the two points of intersection, we have

$$E = \frac{SE^2}{SK^2}.K = \frac{SE^2}{SK^2}.\frac{2a\omega.TK^2}{KK'};$$

and hence

$$F = \frac{\dfrac{\lambda}{SE^3}.\dfrac{SE^2}{SK^2}.\dfrac{2a\omega.TK^2}{KK'}}{EP^2} = \lambda.\frac{2a}{KK'}.\frac{TK^2}{SE.SK^2.EP^2}.\omega.$$

Now, by considering the great circle in which the sphere is cut by a plane through the line SK, we find that

$$\text{(fig. 1)}\quad SK.SE = f^2 - a^2,$$
$$\text{(fig. 2)}\quad KS.SE = a^2 - f^2,$$

and hence $SK.SE = SP.ST$, from which we infer that the triangles KST, PSE are similar; so that $TK : SK :: PE : SP$. Hence

$$\frac{TK^2}{SK^2.PE^2} = \frac{1}{SP^2},$$

and the expression for F becomes

$$F = \lambda . \frac{2a}{KK'} . \frac{1}{SE.SP^2} . \omega.$$

Modifying this by preceding expressions we have

$$\text{(fig. 1)}\quad F = \lambda . \frac{2a}{KK'} . \frac{\omega}{(f^2 - a^2) SP^2} . SK,$$

$$\text{(fig. 2)}\quad F = \lambda . \frac{2a}{KK'} . \frac{\omega}{(a^2 - f^2) SP^2} . KS.$$

Similarly, if F' denote the attraction of E' on P, we have

$$\text{(fig. 1)}\quad F' = \lambda \frac{2a}{KK'} . \frac{\omega}{(f^2 - a^2) SP^2} . SK',$$

$$\text{(fig. 2)}\quad F' = \lambda \frac{2a}{KK'} . \frac{\omega}{(a^2 - f^2) SP^2} . K'S.$$

Now in the triangles which have been shown to be similar, the angles TKS, EPS are equal; and the same may be proved of the angles $K'ST$, PSE'. Hence the two sides SK, SK' of the triangle KSK' are inclined to the third at the same angles as those between the line PS and directions PE, PE' of the two forces on the point P; and the sides SK, SK' are to one another as the forces, F, F', in the directions PE, PE'. It follows, by 'the triangle of forces,' that the resultant of F and F' is along PS, and that it bears to the component forces the same ratios as the side KK' of the triangle bears to the other two sides. Hence the resultant force due to the two elements E and E' on the point P, is towards S, and is equal to

$$\lambda . \frac{2a}{KK} . \frac{\omega}{(f^2 \backsim a^2).SP^2} . KK', \quad \text{or} \quad \frac{\lambda . 2a.\omega}{(f^2 \backsim a^2) SP^2}.$$

The total resultant force will consequently be towards S; and we find, by summation (§ 466) for its magnitude,

$$\frac{\lambda.4\pi a}{(f^2 \backsim a^2) SP^2}.$$

Hence we infer that the resultant force at any point P, separated from S by the spherical surface, is the same as if a quantity of matter equal to $\frac{\lambda.4\pi a}{f^2 \backsim a^2}$ were concentrated at the point S.

492. To find the attraction when S and P are either both without or both within the spherical surface.

Take in CS, or in CS produced through S, a point S_1, such that

$$CS.CS_1=a^2.$$

Then, by a well-known geometrical theorem, if E be any point on the spherical surface, we have

$$\frac{SE}{S_1E}=\frac{f}{a}.$$

Hence we have $$\frac{\lambda}{SE^3}=\frac{\lambda a^3}{f^3.S_1E^3}.$$

Hence, ρ being the electrical density at E, we have

$$\rho=\frac{\frac{\lambda a^3}{f^3}}{S_1E^3}=\frac{\lambda_1}{S_1E^3},$$

if $$\lambda_1=\frac{\lambda a^3}{f^3}.$$

Hence, by the investigation in the preceding section, the attraction on P is towards S_1, and is the same as if a quantity of matter equal to $\frac{\lambda_1.4\pi a}{f_1^2\backsim a^2}$ were concentrated at that point;

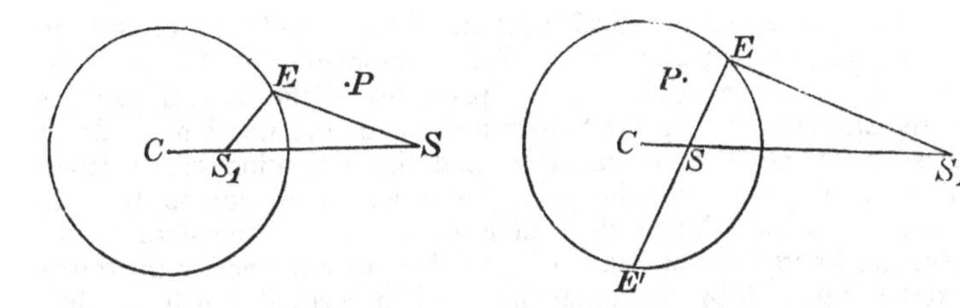

f_1 being taken to denote CS_1. If for f_1 and λ_1 we substitute their values, $\frac{a^2}{f}$ and $\frac{\lambda a^3}{f^3}$, we have the modified expression

$$\frac{\lambda\frac{a}{f}.4\pi a}{a^2\backsim f^2}$$

for the quantity of matter which we must conceive to be collected at S_1.

493. If a spherical surface be electrified in such a way that the electrical density varies inversely as the cube of the distance from an internal point S, or from the corresponding external point S_1, it will attract any external point, as if its whole electricity were concentrated at S, and any internal point, as if a quantity of electricity greater than its own in the ratio of a to f were concentrated at S_1.

Let the density at E be denoted, as before, by $\frac{\lambda}{SE^3}$. Then, if we consider two opposite elements at E and E', which subtend a solid angle ω at the point S, the areas of these elements being $\frac{\omega.2a.SE^2}{EE'}$ and $\frac{\omega.2a.SE'^2}{EE'}$, the quantity of electricity which they possess will be

$$\frac{\lambda.2a.\omega}{EE'}\left(\frac{1}{SE}+\frac{1}{SE'}\right) \text{ or } \frac{\lambda.2a.\omega}{SE.SE'}.$$

Now $SE.SE'$ is constant (Euc. III. 35) and its value is a^2-f^2. Hence, by summation, we find for the total value of electricity on the spherical surface

$$\frac{\lambda.4\pi a}{a^2-f^2}.$$

Hence, if this be denoted by m, the expressions in the preceding paragraphs, for the quantities of electricity which we must suppose to be concentrated at the point S or S_1, according as P is without or within the spherical surface, become respectively

$$m, \text{ and } \frac{a}{f}m.$$

494. The *direct* analytical solution of such problems consists in the expression, by § 408, of the three components of the whole attraction as the sums of its separate parts due to the several particles of the attracting body; the transformation, by the usual methods, of these sums into definite integrals; and the evaluation of the latter. This is, in general, inferior in elegance and simplicity to the less direct mode of solution depending upon the determination of the potential energy of the attracted particle with reference to the forces exerted upon it by the attracting body, a method which we shall presently develop with peculiar care, as it is of incalculable value in the theories of Electricity and Magnetism as well as in that of Gravitation. But before we proceed to it, we give some instances of the direct method.

(*a*) A useful case is that of the attraction of a circular plate of uniform surface density on a point in a line through its centre, and perpendicular to its plane.

All parallel slices, of equal thickness, of any cone attract equally (both in magnitude and direction) a particle at the vertex.

For the proposition is true of a cone of infinitely small angle, the masses of the slices being evidently as the squares of their distances from the vertex. If t be the thickness, ρ the volume density, and ω the angle, the attraction is $\omega t\rho$.

All slices of a cone of infinitely small angle, if of equal thickness and equally inclined to the axis of the cone, exert equal forces on a particle at the vertex. For the area of any inclined section, whatever be its orientation, is greater than that of the corresponding transverse section in the ratio of unity to the cosine of the angle of inclination.

Hence if a plane touch a sphere at a point B, and if the plane and sphere have equal surface density at corresponding points P and p in a line drawn through A, the point diametrically opposite to B, corresponding elements at P and p exert equal attraction on a particle at A.

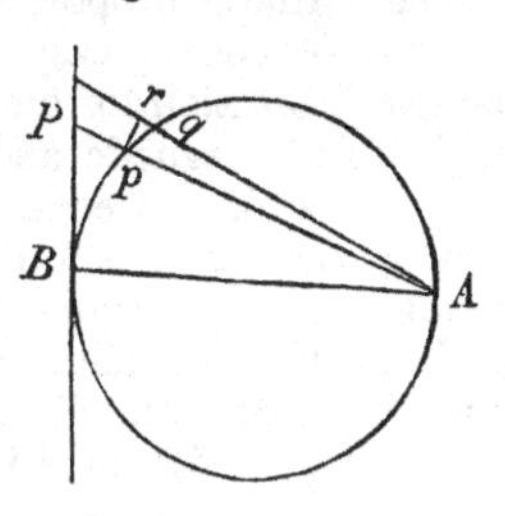

Thus the attraction on A, of any part of the plane, is the same as that of the corresponding part of the sphere, cut out by a cone of infinitely small angle whose vertex is A.

Hence if we resolve along the line AB the attraction of pq on A, the component is equal to the attraction along Ap of the transverse section pr, i. e. $\rho\omega$, where ω is the angle subtended at A by the element pq, and ρ the surface density.

Thus any portion whatever of the sphere attracts A along AB with a force proportional to its spherical opening as seen from A; and the same is, by what was proved above, true of a flat plate.

Hence as a disc of radius a subtends at a point distant h from it, in the direction of the axis of the disc, a spherical angle

$$2\pi\left(1-\frac{h}{\sqrt{h^2+a^2}}\right),$$

the attraction of such a disc is

$$2\pi\rho\left(1-\frac{h}{\sqrt{h^2+a^2}}\right);$$

which for an infinite disc becomes, whatever the distance h,

$$2\pi\rho.$$

From the preceding formula many useful results may easily be deduced: thus,

(*b*) A uniform *cylinder* of length l, and diameter a, attracts a point in its axis at a distance x from the nearest end with a force

$$2\pi\rho\,\{l-\sqrt{(x+l)^2+a^2}+\sqrt{x^2+a^2}\}.$$

When the cylinder is of infinite length (in one direction) the attraction is therefore

$$2\pi\rho\,(\sqrt{x^2+a^2}-x);$$

and, when the attracted particle is in contact with the centre of the end of the infinite cylinder, this is

$$2\pi\rho a.$$

(*c*) A right cone, of semivertical angle α, and length l, attracts a particle at its vertex. Here we have at once for the attraction, the expression

$$2\pi\rho l(1-\cos\alpha),$$

which is simply proportional to the length of the axis.

It is of course easy, when required, to find the necessarily less simple expression for the attraction on any point of the axis.

(*d*) For magnetic and electro-magnetic applications a very useful case is that of two equal uniform discs, each perpendicular to the line joining their centres, on any point in that line—their masses (§ 478) being of opposite sign—that is, one repelling and the other attracting.

Let a be the radius, ρ the mass of a superficial unit, of either, c their distance, x the distance of the attracted point from the nearest disc. The whole force is evidently

$$2\pi\rho\left\{\frac{x+c}{\sqrt{(x+c)^2+a^2}}-\frac{x}{\sqrt{x^2+a^2}}\right\}.$$

In the particular case when c is diminished without limit, this becomes

$$2\pi\rho c\frac{a^2}{(x^2+a^2)^{\frac{3}{2}}}.$$

495. Let P and P' be two points infinitely near one another on two sides of a surface over which matter is distributed; and let ρ be the density of this distribution on the surface in the neighbourhood of these points. Then whatever be the resultant attraction, R, at P, due to all the attracting matter, whether lodging on this surface, or elsewhere, the resultant force, R', on P' is the resultant of a force equal and parallel to R, and a force equal to $4\pi\rho$, in the direction from P' perpendicularly towards the surface. For, suppose PP' to be perpendicular to the surface, which will not limit the generality of the proposition, and consider a circular disc, of the surface, having its centre in PP', and radius infinitely small in comparison with the radii of curvature of the surface but infinitely great in comparison with PP'. This disc will [§ 494] attract P and P' with forces, each equal to $2\pi\rho$ and opposite to one another in the line PP'. Whence the proposition. It is one of much importance in the theory of electricity.

496. It may be shown that at the southern base of a hemispherical hill of radius a and density ρ, the true latitude (as measured by the aid of the plumb-line, or by reflection of starlight in a trough of mercury) is diminished by the attraction of the mountain by the angle

$$\frac{\frac{2}{3}\rho\pi a}{G-\frac{4}{3}\rho a},$$

where G is the attraction of the earth, estimated in the same units.

Hence, if R be the radius and σ the mean density of the earth, the angle is

$$\frac{\frac{2}{3}\pi\rho a}{\frac{4}{3}\pi\sigma R - \frac{4}{3}\rho a}, \quad \text{or} \quad \frac{1}{2}\frac{\rho a}{\sigma R} \text{ approximately.}$$

Hence the latitudes of stations at the base of the hill, north and south of it, differ by $\frac{a}{R}\left(2+\frac{\rho}{\sigma}\right)$; instead of by $\frac{2a}{R}$, as they would do if the hill were removed.

In the same way the latitude of a place at the southern edge of a hemispherical *cavity* is increased on account of the cavity by $\frac{1}{2}\frac{\rho a}{\sigma R}$ where ρ is the density of the superficial strata.

497. As a curious additional example of the class of questions we have just considered, a deep crevasse, extending east and west, increases the latitude of places at its southern edge by (approximately) the angle $\frac{3}{4}\frac{\rho a}{\sigma R}$ where ρ is the density of the crust of the earth, and a is the width of the crevasse. Thus the north edge of the crevasse will have a *lower* latitude than the south edge if $\frac{3}{2}\frac{\rho}{\sigma} > 1$, which might be the case, as there are rocks of density $\frac{2}{3}\times 5{\cdot}5$ or 3·67 times that of water. At a considerable depth in the crevasse, this change of latitudes is nearly *doubled*, and then the southern side has the greater latitude if the density of the crust be not less than 1·83 times that of water.

498. It is interesting, and will be useful later, to consider as a particular case, the attraction of a sphere whose mass is composed of concentric layers, each of uniform density. Let σ be, as above, the mean density of the whole globe, and τ the density of the upper crust.

The attraction at a depth h, small compared with the radius, is

$$\tfrac{4}{3}\pi\sigma_1(R-h) = G_1$$

where σ_1 denotes the mean density of the nucleus remaining when a shell of thickness h is removed from the sphere. Also, evidently,

$$\tfrac{4}{3}\pi\sigma_1(R-h)^3 + 4\pi\tau(R-h)^2h = \tfrac{4}{3}\pi\sigma R^3,$$

or

$$G_1(R-h)^2 + 4\pi\tau(R-h)^2h = GR^2,$$

whence

$$G_1 = G\left(1+\frac{2h}{R}\right) - 4\pi\tau h.$$

The attraction is therefore unaltered at a depth h if

$$\frac{G}{R} = \tfrac{4}{3}\pi\sigma = 2\pi\tau, \quad \text{i. e.} \quad \tau = \tfrac{2}{3}\sigma.$$

499. Some other simple cases may be added here, as their results will be of use to us subsequently.

(*a*) The attraction of a circular arc, AB, of uniform density, on a particle at the centre, C, of the circle, lies evidently in the line CD bisecting the arc. Also the resolved part parallel to CD of the attraction of an element at P is

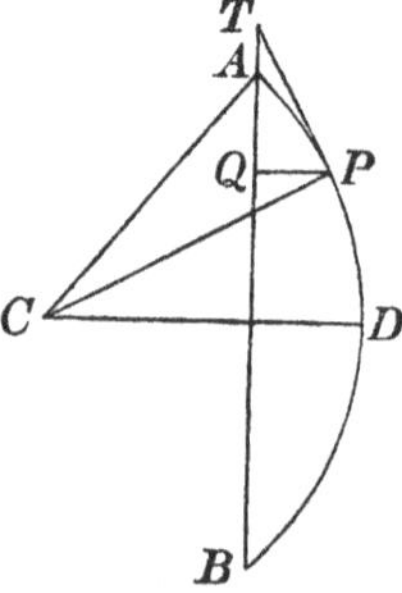

$$\frac{\text{mass of element at } P}{CD^2} \cos. \angle PCD.$$

Now suppose the density of the chord AB to be the same as that of the arc. Then for (mass of element at $P \times \cos \angle PCD$) we may put (mass of projection of element on AB at Q); since, if PT be the tangent at P, $\angle PTQ = \angle PCD$.

Hence attraction along $CD = \dfrac{\text{sum of projected elements}}{CD^2}$

$$= \frac{\rho AB}{CD^2},$$

if ρ be the density of the given arc,

$$= \frac{2\rho \sin \angle ACD}{CD}.$$

It is therefore the same as the attraction of a mass equal to the chord, with the arc's density, concentrated at the point D.

(*b*) Again, a limited straight line of uniform density attracts any external point in the same direction and with the same force as the corresponding arc of a circle of the same density, which has the point for centre, and touches the straight line.

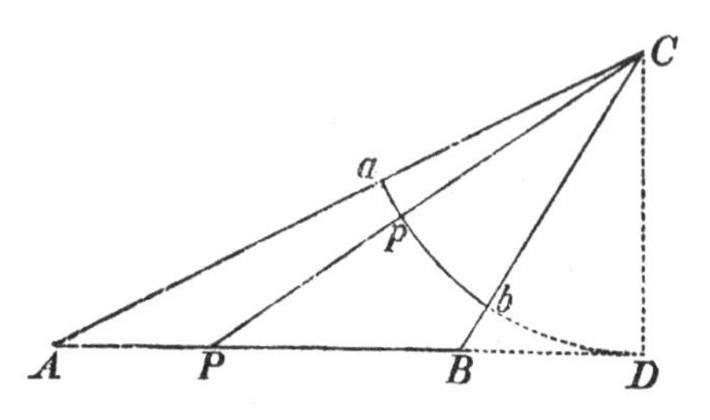

For if CpP be drawn cutting the circle in p and the line in P; element at p : element at $P :: Cp : CP \dfrac{CP}{CD}$; that is, as $Cp^2 : CP^2$. Hence the attractions of these elements on C are equal and in the same line. Thus the arc ab attracts C as the line AB does; and, by the last proposition, the attraction of AB bisects the angle ACB, and is equal to

$$\frac{2\rho}{CD} \sin \tfrac{1}{2} \angle ACB.$$

(*c*) This may be put into other useful forms—thus, let CKF bisect the angle ACB, and let Aa, Bb, EF, be drawn perpendicular to CF from the ends and middle point of AB. We have

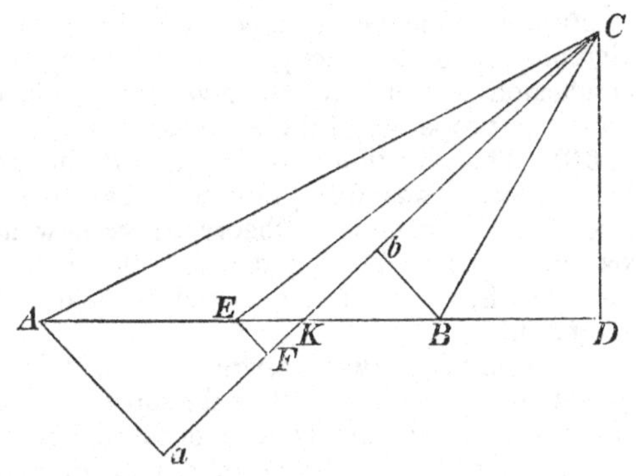

$$\sin \angle KCB = \frac{KB}{BC} \sin \angle CKD = \frac{AB}{AC+CB} \frac{CD}{CK}.$$

Hence the attraction, which is along CK, is

$$\frac{2\rho AB}{(AC+CB)CK} = \frac{\rho AB}{2(AC+CB)(\overline{AC+CB}^2 - AB^2)} \cdot CF. \qquad (1)$$

For, evidently,

$$bK : Ka :: BK : KA :: BC : CA :: bC : Ca,$$

i.e., ab is divided, externally in C, and internally in K, in the same ratio. Hence, by geometry,

$$KC.CF = aC.Cb = \tfrac{1}{4}\{\overline{AC+CB}^2 - AB^2\},$$

which gives the transformation in (1).

(*d*) CF is obviously the tangent at C to a hyperbola, passing through that point, and having A and B as foci. Hence, if in *any* plane through AB any hyperbola be described, with foci A and B, it will be a line of force as regards the attraction of the line AB; that is, as will be more fully explained later, a curve which at every point indicates the direction of attraction.

(*e*) Similarly, if a prolate spheroid be described with foci A and B, and passing through C, CF will evidently be the normal at C; thus the force on a particle at C will be perpendicular to the spheroid; and the particle would evidently rest in equilibrium on the surface, even if it were smooth. This is an instance of (what we shall presently develop at some length) a surface of equilibrium, a level surface, or an equipotential surface.

(*f*) We may further prove, by a simple application of the preceding theorem, that the lines of force due to the attraction of two infinitely long rods in the line AB produced, one of which is attractive and the other repulsive, are the series of ellipses described from the extremities, A and B, as foci, while the surfaces of equilibrium are generated by the revolution of the confocal hyperbolas.

500. As of immense importance, in the theory not only of gravitation but of electricity, of magnetism, of fluid motion, of the conduction of heat, etc., we give here an investigation of the most important properties of the *Potential.*

501. This function was introduced for gravitation by Laplace, but the name was first given to it by Green, who may almost be said to have created the theory, as we now have it. Green's work was neglected till 1846, and before that time most of its important theorems had been re-discovered by Gauss, Chasles, Sturm, and Thomson.

In § 245, the *potential energy* of a conservative system in any configuration was defined. When the forces concerned are forces acting, either really or apparently, at a distance, as attraction of gravitation, or attractions or repulsions of electric or magnetic origin, it is in general most convenient to choose, for the zero configuration, infinite distance between the bodies concerned. We have thus the following definition :—

502. The mutual potential energy of two bodies in any relative position is the amount of work obtainable from their mutual repulsion, by allowing them to separate to an infinite distance asunder. When the bodies attract mutually, as for instance when no other force than gravitation is operative, their mutual potential energy, according to the convention for zero now adopted, is negative, or their *exhaustion of potential energy* is positive.

503. The *Potential* at any point, due to any attracting or repelling body, or distribution of matter, is the mutual potential energy between it and a unit of matter placed at that point. But in the case of gravitation, to avoid defining the potential as a negative quantity, it is convenient to change the sign. Thus the gravitation potential, at any point, due to any mass, is the quantity of work required to remove a unit of matter from that point to an infinite distance.

504. Hence, if V be the potential at any point P, and V_1 that at a proximate point Q, it evidently follows from the above definition that $V-V_1$ is the work required to remove an independent unit of matter from P to Q; and it is useful to note that this is altogether independent of the form of the path chosen between these two points, as it gives us a preliminary idea of the power we acquire by the introduction of this mode of representation.

Suppose Q to be so near to P that the attractive forces exerted on unit of matter at these points, and therefore at any point in the line PQ, may be assumed to be equal and parallel. Then if F represent the resolved part of this force along PQ, $F.PQ$ is the work required to transfer unit of matter from P to Q. Hence

$$V-V_1=F.PQ,$$

or

$$F=\frac{V-V_1}{PQ},$$

that is, the attraction on unit of matter at P in any direction PQ, is the rate at which the potential at P increases per unit of length of PQ.

505. A surface, at every point of which the potential has the same value, and therefore called an *Equipotential Surface*, is such that the attraction is everywhere in the direction of its normal. For in no direction along the surface does the potential change in value, and therefore there is no force in any such direction. Hence if the attracted particle be placed on such a surface (supposed smooth and rigid), it will rest in any position, and the surface is therefore sometimes called a *Surface of Equilibrium*. We shall see later, that the force on a particle of a liquid at the free surface is always in the direction of the normal, hence the term *Level Surface*, which is often used for the other terms above.

506. If a series of equipotential surfaces be constructed for values of the potential increasing by equal small amounts, it is evident from § 504 that the attraction at any point is inversely proportional to the normal distance between two successive surfaces close to that point; since the numerator of the expression for F is, in this case, constant.

507. A line drawn from any origin, so that at every point of its length its tangent is the direction of the attraction at that point, is called a *Line of Force;* and it obviously cuts at right angles every equipotential surface which it meets.

These three last sections are true *whatever* be the law of attraction; in the next we are restricted to the law of the inverse square of the distance.

508. If, through every point of the boundary of an infinitely small portion of an equipotential surface, the corresponding lines of force be drawn, we shall evidently have a tubular surface of infinitely small section. The resultant force, being at every point tangential to the direction of the tube, is inversely as its normal transverse section.

This is an immediate consequence of a most important theorem, which will be proved later. *The surface integral of the attraction exerted by any distribution of matter in the direction of the normal at every point of any closed surface is* $4\pi M$; *where* M *is the amount of matter within the surface, while the attraction is considered positive or negative according as it is inwards or outwards at any point of the surface.*

For in the present case the force perpendicular to the tubular part of the surface vanishes, and we need consider the ends only. When none of the attracting mass is within the portion of the tube considered, we have at once

$$F\varpi - F'\varpi' = 0,$$

F being the force at any point of the section whose area is ϖ.

This is equivalent to the celebrated equation of Laplace.

When the attracting body is symmetrical about a point, the lines of force are obviously straight lines drawn from this point. Hence the tube is in this case a cone, and, by § 486, ϖ is proportional to the square of the distance from the vertex. Hence F is inversely as the square of the distance for points external to the attracting mass.

When the mass is symmetrically disposed about an axis in infinitely long cylindrical shells, the lines of force are evidently perpendicular to the axis. Hence the tube becomes a *wedge*, whose section is proportional to the distance from the axis, and the attraction is therefore inversely as the distance from the axis.

When the mass is arranged in infinite parallel planes, each of uniform density, the lines of force are obviously perpendicular to these planes; the tube becomes a *cylinder*; and, since its section is constant, the force is the same at all distances.

If an infinitely small length l of the portion of the tube considered pass through matter of density ρ, and if ω be the area of the section of the tube in this part, we have

$$F\varpi - F'\varpi' = 4\pi l\omega\rho.$$

This is equivalent to Poisson's extension of Laplace's equation.

509. In estimating work done against a force which varies inversely as the square of the distance from a fixed point, the mean force is to be reckoned as the geometrical mean between the forces at the beginning and end of the path: and, whatever may be the path followed, the effective space is to be reckoned as the difference of distances from the attracting point. Thus the work done in any course is equal to the product of the difference of distances of the extremities from the attracting point, into the geometrical mean of the forces at these distances; or, if O be the attracting point, and m its force on a unit mass at unit distance, the work done in moving a particle, of unit mass, from any position P to any other position P', is

$$(OP' - OP)\sqrt{\frac{m^2}{OP^2\,OP'^2}}, \text{ or } \frac{m}{OP} - \frac{m}{OP'}.$$

To prove this it is only necessary to remark, that for any infinitely small step of the motion, the effective space is clearly the difference of distances from the centre, and the working force may be taken as the force at either end, or of any intermediate value, the geometrical mean for instance: and the preceding expression applied to each infinitely small step shows that the same rule holds for the sum making up the whole work done through any finite range, and by any path.

Hence, by § 503, it is obvious that the potential at P, of a mass m situated at O, is $\frac{m}{OP}$; and thus that the potential of any mass at a point P is to be found by adding the quotients of every portion of the mass, each divided by its distance from P.

510. Let S be any closed surface, and let O be a point, either external or internal, where a mass, m, of matter is collected. Let N be the component of the attraction of m in the direction of the normal drawn inwards from any point P, of S. Then, if $d\sigma$ denotes an element of S, and $\iint$ integration over the whole of it,

$$\iint N d\sigma = 4\pi m, \text{ or } = 0,$$

according as O is internal or external.

Case I, O internal. Let $OP_1P_2P_3\ldots$ be a straight line drawn in any direction from O, cutting S in P_1, P_2, P_3, etc., and therefore passing out at P_1, in at P_2, out again at P_3, in again at P_4, and so on. Let a conical surface be described by lines through O, all infinitely near $OP_1P_2\ldots$, and let ω be its solid angle (§ 482). The portions of $\iint N d\sigma$ corresponding to the elements cut from S by this cone will be clearly each equal in absolute magnitude to ωm, but will be alternately positive and negative. Hence as there is an odd number of them, their sum is $+\omega m$. And the sum of these, for all solid angles round O is (§ 483) equal to $4\pi m$; that is to say, $\iint N d\sigma = 4\pi m$.

Case II, O external. Let $OP_1P_2P_3\ldots$ be a line drawn from O passing across S, inwards at P_1, outwards at P_2, and so on. Drawing, as before, a conical surface of infinitely small solid angle, ω, we have still ωm for the absolute value of each of the portions of $\iint N d\sigma$ corresponding to the elements which it cuts from S; but their signs are alternately negative and positive: and therefore as their number is even, their sum is zero. Hence

$$\iint N d\sigma = 0.$$

From these results it follows immediately that if there be any continuous distribution of matter, partly within and partly without a closed surface S, and N and $d\sigma$ be still used with the same signification, we have

$$\iint N d\sigma = 4\pi M$$

if M denote the whole amount of matter within S.

511. From this it follows that the potential cannot have a maximum or minimum value at a point in free space. For if it were so, a closed surface could be described about the point, and indefinitely near it, so that at every point of it the value of the potential would be less than, or greater than, that at the point; so that N would be negative or positive all over the surface, and therefore $\iint N d\sigma$ would be finite, which is impossible, as the surface contains none of the attracting mass.

512. It is also evident that N must have positive values at some parts of this surface, and negative values at others, unless it is zero all over it. Hence in free space the potential, if not constant round any point, increases in some directions from it, and diminishes in others; and therefore a material particle placed at a point of zero force under the action of any attracting bodies, and free from all constraint, is in unstable equilibrium, a result due to Earnshaw[1].

513. If the potential be constant over a closed surface which contains none of the attracting mass, it has the same constant value throughout the interior. For if not, it must have a maximum or minimum value somewhere within, which is impossible.

514. The mean potential over any spherical surface, due to matter entirely without it, is equal to the potential at its centre; a theorem apparently first given by Gauss. See also Cambridge *Mathematical Journal*, Feb. 1845 (vol. iv. p. 225). This proposition is merely an extension, to any masses, of the converse of the following statement, which is easily seen to follow from the results of §§ 479, 488 expressed in potentials instead of forces. The potential of an uniform spherical shell at an external point is the same as if its mass were condensed at the centre. At all internal points it has the same value as at the surface.

515. If the potential of any masses has a constant value, V, through any finite portion, K, of space, unoccupied by matter, it is equal to V through every part of space which can be reached in any way without passing through any of those masses: a very remarkable proposition, due to Gauss. For, if the potential differ from V in space contiguous to K, it must (§ 513) be greater in some parts and less in others.

From any point C within K, as centre, in the neighbourhood of a place where the potential is greater than V, describe a spherical surface not large enough to contain any part of any of the attracting masses, nor to include any of the space external to K except such as has potential greater than V. But this is impossible, since we have just seen (§ 514) that the mean potential over the spherical surface must be V. Hence the supposition that the potential is greater than V in some places and less in others, contiguous to K and not including masses, is false.

516. Similarly we see that in any case of symmetry round an axis, if the potential is constant through a certain finite distance, however short, along the axis, it is constant throughout the whole space that can be reached from this portion of the axis, without crossing any of the masses.

517. Let S be any finite portion of a surface, or complete closed surface, or infinite surface, and let E be any point on S. (*a*) It is possible to distribute matter over S so as to produce potential equal to $F(E)$, any arbitrary function of the position of E, over the whole of S. (*b*) There is only one whole quantity of matter, and one distribution of it, which can satisfy this condition. For the proof of

[1] Cambridge *Phil. Trans.*, March, 1839.

this and of several succeeding theorems, we refer the reader to our larger work.

518. It is important to remark that, if S consist, in part, of a closed surface, Q, the determination of U within it will be independent of those portions of S, if any, which lie without it; and, *vice versâ*, the determination of U through external space will be independent of those portions of S, if any, which lie within the part Q. Or if S consist, in part, of a surface Q, extending infinitely in all directions, the determination of U through all space on either side of Q, is independent of those portions of S, if any, which lie on the other side.

519. Another remark of extreme importance is this:—If $F(E)$ be the potential at E of any distribution, M, of matter, and if S be such as to separate perfectly any portion or portions of space, H, from all of this matter; that is to say, such that it is impossible to pass into H from any part of M without crossing S; then, throughout H, the value of U will be the potential of M.

520. Thus, for instance, if S consist of three detached surfaces, S_1, S_2, S_3, as in the diagram, of which S_1, S_2 are closed, and S_3 is an open shell, and if $F(E)$ be the potential due to M, at any point, E, of any of these portions of S; then throughout H_1 and H_2, the spaces within S_1 and without S_2, the value of U is simply the potential of M. The value of U through K, the remainder of space, depends, of course, on the character of the composite surface S.

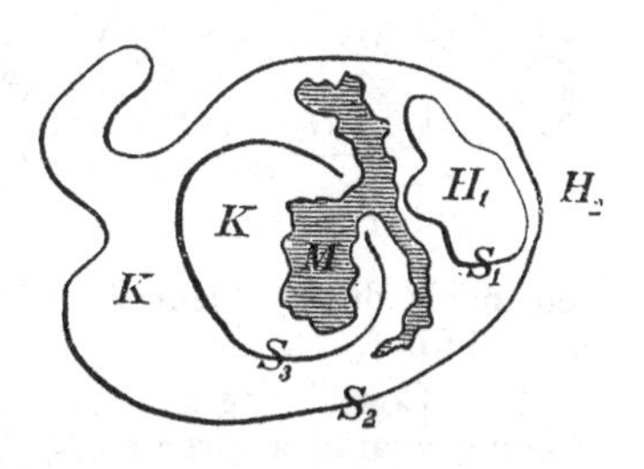

521. From § 518 follows the grand proposition:—*It is possible to find one, but no other than one, distribution of matter over a surface S which shall produce over S, and throughout all space H separated by S from every part of M, the same potential as any given mass M.*

Thus, in the preceding diagram, it is possible to find one, and but one, distribution of matter over S_1, S_2, S_3 which shall produce over S_3 and through H_1 and H_2 the same potential as M.

The statement of this proposition most commonly made is: *It is possible to distribute matter over any surface, S, completely enclosing a mass M, so as to produce the same potential as M through all space outside M;* which, though seemingly more limited, is, when interpreted with proper mathematical comprehensiveness, equivalent to the foregoing.

522. If S consist of several closed or infinite surfaces, S_1, S_2, S_3, respectively separating certain isolated spaces H_1, H_2, H_3, from H, the remainder of all space, and if $F(E)$ be the potential of masses m_1, m_2, m_3, lying in the spaces H_1, H_2, H_3; the portions of U due to

S_1, S_2, S_3, respectively will throughout H be equal respectively to the potentials of m_1, m_2, m_3, separately.

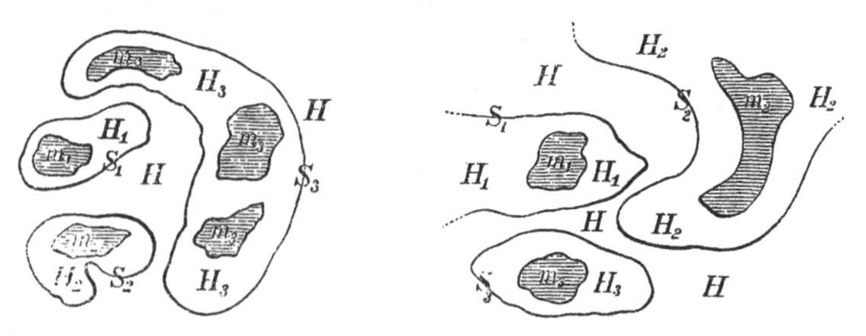

For, as we have just seen, it is possible to find one, but only one, distribution of matter over S, which shall produce the potential of m_1, throughout all the space H, H_2, H_3, etc., and one, but only one, distribution over S_2 which shall produce the potential of m_2 throughout H, H_1, H_3, etc.; and so on. But these distributions on S_1, S_2, etc., jointly constitute a distribution producing the potential $F(E)$ over every part of S, and therefore the sum of the potentials due to them all, at any point, fulfils the conditions presented for U. This is therefore (§ 518) *the* solution of the problem.

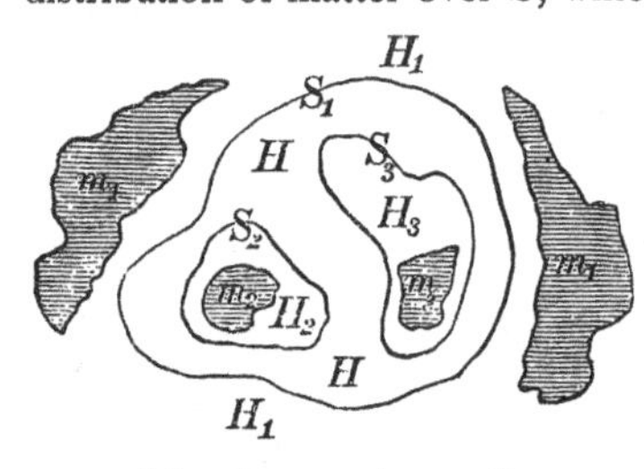

523. Considering still the case in which $F(E)$ is prescribed to be the potential of a given mass, M: let S be an equipotential surface enclosing M, or a group of isolated surfaces enclosing all the parts of M, and each equipotential for the whole of M. The potential due to the supposed distribution over S will be the same as that of M, through all external space, and will be constant (§ 514) through each enclosed portion of space. Its resultant attraction will therefore be the same as that of M on all external points, and zero on all internal points. Hence we see at once that the density of the matter distributed over it, to produce $F(E)$, is equal to $\frac{R}{4\pi}$ where R denotes the resultant force of M, at the point E.

524. When M consists of two portions m_1 and m' separated by an equipotential S_1, and S consists of two portions, S_1 and S', of which the latter separate the former perfectly from m'; we see, by § 522, that the distribution over S_1 produces through all space on the side of it on which S' lies, the same potential, V_1, as m_1, and the distribution on S' produces through space on the side of it on which S_1 lies, the same potential, V', as m'. But the supposed distribution on the whole of S is such as to produce a constant potential, C_1 over S_1, and consequently the same at every point within S_1. Hence the internal potential, due to S_1 alone, is $C_1 - V'$.

Thus, passing from potentials to attractions, we see that the resultant attraction of S_1 alone, on all points on one side of it is the same as that of m_1; and on the other side is equal and opposite to that of the remainder m' of the whole mass. The most direct and simple complete statement of this result is as follows:—

If masses m, m', in portions of space, H, H', completely separated from one another by one continuous surface S, whether closed or infinite, are known to produce tangential forces equal and in the same direction at each point of S, one and the same distribution of matter over S will produce the force of m throughout H', and that of m' throughout H. The density of this distribution is equal to $\frac{R}{4\pi}$, if R denote the resultant force due to one of the masses, and the other with its *sign* changed. And it is to be remarked that the direction of this resultant force is, at every point, E, of S, perpendicular to S, since the potential due to one mass, and the other with its sign changed, is constant over the whole of S.

525. Green, in first publishing his discovery of the result stated in § 523, remarked that it shows a way to find an infinite variety of closed surfaces for any one of which we can solve the problem of determining the distribution of matter over it which shall produce a given uniform potential at each point of its surface, and consequently the same also throughout its interior. Thus, an example which Green himself gives, let M be a uniform bar of matter, AA'. The equipotential surfaces round it are, as we have seen above (§ 499 (*e*)), prolate ellipsoids of revolution, each having A and A' for its foci; and the resultant force at C was found to be

$$\frac{m}{l(l^2-a^2)} \cdot CF,$$

the whole mass of the bar being denoted by m, its length by $2a$, and $A'C+AC$ by $2l$. We conclude that a distribution of matter over the surface of the ellipsoid, having

$$\frac{1}{4\pi}\frac{m.CF}{l(l^2-a^2)}$$

for density at C, produces on all external space the same resultant force as the bar, and zero force or a constant potential through the internal space. This is a particular case of the general result regarding ellipsoidal shells, proved below, in §§ 536, 537.

526. As a second example, let M consist of two equal particles, at points I, I'. If we take the mass of each as unity, the potential at P is $\frac{1}{IP}+\frac{1}{I'P}$; and therefore

$$\frac{1}{IP}+\frac{1}{I'P}=C$$

is the equation of an equipotential surface; it being understood that negative values of IP and $I'P$ are inadmissible, and that any constant value, from ∞ to 0, may be given to C. The curves in the annexed diagram have been drawn, from this equation, for the cases of C equal respectively to 10, 9, 8, 7, 6, 5, 4·5, 4·3, 4·2, 4·1, 4, 3·9, 3·8, 3·7, 3·5, 3, 2·5, 2; the value of II' being unity.

The corresponding equipotential surfaces are the surfaces traced by these curves, if the whole diagram is made to rotate round II' as

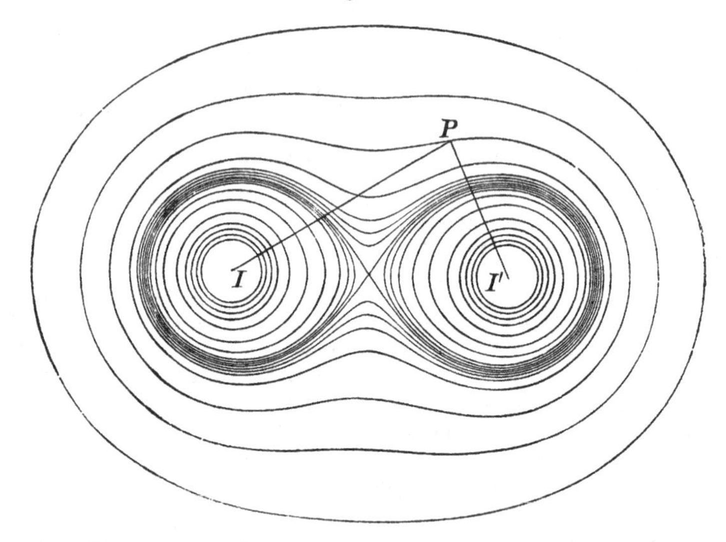

axis. Thus we see that for any values of C less than 4 the equipotential surface is one closed surface. Choosing any one of these surfaces, let R denote the resultant of forces equal to $\frac{1}{IP^2}$ and $\frac{1}{I'P^2}$ in the lines PI and PI'. Then if matter be distributed over this surface, with density at P equal to $\frac{R}{4\pi}$, its attraction on any internal point will be zero; and on any external point, will be the same as that of I and I'.

527. For each value of C greater than 4, the equipotential surface consists of two detached ovals approximating (the last three or four in the diagram, very closely) to spherical surfaces, with centres lying between the points I and I', but approximating more and more closely to these points, for larger and larger values of C.

Considering one of these ovals alone, one of the series enclosing I', for instance, and distributing matter over it according to the same law of density, $\frac{R}{4\pi}$, we have a shell of matter which exerts (§ 525) on external points the same force as I'; and on internal points a force equal and opposite to that of I.

528. As an example of exceedingly great importance in the theory of electricity, let M consist of a positive mass, m, concentrated at a point I, and a negative mass, $-m'$, at I'; and let S be a spherical surface cutting II', and II' produced in points A, $A_{,}$, such that

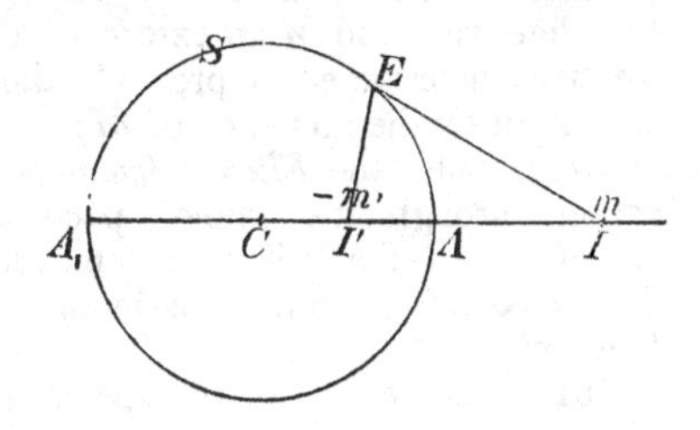

$IA : AI' :: IA_{,} : I'A_{,} :: m : m'$.

Then, by a well-known geometrical proposition, we shall have $IE : I'E :: m : m'$; and therefore

$$\frac{m}{IE} = \frac{m'}{I'E}.$$

Hence, by what we have just seen, one and the same distribution of matter over S will produce the same force as m' through all external space, and the same as m through all the space within S. And, finding the resultant of the forces $\frac{m}{IE^2}$ in EI, and $\frac{m'}{I'E^2}$ in $I'E$, produced, which, as these forces are inversely as IE to $I'E$, is (§ 222) equal to

$$\frac{m}{IE^2 . I'E} II', \text{ or } \frac{m^2 II'}{m'} \frac{1}{IE^3},$$

we conclude that the density in the shell at E is

$$\frac{m^2 II'}{4\pi m'} \cdot \frac{1}{IE^3}.$$

That the shell thus constituted does attract external points as if its mass were collected at I', and internal points as a certain mass collected at I, was proved geometrically in § 491 above.

529. If the spherical surface is given, and one of the points, I, I', for instance I, the other is found by taking $CI' = \frac{CA^2}{CI}$; and for the mass to be placed at it we have

$$m' = m\frac{I'A}{AI} = m\frac{CA}{CI} = m\frac{CI'}{CA}.$$

Hence, if we have any number of particles m_1, m_2, etc., at points I_1, I_2, etc., situated without S, we may find in the same way corresponding internal points I'_1, I'_2, etc., and masses m'_1, m'_2, etc.; and, by adding the expressions for the density at E given for each pair by the preceding formula, we get a spherical shell of matter which has the property of acting on all external space with the same force as $-m'_1$, $-m'_2$, etc., and on all internal points with a force equal and opposite to that of m_1, m_2, etc.

530. An infinite number of such particles may be given, constituting a continuous mass M; when of course the corresponding internal particles will constitute a continuous mass, $-M'$, of the opposite kind of matter; and the same conclusion will hold. If S is the surface of a solid or hollow metal ball connected with the earth by a fine wire, and M an external influencing body, the shell of matter we have determined is precisely the distribution of electricity on S called out by the influence of M: and the mass $-M'$, determined as above, is called the *Electric Image* of M in the ball, since the electric action through the whole space external to the ball would be unchanged if the ball were removed and $-M'$ properly placed in the space left vacant. We intend to return to this subject under Electricity.

531. Irrespectively of the special electric application, this method of images gives a remarkable kind of transformation which is often useful. It suggests for mere geometry what has been called the transformation by reciprocal radius-vectors; that is to say, the substitution for any set of points, or for any diagram of lines or surfaces, another obtained by drawing radii to them from a certain fixed point or origin, and measuring off lengths inversely proportional to these radii along their directions. We see in a moment by elementary geometry that any line thus obtained cuts the radius-vector through any point of it at the same angle and in the same plane as the line from which it is derived. Hence any two lines or surfaces that cut one another give two transformed lines or surfaces cutting at the same angle: and infinitely small lengths, areas, and volumes transform into others whose magnitudes are altered respectively in the ratios of the first, second, and third powers of the distances of the latter from the origin, to the same powers of the distances of the former from the same. Hence the lengths, areas, and volumes in the transformed diagram, corresponding to a set of given equal infinitely small lengths, areas, and volumes, however situated, at different distances from the origin, are inversely as the squares, the fourth powers and the sixth powers of these distances. Further, it is easily proved that a straight line and a plane transform into a circle and a spherical surface, each passing through the origin; and that, generally, circles and spheres transform into circles and spheres.

532. In the theory of attraction, the transformation of masses, densities, and potentials has also to be considered. Thus, according to the foundation of the method (§ 530), equal masses, of infinitely small dimensions at different distances from the origin, transform into masses inversely as these distances, or directly as the transformed distances: and, therefore, equal densities of lines, of surfaces, and of solids, given at any stated distances from the origin, transform into densities directly as the first, the third, and the fifth powers of those distances; or inversely as the same powers of the distances, from the origin, of the corresponding points in the transformed system. The usefulness of this transformation in the theory of electricity, and of attraction in general, depends entirely on the following theorem:—

Let ϕ denote the potential at P due to the given distribution, and ϕ' the potential at P' due to the transformed distribution: then shall

$$\phi' = \frac{r}{a}\phi = \frac{a}{r'}\phi.$$

Let a mass m collected at I be any part of the given distribution, and let m' at I' be the corresponding part in the transformed distribution. We have

$$a^2 = OI'.OI = OP'.OP,$$

and therefore

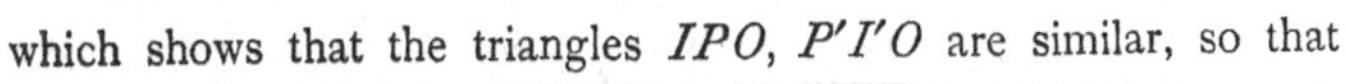

$$OI : OP :: OP' : OI';$$

which shows that the triangles IPO, $P'I'O$ are similar, so that

$$IP : P'I' :: \sqrt{OI.OP} : \sqrt{OP'.OI'} :: OI.OP : a^2.$$

We have besides

$$m : m' :: OI : a,$$

and therefore

$$\frac{m}{IP} : \frac{m'}{I'P'} :: OP : a.$$

Hence each term of ϕ bears to the corresponding term of ϕ' the same ratio; and therefore the sum, ϕ, must be to the sum, ϕ', in that ratio, as was to be proved.

533. As an example, let the given distribution be confined to a spherical surface, and let O be its centre and a its own radius. The transformed distribution is the same. But the space within it becomes transformed into the space without it. Hence if ϕ be the potential due to any spherical shell at a point P, within it, the potential due to the same shell at the point P' in OP produced till $OP' = \frac{a^2}{OP}$, is equal to $\frac{a}{OP'}\phi$ (which is an elementary proposition in the spherical harmonic treatment of potentials, as we shall see presently). Thus, for instance, let the distribution be uniform. Then, as we know there is no force on an interior point, ϕ must be constant; and therefore the potential at P', any external point, is inversely proportional to its distance from the centre.

Or let the given distribution be a uniform shell, S, and let O be any eccentric or any external point. The transformed distribution becomes (§§ 531, 532) a spherical shell, S', with density varying inversely as the cube of the distance from O. If O is within S, it is also enclosed by S', and the whole space within S transforms into the whole space without S'. Hence (§ 532) the potential of S' at any point without it is inversely as the distance from O, and is therefore that of a certain quantity of matter collected at O. Or if O is external to S, and consequently also external to S', the space within S transforms into the space within S'. Hence the potential of S' at any point within it is the same as that of a certain quantity of matter collected at O, which is now a point external to it. Thus, without taking advantage of the general theorems (§§ 517, 524), we fall back on the same results as we inferred from them in § 528, and as we proved synthetically earlier (§§ 488, 491, 492). It may be remarked that those synthetical demonstrations consist merely of transformations of Newton's demonstration, that attractions balance on a point within a uniform shell. Thus the first of them (§ 488) is the image of Newton's in a concentric spherical surface; and the second is its image in a spherical surface having its centre external to the shell, or internal but eccentric, according as the first or the second diagram is used.

534. We shall give just one other application of the theorem of § 532 at present, but much use of it will be made later in the theory of Electricity.

Let the given distribution of matter be a uniform solid sphere, B, and let O be external to it. The transformed system will be a solid sphere, B', with density varying inversely as the fifth power of the distance from O, a point external to it. The potential of S is the same throughout external space as that due to its mass, m, collected at its centre, C. Hence the potential of S' through space external to it is the same as that of the corresponding quantity of matter collected at C', the transformed position of C. This quantity is of course equal to the mass of B'. And it is easily proved that C' is the position of the image of O in the spherical surface of B'. We conclude that a solid sphere with density varying inversely as the fifth power of the distance from an external point, O, attracts any external point as if its mass were condensed at the image of O in its external surface. It is easy to verify this for points of the axis by direct integration, and thence the general conclusion follows according to § 508.

535. The determination of the attraction of an ellipsoid, or of an ellipsoidal shell, is a problem of great interest, and its results will be of great use to us afterwards, especially in Magnetism. We have left it till now, in order that we may be prepared to apply the properties of the potential, as they afford an extremely elegant method of treatment. A few definitions and lemmas are necessary.

Corresponding points on two confocal ellipsoids are such as coincide when either ellipsoid by a pure strain is deformed so as to coincide with the other.

And it is easily shown, that if any two points, P, Q, be assumed on one shell, and their *corresponding* points, p, q, on the other, we have $Pq = Qp$.

The species of shell which it is most convenient to employ in the subdivision of a homogeneous ellipsoid is bounded by similar, similarly situated, and concentric ellipsoidal surfaces; and it is evident from the properties of pure strain (§ 141) that such a shell may be produced from a spherical shell of uniform thickness by uniform extensions and compressions in three rectangular directions. Unless the contrary be specified, the word 'shell' in connection with this subject will always signify an infinitely thin shell of the kind now described.

536. Since, by § 479, a homogeneous spherical shell exerts no attraction on an internal point, a homogeneous shell (which need not be infinitely thin) bounded by similar, and similarly situated, and concentric ellipsoids, exerts no attraction on an internal point.

For suppose the spherical shell of § 479, by simple extensions and compressions in three rectangular directions, to be transformed into an ellipsoidal shell. In this distorted form the masses of all parts are reduced or increased in the proportion of the mass of the ellipsoid to that of the sphere. Also the ratio of the lines HP, PK is unaltered, § 139. Hence the elements IH, KL still attract P equally, and the proposition follows as in § 479.

Hence inside the shell the potential is constant.

537. Two confocal shells (§ 535) being given, the potential of the first at any point, P, of the surface of the second, is to that of the second at the corresponding point, p, on the surface of the first, as the mass of the first is to the mass of the second. This beautiful proposition is due to Chasles.

To any element of the mass of the outer shell at Q corresponds an element of mass of the inner at q, and these bear the same ratio to the whole masses of their respective shells, that the corresponding element of the spherical shell from which either may be derived bears to its whole mass. Whence, since $Pq=Qp$, the proposition is true for the corresponding elements at Q and q, and therefore for the entire shells.

Also, as the potential of a shell on an internal point is constant, and as one of two confocal ellipsoids is wholly within the other: it follows that the external equipotential surfaces for any such shell are confocal ellipsoids, and therefore that the attraction of the shell on an external point is normal to a confocal ellipsoid passing through the point.

538. Now it has been shown (§ 495) that the attraction of a shell on an external point near its surface exceeds that on an internal point infinitely near it by $4\pi\rho$ where ρ is the surface-density of the shell at that point. Hence, as (§ 536) there is no attraction on an internal point, the attraction of a shell on a point at its exterior surface is $4\pi\rho$: or $4\pi\rho t$ if ρ be now put for the volume-density, and t for the (infinitely small) thickness of the shell, § 495. From this it is easy to obtain by integration the determination of the whole attraction of a homogeneous ellipsoid on an external particle.

539. The following splendid theorem is due to Maclaurin:—

The attractions exerted by two homogeneous and confocal ellipsoids on the same point external to each, or external to one and on the surface of the other, are in the same direction and proportional to their masses.

540. Ivory's theorem is as follows:—

Let corresponding points P, p, be taken on the surfaces of two homogeneous confocal ellipsoids, E, e. The x component of the attraction of E on p, is to that of e on P as the area of the section of E by the plane of yz is to that of the coplanar section of e.

Poisson showed that this theorem is true for any law of force whatever. This is easily proved by employing in the general expressions for the components of the attraction of any body, after *one* integration, the properties of corresponding points upon confocal ellipsoids (§ 535).

541. An ingenious application of Ivory's theorem, by Duhamel, must not be omitted here. Concentric spheres are a particular case of confocal ellipsoids, and therefore the attraction of any sphere on a point on the surface of an internal concentric sphere, is to that of the latter upon a point in the surface of the former as the squares of the radii of the spheres. Now *if the law of attraction be such that a homogeneous spherical shell of uniform thickness exerts no attraction on an internal point*, the action of the larger sphere on the internal point is reducible to that of the smaller. Hence *the law is that of the inverse square of the distance*, as is easily seen by making the smaller sphere less and less till it becomes a mere particle. This theorem is due originally to Cavendish.

542. (*Definition.*) If the action of terrestrial or other gravity on a rigid body is reducible to a single force in a line passing always through one point fixed relatively to the body, whatever be its position relatively to the earth or other attracting mass, that point is called its *centre of gravity*, and the body is called a *centrobaric body.*

543. One of the most startling results of Green's wonderful theory of the potential is its establishment of the existence of centrobaric bodies; and the discovery of their properties is not the least curious and interesting among its very various applications.

544. If a body (B) is centrobaric relatively to any one attracting mass (A), it is centrobaric relatively to every other: and it attracts all matter external to itself as if its own mass were collected in its centre of gravity.[1]

545. Hence §§ 510, 515 show that—

(*a*) *The centre of gravity of a centrobaric body necessarily lies in its interior;* or in other words, *can only be reached from external space by a path cutting through some of its mass.* And

(*b*) *No centrobaric body can consist of parts isolated from one another,*

[1] Thomson. *Proc. R.S.E.*, Feb. 1864.

each in space external to all: in other words, *the outer boundary of every centrobaric body is a single closed surface.*

Thus we see, by (a), that no symmetrical ring, or hollow cylinder with open ends, can have a centre of gravity; for its centre of gravity, if it had one, would be in its axis, and therefore external to its mass.

546. *If any mass whatever, M, and any single surface, S, completely enclosing it be given, a distribution of any given amount, M', of matter on this surface may be found which shall make the whole centrobaric with its centre of gravity in any given position (G) within that surface.*

The condition here to be fulfilled is to distribute M' over S, so as by it to produce the potential

$$\frac{M+M'}{EG} - V,$$

any point, E, of S; V denoting the potential of M at this point. The possibility and singleness of the solution of this problem were stated above (§ 517). It is to be remarked, however, that if M' be not given in sufficient amount, an extra quantity must be taken, but neutralized by an equal quantity of negative matter, to constitute the required distribution on S.

The case in which there is no given body M to begin with is important; and yields the following:—

547. *A given quantity of matter may be distributed in one way, but in only one way, over any given closed surface, so as to constitute a centrobaric body with its centre of gravity at any given point within it.*

Thus we have already seen that the condition is fulfilled by making the density inversely as the distance from the given point, if the surface be spherical. From what was proved in §§ 519, 524 above, it appears also that a centrobaric shell may be made of either half of the lemniscate in the diagram of § 526, or of any of the ovals within it, by distributing matter with density proportional to the resultant force of m at I and m' at I'; and that the one of these points which is within it is its centre of gravity. And generally, by drawing the equipotential surfaces relatively to a mass m collected at a point I, and any other distribution of matter whatever not surrounding this point; and by taking one of these surfaces which encloses I but no other part of the mass, we learn, by Green's general theorem, and the special proposition of § 524, how to distribute matter over it so as to make it a centrobaric shell with I for centre of gravity.

548. Under *hydrokinetics* the same problem will be solved for a cube, or a rectangular parallelepiped in general, in terms of converging series; and under *electricity* (in a subsequent volume) it will be solved in finite algebraic terms for the surface of a lense bounded by two spherical surfaces cutting one another at any sub-multiple of two right angles, and for either part obtained by dividing this surface in two by a third spherical surface cutting each of its sides at right angles.

549. *Matter may be distributed in an infinite number of ways throughout a given closed space, to constitute a centrobaric body with its centre of gravity at any given point within it.*

For by an infinite number of surfaces, each enclosing the given point, the whole space between this point and the given closed surface may be divided into infinitely thin shells; and matter may be distributed on each of these so as to make it centrobaric with its centre of gravity at the given point. Both the forms of these shells and the quantities of matter distributed on them, may be arbitrarily varied in an infinite variety of ways.

Thus, for example, if the given closed surface be the pointed oval constituted by either half of the lemniscate of the diagram of § 526, and if the given point be the point I within it, a centrobaric solid may be built up of the interior ovals with matter distributed over them to make them centrobaric shells as above (§ 547). From what was proved in § 534, we see that a solid sphere with its density varying inversely as the fifth power of the distance from an external point, is centrobaric, and that its centre of gravity is the *image* (§ 530) of this point relatively to its surface.

550. The centre of gravity of a centrobaric body composed of true gravitating matter is its centre of inertia. For a centrobaric body, if attracted only by another infinitely distant body, or by matter so distributed round itself as to produce (§ 517) uniform force in parallel lines throughout the space occupied by it, experiences (§ 544) a resultant force always through its centre of gravity. But in this case this force is the resultant of parallel forces on all the particles of the body, which (see *Properties of Matter*, below) are rigorously proportional to their masses: and it is proved that the resultant of such a system of parallel forces passes through the point defined in § 195, as the centre of inertia.

551. The moments of inertia of a centrobaric body are equal round all axes through its centre of inertia. In other words (§ 239), all these axes are principal axes, and the body is kinetically symmetrical round its centre of inertia.

CHAPTER VII.

STATICS OF SOLIDS AND FLUIDS.

552. FORCES whose lines meet. Let ABC be a rigid body acted on by two forces, P and Q, applied to it at different points, D and E respectively, in lines in the same plane.

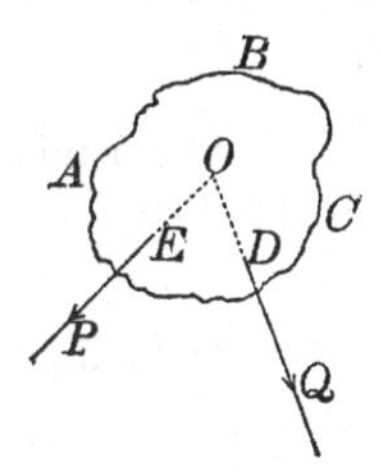

Since the lines are not parallel, they will meet if produced; let them be produced and meet in O. Transmit the forces to act on that point; and the result is that we have simply the case of two forces acting on a material point, which has been already considered.

553. The preceding solution is applicable to every case of non-parallel forces in a plane, however far removed the point may be in which their lines of action meet, and the resultant will of course be found by the parallelogram of forces. The limiting case of parallel forces, or forces whose lines of action, however far produced, do not meet, was considered above, and the position and magnitude of the resultant were investigated. The following is an independent demonstration of the conclusion arrived at.

554. Parallel forces in a plane. The resultant of two parallel forces is equal to their sum, and is in the parallel line which divides any line drawn across their lines of action into parts inversely as their magnitudes.

1°. Let P and Q be two parallel forces acting on a rigid body in similar directions in lines AB and CD. Draw any line AC across their lines. In it introduce any pair of balancing forces, S in AG and S in CH. These forces will not disturb the equilibrium of the body. Suppose the forces P and S in AG, and Q and S in CH to act respectively on the points A and C of the rigid body. The forces P and S, in AB and AG, have a single resultant in some line AM, within the angle

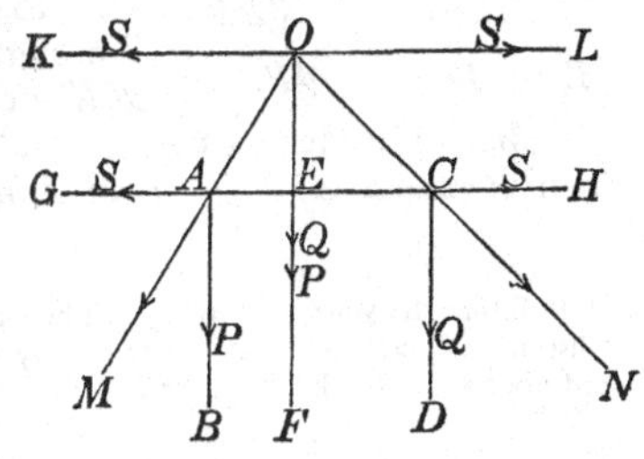

GAB; and Q and S in CD and CH have a resultant in some line CN, within the angle DCH.

The angles MAC, NCA are together greater than two right angles, hence the lines MA, NC will meet if produced. Let them meet in O. Now the two forces P and S may be transferred to parallel lines through O. Similarly the forces Q and S may be also transferred. Then there are four forces acting on O, two of which, S in OK and S in OL, are equal and directly opposed. They may, therefore, be removed, and there are left two forces equal to P and Q in one line on O, which are equivalent to a single force $P+Q$ in the same line.

2°. If, for a moment, we suppose OE to represent the force P, then the force representing S must be equal and parallel to EA, since the resultant of the two is in the direction OA. That is to say,

$$S : P :: EA : OE;$$

and in like manner, by considering the forces S in OL and Q in OE, we find that

$$Q : S :: OE : EC.$$

Compounding these analogies, we get at once

$$Q : P :: EA : EC,$$

that is, the parts into which the line is divided by the resultant are inversely as the forces.

555. Forces in dissimilar directions. The resultant of two parallel forces in dissimilar directions[1], of which one is greater than the other, is found by the following rule: Draw any line across the lines of the forces and produce it across the line of the greater, until the whole line is to the part produced as the greater force is to the less; a force equal to the excess of the greater force above the less, applied at the extremity of this line in a parallel line and in the direction similar to that of the greater, is the resultant of the system.

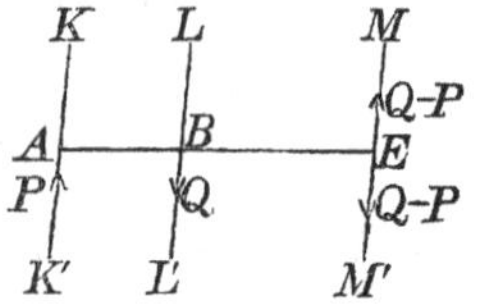

Let P and Q in KK' and LL', be the contrary forces. From any point A, in the line of P, draw a line AB across the line of Q cutting it in B, and produce the line to E, so that $AE : BE :: Q : P$. Through E draw a line MM' parallel to KK' or LL'.

In MM' introduce a pair of balancing forces each equal to $Q-P$. Then P in AK' and $Q-P$ in EM have a resultant equal to their

[1] In future the word 'contrary' will be employed instead of the phrase 'parallel and in dissimilar directions' to designate merely *directional opposition*, while the unqualified word 'opposite' will be understood to signify *contrary and in one line.*

sum, or Q. This resultant is in the line LL'; for, from the analogy,

$$AE : BE :: Q : P,$$

we have
$$AE-BE : BE :: Q-P : P,$$

or
$$AB : BE :: Q-P : P.$$

Hence P in AK', Q in BL', and $Q-P$ in EM are in equilibrium and may be removed. There remains only $Q-P$ in EM', which is therefore the resultant of the two given forces. This fails when the forces are equal.

556. Any number of parallel forces in a plane. Let P_1, P_2, P_3, &c., be any number of parallel forces acting on a rigid body in one plane. To find their resultant in position and magnitude, draw any line across their lines of action, cutting them in points, denoted respectively by A_1, A_2, A_3, &c., and in it choose a point of reference O. Let the distances of the lines of the forces from this point be denoted by a_1, a_2, a_3, &c.; as $OA_1=a_1$, $OA_2=a_2$, &c. Also let R denote the resultant, and x its distance from O.

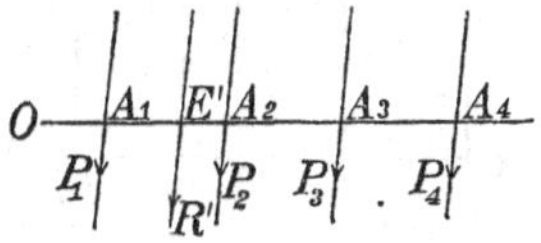

Find the resultant of any two of the forces, as P_1 and P_2, by § 554. Then if we denote this resultant by R', we have

$$R'=P_1+P_2.$$

Divide $A_1 A_2$ in E' into parts inversely as the forces, so that

$$P_1 \times A_1E'=P_2 \times E'A_2.$$

Hence if we denote OE' by x' we have

$$P_1 \times (x'-a_1)=P_2 \times (a_2-x')$$

or
$$(P_1+P_2)\,x'=P_1\,a_1+P_2\,a_2,$$

that is
$$R'x'=P_1\,a_1+P_2\,a_2.$$

Similarly we shall find the resultant of R' and P_3 to be

$$R''=R'+P_3=P_1+P_2+P_3;$$

and
$$R''x''=R'x'+P_3\,a_3=P_1\,a_1+P_2\,a_2+P_3\,a_3.$$

Hence, finally we have

$$R=P_1+P_2+P_3+\dots\dots\dots\dots+P_n \qquad (1)$$

and
$$Rx=P_1\,a_1+P_2\,a_2+P_3\,a_3+\dots\dots\dots+P_n\,a_n. \qquad (2)$$

In this method negative forces or negative values of any of the quantities a_1, a_2, ..., may be included, provided the generalized rules of multiplication and addition in algebra are followed.

557. Any number of parallel forces not in one plane. To find the resultant, let a plane cut the lines of all the forces, and let the points in which they are cut be specified by reference to two rectangular axes in the plane. Let the plane be YOX: OX, OY, the axes of reference, O the origin of co-ordinates, and A_1, A_2, A_3, &c., the points in which the plane cuts the lines of the forces, P_1, P_2, P_3, &c. Thus each of these points will be specified by perpendiculars drawn from it to the axis. Let the co-ordinates of the point A_1 be denoted by $x_1\, y_1$; of A_2, by $x_2\, y_2$; and so on; that is, $ON_1 = x_1$, $N_1 A_1 = y_1$; $ON_2 = x_2$, $N_2 A_2 = y_2$, &c.; let also the final resultant be denoted by R, and its co-ordinates by x and y.

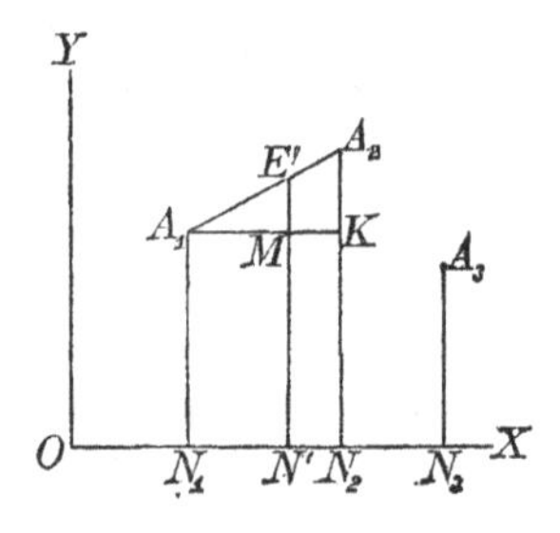

Find the resultant of P_1 and P_2 by joining $A_1\, A_2$, and dividing the line inversely as the forces. Suppose E' the point in which this resultant cuts the plane of reference. Then

$$P_1 \times A_1E' = P_2 \times E'A_2.$$

To find the co-ordinates, which may be denoted by $x'y'$, of the point E' with reference to OX and OY; draw $E'N'$ perpendicular to OX and cutting it in N', and from A_1 draw A_1K parallel to OX, or perpendicular to A_2N_2, and cutting it in K and $E'N'$ in M. Then (Euclid VI. 2)

$$A_1E' : E'A_2 :: A_1M : MK.$$

Hence $$P_1 \times A_1M = P_2 \times MK,$$

or $$P_1(x' - x_1) = P_2(x_2 - x'),$$

whence we get $$(P_1 + P_2)x' = P_1x_1 + P_2x_2;$$

and since $$R' = P_1 + P_2,$$

we have $$R'x' = P_1x_1 + P_2x_2,$$

and similarly, $$R'y' = P_1y + P_2y_2.$$

We may find the resultant of R' and P_3 in like manner, and so with all the forces. Hence we have for the final resultant,

$$R = P_1 + P_2 + P_3 + \ldots\ldots\ldots\ldots + P_n. \quad (3)$$

$$Rx = P_1x_1 + P_2x_2 + P_3x_3 + \ldots + P_nx_n. \quad (4)$$

$$Ry = P_1y_1 + P_2y_2 + P_3y_3 + \ldots + P_ny_n. \quad (5)$$

These equations may include negative forces, or negative co-ordinates.

558. Conditions of equilibrium of any number of parallel forces. In order that any given parallel forces may be in equilibrium, it is not sufficient alone, that their algebraic sum be equal to zero.

For, let $$R = P_1 + P_2 + \&c. = 0.$$

From this equation it follows that if the forces be divided into two groups, one consisting of the forces reckoned positive, the other of those reckoned negative, the sum, or resultant (§ 556) of the former is equal to the resultant of the latter; that is, if ${}_{,}R$ and ${}'R$ denote the resultants of the positive and negative groups respectively,

$$ {}_{,}R = {}'R. $$

But unless these resultants are directly opposed they do not balance one another; wherefore, if $({}_{,}x, {}_{,}y)$ and $({}'x{}'y)$ be the co-ordinates of ${}_{,}R$ and ${}'R$ respectively, we must have for equilibrium

$$ {}_{,}x = {}'x $$

and $$ {}_{,}y = {}'y; $$

whence we get $$ {}_{,}R\,{}_{,}x - {}'R{}'x = 0 $$

and $$ {}_{,}R{}_{,}y - {}'R{}'y = 0. $$

But ${}_{,}R{}_{,}x$ is equal to the sum of those of the terms P_1x_1, P_2x_2, &c., which are positive, and ${}'R{}'x$ is equal to the sum of the others each with its sign changed: and so for ${}_{,}R{}_{,}y$ and ${}'R{}'y$. Hence the preceding equations are equivalent to

$$ P_1x_1 + P_2x_2 + \dots\dots\dots\dots + P_nx_n = 0. $$
$$ P_1y_1 + P_2y_2 + \dots\dots\dots\dots + P_ny_n = 0. $$

We conclude that, for equilibrium, it is necessary and sufficient that each of the following three equations be satisfied:—

$$ P_1 + P_2 + P_3 + \dots\dots\dots\dots + P_n = 0. \qquad (6) $$
$$ P_1x_1 + P_2x_2 + P_3x_3 + \dots\dots + P_nx_n = 0. \qquad (7) $$
$$ P_1y + P_2y_2 + P_3y_3 + \dots\dots + P_ny_n = 0. \qquad (8) $$

559. If equation (6) do not hold, but equations (7) and (8) do, the forces have a single resultant through the origin of co-ordinates. If equation (6) and either of the other two do not hold, there will be a single resultant in a line through the corresponding axis of reference, the co-ordinates of the other vanishing. If equation (6) and either of the other two do hold, the system is reducible to a single couple in a plane through that line of reference for which the sum of the products is not equal to nothing. If the plane of reference is perpendicular to the lines of the forces, the moment of this couple is equal to the sum of the products not equal to nothing.

560. In finding the resultant of two contrary forces in any case in which the forces are unequal—the smaller the difference of magnitude between them, the farther removed is the point of application of the resultant. When the difference is nothing, the point is removed to an infinite distance, and the construction (§ 555) is thus rendered nugatory. The general solution gives in this case $R=0$; yet the forces are not in equilibrium, since they are not directly opposed. Hence two equal contrary forces neither balance, nor have a single resultant. It is clear that they have a tendency to turn the body to which they are applied. This system was by Poinsot denominated a couple.

In actual cases the direction of a couple is generally reckoned positive if the couple tends to turn contrary to the hands of a watch as seen by a person looking at its face, negative when it tends to turn with the hands. Hence the axis, which may be taken to represent a couple, will show, if drawn according to the rule given in § 201, whether the couple is positive or negative, according to the side of its plane from which it is regarded.

561. Proposition I. Any two couples in the same or in parallel planes are in equilibrium if their moments are equal and they tend to turn in contrary directions.

1°. Let the forces of the first couple be parallel to those of the second, and let all four forces be in one plane.

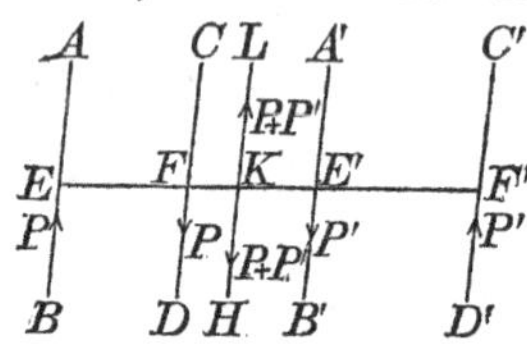

Let the forces of the first couple be P in AB and CD, and of the second P' in $A'B'$ and $C'D'$. Draw any line EF' across the lines of the forces, cutting them respectively in points E, F, E' and F'; then the moment of the first couple is $P.EF$ and of the second $P'.E'F'$; and since the moments are equal we have

$$P.EF = P'.E'F'.$$

Of the four forces, P in AB and P' in $C'D'$ act in similar directions, and P in CD and P' in $A'B'$ also act in similar directions; and their resultants respectively can be determined by the general method (§ 556). The resultant of P in AB, and P' in $C'D'$, is thus found to be equal to $P+P'$, and if HL is the line in which it acts,

$$P.EK = P'.KF'.$$

Again, we have
$$P.EF = P'.E'F'.$$

Subtract the first member of the latter equation from the first member of the former, and the second member of the latter from the second member of the former: there remains

$$P.FK = P'.KE',$$

from which we conclude, that the resultant of P in CD and P' in $A'B'$ is in the line LH. Its magnitude is $P+P'$. Thus the given system is reduced to two equal resultants acting in opposite directions in the same straight line. These balance one another, and therefore the given system is in equilibrium.

Corollary. A couple may be transferred from its own arm to any other arm in the same line, if its moment be not altered.

562. Proposition I. 2°. All four forces in one plane, but those of one couple not parallel to those of the other.

Produce their lines to meet in four points; and consider the parallelogram thus formed. The products of the sides, each into its perpendicular distance from the side parallel to it, are equal, each product being the area of the parallelogram. Hence, since the moments of the two couples are equal, their forces are proportional to the sides of the parallelogram along which they act. And, since the couples tend to turn in opposite directions, the four forces represented by the sides of a parallelogram act in similar directions relatively to the angles, and dissimilar directions in the parallels, and therefore balance one another.

Corollary. The statical effect of a couple is not altered, if its arm be turned round any point in the plane of the couple.

563. Proposition I. 3°. The two couples not in the same plane, but the forces equal and parallel.

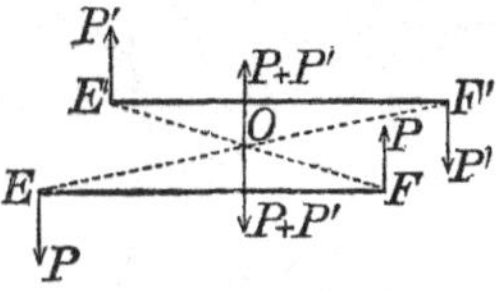

Let there be two couples, acting respectively on arms EF and $E'F'$, which are parallel but not in the same plane. Join EF' and $E'F$. These lines bisect one another in O.

Of the four forces, P on F and P' on E', act in similar directions, and their resultant equal to $P+P'$, may be substituted for them. It acts in a parallel line through O. Similarly P on E and P' on F' have also a resultant equal to $P+P'$ through O; but these resultants being equal and opposite, balance, and therefore the given system is in equilibrium.

Remark 1.—A corresponding demonstration may be applied to every case of two couples, the moments of which are equal, though the forces and arms may be unequal. When the forces and arms are unequal, the lines EF', $E'F$ cut one another in O into parts inversely as the forces.

Remark 2.—Hence as an extreme case, Proposition I, 1°, may be brought under this head. Let EF be the arm of one couple, $E'F'$ of the other, both in one straight line. Join FE', and divide it inversely as the forces. Then $FK : KE' :: EF : E'F'$ and EF' is divided in the same ratio.

Corollary. Transposition of couples. Any two couples in the same or in parallel planes, are equivalent, provided their moments are equal, and they tend to turn in similar directions.

564. Proposition II. Any number of couples in the same or in parallel planes, may be reduced to a single resultant couple, whose moment is equal to the algebraic sum of their moments, and whose plane is parallel to their planes.

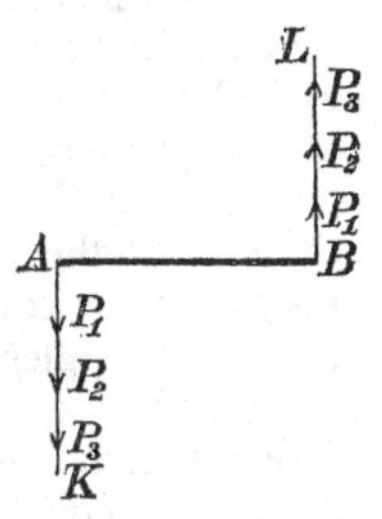

Reduce all the couples to forces acting on one arm AB, which may be denoted by a. Then if P_1, P_2, P_3, &c., be the forces, the moments of the couples will be P_1a, P_2a, P_3a, &c. Thus we have P_1, P_2, P_3, &c., in AK, reducible to a single force, their sum, and similarly, a single force P_1+P_2+ &c., in BL.

These two forces constitute a couple whose moment is $(P_1+P_2+P_3+$ &c.$)\,a$. But this product is equal to $P_1a+P_2a+P_3a+$ &c., the sum of the moments of the given couples, and therefore any number of couples, &c. If any of the couples act in the direction opposite to that reckoned positive, their moments must be reckoned as negative in the sum.

565. Proposition III. Any two couples not in parallel planes may be reduced to a single resultant couple, whose axis is the diagonal through the point of reference of the parallelogram described upon their axes.

1°. Let the planes of the two couples cut the plane of the diagram perpendicularly in the lines AA' and BB' respectively; let the planes of the couples also cut each other in a line cutting the plane of the diagram in O. Through O, as a point of reference, draw OK the axis of the first couple, and OL the axis of the second. On OK and OL construct the parallelogram $OKML$. Its diagonal OM is the axis of the resultant couple.

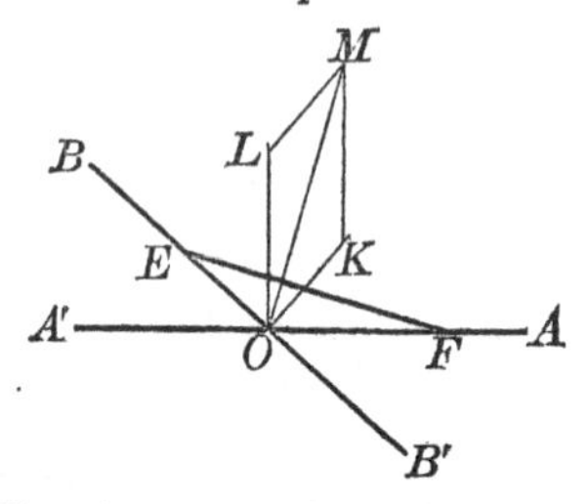

Let the moment of the couple acting in the plane BB', be denoted by G, and of that in AA', by H. For the given couples, substitute two others, with arms equal respectively to G and H, and therefore with forces equal to unity.

From OB and OA measure off $OE=G$, and $OF=H$, and let these lines be taken as the arms of the two couples respectively. The forces of the couples will thus be perpendicular to the plane of the diagram: those of the first, acting outwards at E, and inwards at O; and those of the second, outwards at O, and inwards at F. Thus, of the four equal forces which we have in all, there are two equal and opposite at O, which therefore balance one another, and may be removed; and there remain two equal parallel forces, one acting outwards at E, and the other inwards at F, which constitute a couple on an arm EF.

This single couple is therefore equivalent to the two given couples.

2°. It remains to be proved that its axis is OM. Join EF. As, by construction, OL and OK are respectively perpendicular to OA, and OB, the angle KOL is equal to the angle AOB'. Hence, MLO the supplement of the former is equal to EOF, the supplement of the latter. But OK is equal to OE; each being equal to the moment of the first of the given couples; and therefore LM, which is equal to the former, is equal to OE. Similarly OL is equal to OF. Thus there are two triangles, MLO and EOF, with two sides of one respectively equal to two sides of the other, and the contained angles equal: therefore the remaining sides OM, EF are equal, and the angles LOM, OFE are equal. But since OL is perpendicular to OF, OM is perpendicular to EF. Hence OM is the axis of the resultant couple.

566. Proposition IV. Any number of couples whatever are either in equilibrium with one another, or may be reduced to a single couple, under precisely the same conditions as those already investigated for forces acting on one point, the axes of the couples being now taken everywhere instead of the lines formerly used to represent the forces.

1°. Resolve each couple into three components having their axes along three rectangular lines of reference, OX, OY, OZ. Add all the components corresponding to each of these three lines. Then if the resultant of all the couples whose axes are along the line

OX, be denoted by L,
OY, " " M,
OZ, " " N,

and if G be the resultant of these three, we have

$$G=\sqrt{(L^2+M^2+N^2)}:$$

and if ζ, η, θ, be the angles which the axis of this couple G, makes with the three axes OX, OY, OZ, respectively, we have

$$\cos\zeta=\frac{L}{G};\ \cos\eta=\frac{M}{G};\ \cos\theta=\frac{N}{G}.$$

567. 2°. Conditions of equilibrium of any number of couples. For equilibrium the resultant couple must be equal to nothing: but as it is compounded of three subsidiary resultant couples in planes at right angles to one another, they also must each be equal to nothing. The remarks already made, and the equations already given in §§ 471, 472, apply with the necessary modification to couples also. Thus, for instance, the equations of equilibrium are

$$G_1\cos\zeta_1+G_2\cos\zeta_2+G_3\cos\zeta_3+\&c.,=0,$$
$$G_1\cos\eta_1+G_2\cos\eta_2+G_3\cos\eta_3+\&c.,=0,$$
$$G_1\cos\theta_1+G_2\cos\theta_2+G_3\cos\theta_3+\&c.,=0.$$

568. Before investigating the conditions of equilibrium of any number of forces acting on a rigid body, we shall establish some preliminary propositions.

1°. A force and a couple in the same or in parallel planes may be reduced to a single force. Let the plane of the couple be the plane of the diagram, and let its moment be denoted by G. Let R, acting in the line OA in the same plane, be the force. Transfer the couple to an arm (which may be denoted by a) through the point O, such that each force shall be equal to R; and let its position be so chosen, that one of the forces shall act in the same straight line with R in OA, but in the opposite direction to it.

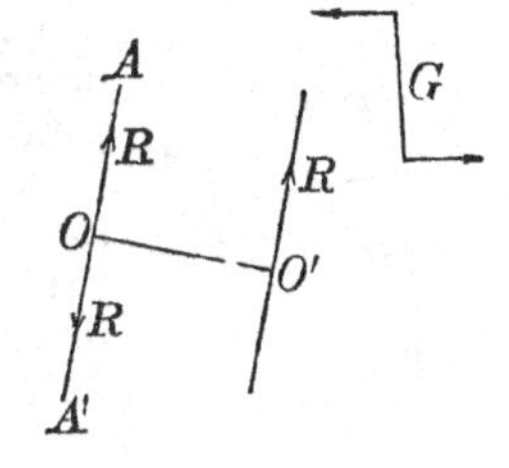

R and G being known, the length of this arm can be found, for since the moment of the transposed couple is

$$Ra = G$$

we have

$$a = \frac{G}{R}.$$

Through O then, draw a line OO' perpendicular to OA, making it equal to a. On this arm apply the couple, a force equal to R, acting on O' in a line perpendicular to OO', and another in the opposite direction at the other extremity. There are now three forces, two of which, being equal and opposite to one another, in the line AA', may be removed. One, acting on the point O', remains, which is therefore equivalent to the given system.

569. 2°. A couple and a force in a given line inclined to its plane may be reduced to a smaller couple in a plane perpendicular to the force, and a force equal and parallel to the given force.

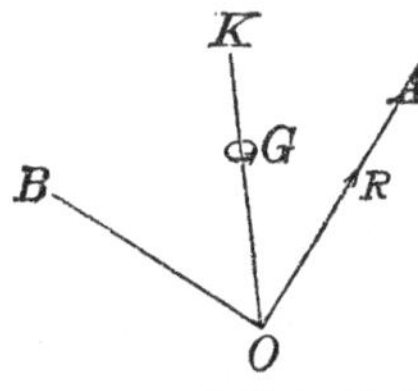

Let OA be the line of action of the force R, and let OK be the axis of the couple. Let the moment be denoted by G: and let AOK, the inclination of its axis to the line of the force, be θ. Draw OB perpendicular to OA. By Prop. IV (§ 566) resolve the couple into two components, one acting round OA as axis, and one round OB. Thus the component round OA will be

$$G \cos\theta,$$

and the component round OB,

$$G \sin\theta.$$

Now as $G \sin\theta$ acts in the same plane as the given force R, this component together with R may be reduced by § 568 to one force. This force which is equal to R, will act not at O in the line OA, but in a parallel line through a point O' out of the plane of the diagram. Thus the given system is reduced to a smaller couple $G \cos\theta$, and to a force in a line which, by Poinsot, was denominated the central axis of the system.

570. 3°. Any number of forces may be reduced to a force and a couple.

Let P_1 acting on M_1 be one of a number of forces acting in different directions on different points of a rigid body. Choose any point of reference O, for the different forces, and through it draw a line AA' parallel to the line of the first force P_1. Through O, draw OO' perpendicular to AA' or the line of the force P_1. In the line AA' introduce two equal opposite forces, each equal to P_1. There are now three forces, producing the same effect as the given force, and they may be grouped differently: P_1 acting

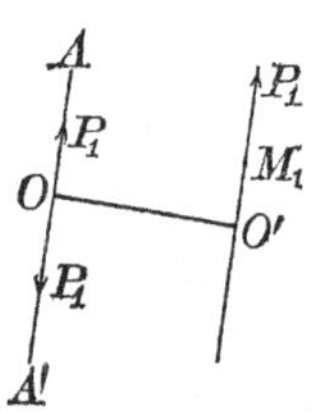

in O in the line OA, and a couple, P_1 acting at O', and P_1 at O in the line OA', on an arm OO'. Reduce similarly all the other forces, each to a force acting on O, and to a couple. But all the couples thus obtained are equivalent to a single couple, and all the forces are equivalent to one force. Hence, &c.

571. Reduction of any number of forces to their simplest equivalent system.

Suppose any number of forces acting in any directions on different points of a rigid body. Choose three rectangular planes of reference meeting in a point O, the origin of co-ordinates. In order to effect the reduction it is necessary to bring in all the forces to the point O. This may be done in two different ways—either in two steps, or directly.

572. 1°. Let the magnitudes of the forces be P_1, P_2, &c., and the co-ordinates, with reference to the rectangular planes, of the points at which they act respectively, be (x_1, y_1, z_1), (x_2, y_2, z_2), &c. Let also the direction cosines be (l_1, m_1, n_1), (l_2, m_2, n_2), &c. Resolve each force into three components, parallel to OX, OY, OZ, respectively. Thus, if (X_1, Y_1, Z_1), &c., be the components of P_1, &c., we shall have

$$X_1 = P_1 l_1;\ X_2 = P_2 l_2;\ \text{\&c.} \qquad (1)$$
$$Y_1 = P_1 m_1;\ Y_2 = P_2 m_2;\ \text{\&c.} \qquad (2)$$
$$Z_1 = P_1 n_1;\ Z_2 = P_2 n_2;\ \text{\&c.} \qquad (3)$$

To transfer these components to the point O. Let X_1, in MK, be the component, parallel to OX, of the force P_1 acting on the point M. From M transmit it along its line to a point N in the plane ZOY: the co-ordinates of this point will be y_1, z_1. From N draw a perpendicular NB to OY, and through B draw a line parallel to MK or OX. Introducing in this line a pair of balancing forces each equal to X_1, we have a couple acting on an arm z_1 in a plane parallel to XOZ, and a single force X_1 parallel to OX in the plane XOY. The moment of this couple is $X_1 z_1$, and its axis is along OY.

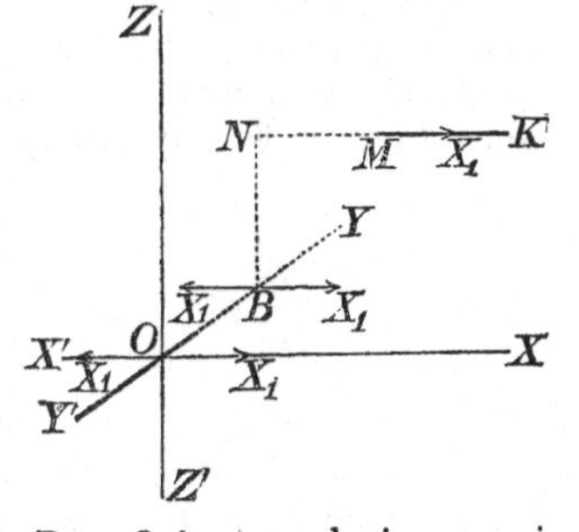

Next transfer the force X_1 from B to O, by introducing a pair of balancing forces in $X'OX$, one of which, with the force X_1 in the line through B parallel to $X'X$ and the direction similar to OX, form a couple acting on an arm y_1. This couple, when Z and X are both positive, tends to turn in the plane XOY from OY to OX. Therefore by the rule, § 201, its axis must be drawn from O in the direction OZ'. Hence its moment is to be reckoned as $Z_1 x_1$. Besides this couple there remains a single force equal to X_1, in the direction OX, through the point O. Similarly by successive steps transfer the forces Y_1, Z_1, to the origin of co-ordinates. In this way six couples of transference are got, three tending to turn in one direction round the axes respectively, and three in the opposite direction; and three single forces at right angles to one another, acting at the point O. Thus for the force P_1, at the point (x_1, y_1, z_1), we have as equivalent to it at the point O, three forces X_1, Y_1, Z_1, and three couples;

$$Z_1 y_1 - Y_1 z_1; \text{ moment of the couple round } OX; \tag{4}$$
$$X_1 z_1 - Z_1 x_1; \text{ moment of the couple round } OY; \tag{5}$$
$$Y_1 x_1 - X_1 y_1; \text{ moment of the couple round } OZ. \tag{6}$$

All the forces may be brought in to the origin of co-ordinates in a similar way.

573. 2°. Otherwise: Let P be one of the forces acting in the line MT on a point M of a rigid body. Let O be the origin of co-ordinates; OX, OY, OZ, three rectangular lines of reference. Join OM and produce the line to S. From O draw ON, cutting at right angles in the point N, the line MT produced through M. Let ON be denoted by p, and the angle TMS by κ. In a line through O parallel to MT (not shown in diagram) suppose introduced a pair of balancing forces each equal to P. We have thus a single force equal to P acting at O, and a couple, whose moment is Pp, in the plane ONM. The direction cosines of this plane, or, which is the same thing, the direction cosines of a perpendicular to it, that is, the axis of the couple are (§ 464), if we denote them by ϕ, χ, ψ, respectively,

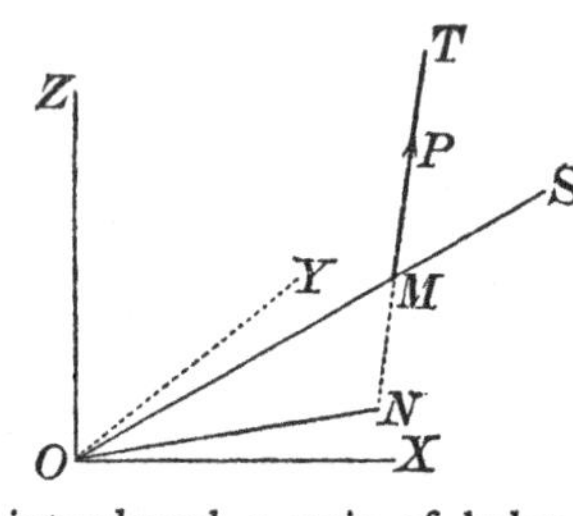

$$\phi = \frac{\frac{y}{r} n - \frac{z}{r} m}{\sin \kappa},$$

$$\chi = \frac{\frac{z}{r} l - \frac{x}{r} n}{\sin \kappa},$$

$$\psi = \frac{\frac{x}{r} m - \frac{y}{r} l}{\sin \kappa}.$$

Now in the triangle ONM,

$$ON = OM \sin OMN,$$

that is

$$p = r \sin \kappa.$$

Hence, if we substitute p for its value in the three preceding equations, the expression for the direction cosines are reduced to

$$\phi=\frac{ny-mz}{p}, \quad (7)$$

$$\chi=\frac{lz-nx}{p}, \quad (8)$$

$$\psi=\frac{mx-ly}{p}. \quad (9)$$

To find the component couples round OX, OY, OZ, multiply these direction cosines respectively by Pp; whence we get

$$Pp.\phi=P(ny-mz), \text{ moment of couple round } OX, \quad (10)$$

$$Pp.\chi=P(lz-nx), \text{ moment of couple round } OY, \quad (11)$$

$$Pp.\psi=P(mx-ly), \text{ moment of couple round } OZ. \quad (12)$$

That this result is the same as that got by the other method will be evident, by considering that (equations 1, 2, 3),

$$Pl=X; \ Pm=Y; \ Pn=Z.$$

574. When by either of the methods all the forces have been referred to O, there is obtained a set of couples acting round OX, OY, OZ; and a set of forces acting along OX, OY, OZ. Find then the resultant moments of all the couples; and the sums of all the forces: if L, M, N be the resultant moments round OX, OY, OZ respectively, we have

$$L=(Z_1y_1-Y_1z_1)+(Z_2y_2-Y_2z_2)+\&c. \quad (13)$$

$$M=(X_1z_1-Z_1x_1)+(X_2z_2-Z_2x_2)+\&c. \quad (14)$$

$$N=(Y_1x_1-X_1y_1)+(Y_2x_2-X_2y_2)+\&c. \quad (15)$$

and if X, Y, Z be the resultant forces,

$$X=X_1+X_2+X_3+\&c. \quad (16)$$

$$Y=Y_1+Y_2+Y_3+\&c. \quad (17)$$

$$Z=Z_1+Z_2+Z_3+\&c. \quad (18)$$

575. Finally, find the resultant of the three forces by the formulae of Chap. VI, and the resultant of the three couples by Prop. IV (§ 566). Thus, if l, m, n be the direction cosines of the resultant force R, we have (§§ 463, 467)

$$l=\frac{X}{R}; \ m=\frac{Y}{R}; \ n=\frac{Z}{R}, \quad (19)$$

and if λ, μ, ν be the direction cosines of the axis of the resultant couple, we have (§ 566)

$$\lambda=\frac{L}{G}; \ \mu=\frac{M}{G}; \ \nu=\frac{N}{G}. \quad (20)$$

576. Conditions of Equilibrium. The conditions of equilibrium of three forces at right angles to one another have been already stated in § 470; and the conditions for three rectangular couples in § 567.

If a body be acted on by three forces and three couples simultaneously, all the conditions applicable when they act separately, must also be satisfied when they act conjointly, since a force cannot balance a couple. Six Equations of Equilibrium therefore are necessary and sufficient for a rigid body acted on by any number of forces. These are

$$P_1 \cos \alpha_1 + P_2 \cos \alpha_2 + \&c. = 0,$$
$$P_1 \cos \beta_1 + P_2 \cos \beta_2 + \&c. = 0,$$
$$P_1 \cos \gamma_1 + P_2 \cos \gamma_2 + \&c. = 0,$$
$$G_1 \cos \zeta_1 + G_2 \cos \zeta_2 + \&c. = 0,$$
$$G_1 \cos \eta_1 + G_2 \cos \eta_2 + \&c. = 0,$$
$$G_1 \cos \theta_1 + G_2 \cos \theta_2 + \&c. = 0.$$

577. If the line of the resultant found by § 575, is perpendicular to the plane of the couple, that is, if

$$\lambda = l, \ \mu = m, \ \nu = n;$$

or

$$\frac{L}{X} = \frac{M}{Y} = \frac{N}{Z}, \tag{21}$$

the system cannot be reduced to another with a force and a smaller couple, and in this case the line found for the resultant force is the central axis of the system.

578. If, on the other hand, the plane of the couple is parallel to the line of the force, or the axis of the couple perpendicular to the line of the force, that is, if

$$l\lambda + m\mu + n\nu = 0,$$

or

$$LX + MY + NZ = 0, \tag{22}$$

the force and couple may (§ 568) be reduced to one force: and this force is parallel to the former, at a distance from it equal to $\frac{G}{R}$, in the plane of it and the couple. Thus, OO' will be perpendicular to the line of the resultant force, and to the axis of the resultant couple, and therefore its direction cosines are (§ 464, b);

$$m\nu - n\mu, \ n\lambda - l\nu, \ l\mu - m\lambda, \tag{23}$$

each of which will be positive when O' lies within the solid angle edged by OX, OY, OZ. Hence, remembering that $OO' = \frac{G}{R}$, and using the expressions (19) and (20), we find for the co-ordinates of O'

$$\frac{YN - ZM}{R^2}, \ \frac{ZL - XN}{R^2}, \ \frac{XM - YL}{R^2}, \tag{24}$$

and we thus complete the specification of the single force to which the system is reduced when (22) holds.

579. If the line of the force is inclined at any angle to the plane of the couple, the resultant system can be further reduced by § 569, to a smaller couple and a force in a determinate line, the 'central axis.' This couple is $G\cos\theta$, and according to the notation, may be thus expressed by § 464, (7), if we substitute the values given in (19) and (20),

$$G\cos\theta = \frac{XL+YM+ZN}{R}. \tag{25}$$

The other component couple, $G\sin\theta$, lies in the same plane as R, and with it may be reduced by § 568 to one force, which will be parallel to R, that is, in the direction (l, m, n), at a distance from it equal to $\frac{G\sin\theta}{R}$. Hence the direction cosines of OO' will be

$$\frac{m\nu - n\mu}{\sin\theta}, \quad \frac{n\lambda - l\nu}{\sin\theta}, \quad \frac{l\mu - m\lambda}{\sin\theta}. \tag{26}$$

Substituting in each of these for l, λ, &c., their respective values, and multiplying each member by $\frac{G\sin\theta}{R}$, we have for the co-ordinates of the point O', as in § 578,

$$\frac{YN-ZM}{R^2}, \quad \frac{ZL-XN}{R^2}, \quad \frac{XM-YL}{R^2}. \tag{27}$$

A single force, R, through the point thus specified in the direction (l, m, n), with a couple in a plane perpendicular to it, and having

$$\frac{XL+YM+ZN}{R}$$

for its moment, is consequently the system of *force along central axis* and *minimum couple*, to which the given set of forces is determinately reducible by Poinsot's beautiful method.

580. The position of the central axis may be determined otherwise; thus, instead of in the first place bringing the forces to O, bring them to any point T, of which let (x, y, z) be the co-ordinates. Then instead of $Y_1 z_1 + Y_2 z_2 + \text{\&c.}$, which we had before (§ 574), we have now

$$Y_1(z_1 - z) + Y_2(z_2 - z) + \text{\&c.},$$

or

$$Y_1 z_1 + Y_2 z_2 + \text{\&c.} - (Y_1 + Y_2 + \text{\&c.})\, z,$$

and so for the others. Then for the moments of the couples of transference we have

$$\mathfrak{L} = L - (Zy - Yz),$$
$$\mathfrak{M} = M - (Xz - Zx),$$
$$\mathfrak{N} = N - (Yx - Xy).$$

Now, let T be chosen, if possible, so as to make the resultant couple lie in a plane perpendicular to it. The condition to be fulfilled in this case is

$$\frac{\mathfrak{L}}{X}=\frac{\mathfrak{M}}{Y}=\frac{\mathfrak{N}}{Z},$$

which, when for $\mathfrak{L}$, &c., we substitute their values, becomes,

$$\frac{L-(Zy-Yz)}{X}=\frac{M-(Xz-Zx)}{Y}=\frac{N-(Yx-Xy)}{Z},$$

which is the equation of the central axis of the system.

To show that O', the point determined in §§ 578, 579, is in the central axis thus found; we have, substituting for x, y, z, the values given in (24),

$$\frac{L-\dfrac{Z(ZL-XN)+Y(XM-YL)}{R^2}}{X}=\&c.$$

Reducing, and remarking that

$$LR^2-LY^2-LZ^2=LX^2,$$

we find that the first member becomes

$$\frac{(LX+MY+NZ)X}{X}=LX+MY+NZ,$$

and is therefore equal to each of the two others. Thus is verified the comparison of the two methods.

581. In one respect, this reduction of a system of forces to a couple, and a force perpendicular to its plane, is the best and simplest, especially in having the advantage of being determinate, and it gives very clear and useful conceptions regarding the effect of force on a rigid body. The system may, however, be farther reduced to two equal forces acting symmetrically on the rigid body, but whose position is indeterminate. Thus, supposing the central axis of the system has been found, draw a line AA', at right angles through any point C in it, so that CA may be equal to CA'. For R, acting along the central axis, substitute $\frac{1}{2}R$ at each end of AA'. Thus, choosing this line AA' as the arm of the couple, and calling it a, we have at each extremity of it two forces, $\frac{G}{a}$ perpendicular to the central axis, and $\frac{1}{2}R$ parallel to the central axis. Compounding these, we get two forces, each equal to $\left(\frac{1}{4}R^2+\frac{G^2}{a^2}\right)^{\frac{1}{2}}$, through A and A' respectively, perpendicular to AA', and equally inclined at the angle $\tan^{-1}\frac{2G}{aR}$ on the two sides of the plane through AA' and the central axis.

582. It is obvious, from the formulae of § 195, that if masses proportional to the forces be placed at the several points of application of these forces, the centre of inertia of these masses will be the same point in the body as the centre of parallel forces. Hence the reactions of the different parts of a rigid body against acceleration in parallel lines are rigorously reducible to one force, acting at the centre of inertia. The same is true approximately of the action of gravity on a rigid body of small dimensions relatively to the earth, and hence the centre of inertia is sometimes (§ 195) called the *Centre of Gravity*. But, except on a centrobaric body (§ 543), gravity is not in general reducible to a single force: and when it is so, this force does not pass through a point fixed relatively to the body in all positions.

583. The resultant of a system of parallel forces is not a single force when the algebraic sum of the given forces vanishes. In this case the resultant is a couple whose plane is parallel to the common direction of the forces. A good example of this is furnished by a magnetized mass of steel, of moderate dimensions, subject to the influence of the earth's magnetism only. As will be shown later, the amounts of the so-called north and south magnetisms in each element of the mass are equal, and are therefore subject to equal and opposite forces, all parallel to the line of dip. Thus a compass-needle experiences from the earth's magnetism merely a couple or *directive* action, and is not attracted or repelled as a whole.

584. If three forces, acting on a rigid body, produce equilibrium, their directions must lie in one plane; and must all meet in one point, or be parallel. For the proof, we may introduce a consideration which will be very useful to us in investigations connected with the statics of flexible bodies and fluids.

If any forces, acting on a solid or fluid body, produce equilibrium, we may suppose any portions of the body to become fixed, or rigid, or rigid and fixed, without destroying the equilibrium.

Applying this principle to the case above, suppose any two points of the body, respectively in the lines of action of two of the forces, to be fixed—the third force must have no moment about the line joining these points; that is, its direction must pass through the line joining them. As any two points in the lines of action may be taken, it follows that the three forces are coplanar. And three forces in one plane cannot equilibrate, unless their directions are parallel or pass through a point.

585. It is easy and useful to consider various cases of equilibrium when no forces act on a rigid body but gravity and the pressures, normal or tangential, between it and fixed supports. Thus, if one given point only of the body be fixed, it is evident that the centre of gravity must be in the vertical line through this point—else the weight and the reaction of the support would form an unbalanced couple. Also for *stable* equilibrium the centre of gravity must be *below* the point of suspension. Thus a body of any form may be made to stand in stable equilibrium on the point of a needle if we rigidly attach to it such a mass as to cause the joint centre of gravity to be below the point of the needle.

586. An interesting case of equilibrium is suggested by what are called Rocking Stones, where, whether by natural or by artificial processes, the lower surface of a loose mass of rock is worn into a convex form which may be approximately spherical, while the bed of rock on which it rests in equilibrium is, whether convex or concave, also approximately spherical, if not plane. A loaded sphere resting on a spherical surface is therefore a type of such cases.

Let O, O' be the centres of curvature of the fixed and rocking bodies respectively, when in the position of equilibrium. Take any two infinitely small equal arcs PQ, Pp; and at Q make the angle $O'QR$ equal to POp. When, by displacement, Q and p become the points in contact, QR will evidently be vertical; and, if the centre of gravity G, which must be OPO' when the movable body is in its position of equilibrium, be to the left of QR, the equilibrium will obviously be stable. Hence, if it be below R, the equilibrium is stable, and not unless.

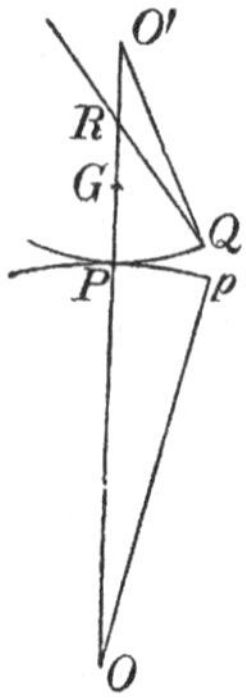

Now if ρ and σ be the radii of curvature OP, $O'P$ of the two surfaces, and θ the angle POp, the angle $QO'R$ will be equal to $\frac{\rho\theta}{\sigma}$; and we have in the triangle $QO'R$ (§ 119)

$$RO' : \sigma :: \sin\theta : \sin\left(\theta + \frac{\rho\theta}{\sigma}\right)$$

$$:: \sigma : \sigma + \rho \text{ (approximately).}$$

Hence
$$PR = \sigma - \frac{\sigma^2}{\sigma+\rho} = \frac{\rho\sigma}{\rho+\sigma};$$

and therefore, for stable equilibrium,

$$PG < \frac{\rho\sigma}{\rho+\sigma}.$$

If the lower surface be plane, ρ is infinite, and the condition becomes (as in § 256)
$$PG < \sigma.$$

If the lower surface be concave, the sign of ρ must be changed, and the condition becomes
$$PG < \frac{\rho\sigma}{\rho-\sigma},$$

which cannot be negative, since ρ *must* be numerically greater than σ in this case.

587. If two points be fixed, the only motion of which the system is capable is one of rotation about a fixed axis. The centre of gravity must then be in the vertical plane passing through those points, and *below* the line adjoining them.

588. If a rigid body rest on a fixed surface, there will in general be only *three* points of contact, § 380; and the body will be in stable equilibrium if the vertical line drawn from its centre of gravity cuts the plane of these three points *within* the triangle of which they form the corners. For if one of these supports be removed, the body will obviously tend to fall towards that support. Hence each of the three prevents the body from rotating about the line joining the other two. Thus, for instance, a body stands stably on an inclined plane (if the friction be sufficient to prevent it from sliding down) when the vertical line drawn through its centre of gravity falls within the base, or area bounded by the shortest line which can be drawn round the portion in contact with the plane. Hence a body, which cannot stand on a horizontal plane, may stand on an inclined plane.

589. A curious theorem, due to Pappus, but commonly attributed to Guldinus, may be mentioned here, as it is employed with advantage in some cases in finding the centre of gravity of a body—though it is really one of the geometrical properties of the Centre of Inertia. It is obvious from § 195. *If a plane closed curve revolve through any angle about an axis in its plane, the solid content of the surface generated is equal to the product of the area of either end into the length of the path described by its centre of gravity; and the area of the curved surface is equal to the product of the length of the curve into the length of the path described by its centre of gravity.*

590. The general principles upon which forces of constraint and friction are to be treated have been stated above (§§ 258, 405). We add here a few examples, for the sake of illustrating the application of these principles to the equilibrium of a rigid body in some of the more important practical cases of constraint.

591. The application of statical principles to the *Mechanical Powers*, or elementary machines, and to their combinations, however complex, requires merely a statement of their kinematical relations (as in §§ 91, 97, 113, &c.) and an immediate translation into Dynamics by Newton's principle (§ 241); or by Lagrange's Virtual Velocities (§ 254), with special attention to the introduction of forces of friction, as in § 405. In no case can this process involve further difficulties than are implied in seeking the geometrical circumstances of any infinitely small disturbance, and in the subsequent solution of the equations to which the translation into dynamics leads us. We will not, therefore, stop to discuss any of these questions; but will take a few examples of no very great difficulty, before for a time quitting this part of the subject. The principles already developed will be of constant use to us in the remainder of the work, which will furnish us with ever-recurring opportunities of exemplifying their use and mode of application.

Let us begin with the case of the Balance, of which we promised (§ 384) to give an investigation.

592. *Ex.* I. We will assume the line joining the points of attachment of the scale-pans to the arms to be at right angles to the line joining the centre of gravity of the beam with the fulcrum. It is obvious that the centre of gravity of the beam must not coincide with the knife-edge, else the beam would rest indifferently in any position. We will suppose, in the first place, that the arms are not of equal length.

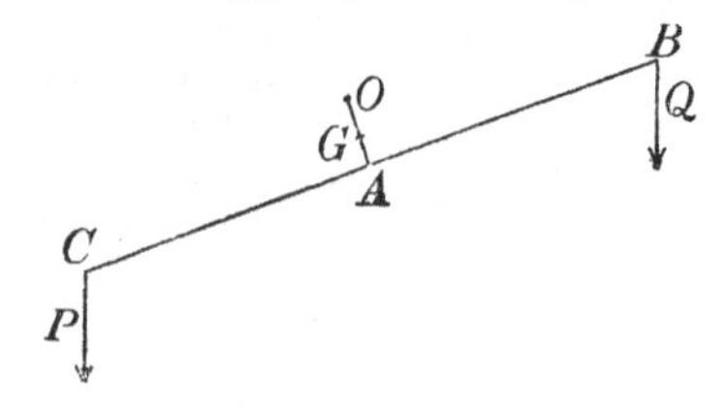

Let O be the fulcrum, G the centre of gravity of the beam, M its mass; and suppose that with loads P and Q in the pans the beam rests (as drawn) in a position making an angle θ with the horizontal line.

Taking moments about O, and, for convenience (see § 185), using gravitation measurement of the forces, we have

$$Q(AB\cos\theta+OA\sin\theta)+M.OG\sin\theta=P(AC\cos\theta-OA\sin\theta).$$

From this we find
$$\tan\theta=\frac{P.AC-Q.AB}{(P+Q)\,OA+M.OG}.$$

If the arms be equal we have

$$\tan\theta=\frac{(P-Q)\,AB}{(P+Q)\,OA+M.OG}.$$

Hence the Sensibility (§ 384) is greater, (1) as the arms are longer, (2) as the mass of the beam is less, (3) as the fulcrum is nearer to the line joining the points of attachment of the pans, (4) as the fulcrum is nearer to the centre of gravity of the beam. If the fulcrum be *in* the line joining the points of attachment of the pans, the sensibility is the same for the same *difference* of loads in the pan.

To determine the Stability we must investigate the time of oscillation of the balance when slightly disturbed. It will be seen, by reference to a future chapter, that the equation of motion is approximately

$$\{Mk^2+(P+Q)\,AB^2\}\ddot\theta+Qg(AB\cos\theta+OA\sin\theta)$$
$$+MgOG\sin\theta-Pg(AC\cos\theta-OA\sin\theta)=0,$$

k being the radius of gyration (§ 235) of the beam. If we suppose the arms and their loads equal, we have for the time of an infinitely small oscillation

$$\pi\sqrt{\frac{Mk^2+2P.AB^2}{(2P.OA+M.OG)g}}.$$

Thus the stability is greater for a given load, (1) the less the length of the beam, (2) the less its mass, (3) the less its radius of gyration, (4) the further the fulcrum from the beam, and from its centre of gravity. With the exception of the second, these adjustments are the very opposite of those required for sensibility. Hence all we can do is to effect a judicious compromise; but the less the mass of the beam, the better will the balance be, in *both* respects.

The general equation, above written, shows that if the length, and the radius of gyration, of one arm be diminished, the corresponding load being increased so as to maintain equilibrium—a form of balance occasionally useful—the sensibility is increased.

Ex. II. Find the position of equilibrium of a rod AB resting on a smooth horizontal rail D, its lower end pressing against a smooth vertical wall AC parallel to the rail.

The figure represents a vertical section through the rod, which must evidently be in a plane perpendicular to the wall and rail.

The only forces acting are three, R the pressure of the wall on the rod, horizontal; S that of the rail on the rod, perpendicular to the rod; W the weight of the rod, acting vertically downwards at its centre of gravity. If the half-length of the rod be a, and the distance of the rail from the wall b, these are given —and all that is wanted to fix the position of equilibrium is the angle the rod makes with the wall.

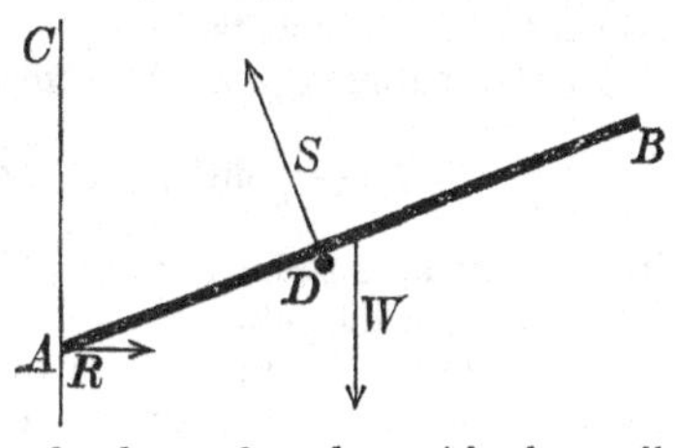

Call $\widehat{CAB}$, θ. Then we see at once that $AD=\dfrac{b}{\sin\theta}$.

Resolving horizontally $\quad R-S\cos\theta=0, \qquad (1)$

vertically $\quad W-S\sin\theta=0. \qquad (2)$

Taking moments about A,

$$S.AD-Wa\sin\theta=0,$$

or

$$Sb-Wa\sin^2\theta=0. \qquad (3)$$

As there are only three unknown quantities, R, S, and θ, these three equations contain the complete solution of the problem. By (2) and (3)

$$\sin^3\theta=\frac{b}{a}, \text{ which gives } \theta.$$

Hence by (2)

$$S=\frac{W}{\sin\theta},$$

and by (1)

$$R=S\cos\theta=W\cot\theta.$$

Ex. III. As an additional example, suppose the wall and rail to be rough, and μ to be the co-efficient of statical friction for both. If the rod be placed in the position of equilibrium just investigated for the case of no friction, none will be called into play, for there will be no tendency to motion to be overcome. If the end A be brought lower and lower, more and more friction will be called into play to overcome the tendency of the rod to fall between the wall and the rail, until we come to a limiting position in which motion is about to commence. At that instant the friction at A is μ times the pressure on the wall, and acts *upwards*. That at D is μ times the pressure on the rod, and acts in the direction DB. Calling $CAD=\theta_1$ in this case, our three equations become

$$R_1+\mu S_1 \sin\theta_1 - S_1\cos\theta_1 \qquad =0, \qquad (1_1)$$

$$W-\mu R_1 - S_1\sin\theta_1 - \mu S_1\cos\theta_1=0, \qquad (2_1)$$

$$S_1 b - Wa\sin^2\theta_1 \qquad =0. \qquad (3_1)$$

The directions of both the friction-forces passing through A, neither appears in (3_1). This is why A is preferable to any other point about which to take moments.

By eliminating R_1 and S_1 from these equations we get

$$1-\frac{a}{b}\sin^3\theta_1=\mu\frac{a}{b}\sin^2\theta_1(2\cos\theta_1-\mu\sin\theta_1), \qquad (4_1)$$

from which θ_1 is to be found. Then S_1 is known from (3_1), and R_1 from either of the others.

If the end A be raised above the position of equilibrium without friction, the tendency is for the rod to fall *outside* the rail; more and more friction will be called into play, till the position of the rod (θ_2) is such that the friction reaches its greatest value, μ times the pressure. We may thus find another *limiting* position for stability; and between these the rod is in equilibrium in any position.

It is useful to observe that in this second case the direction of each friction is the opposite to that in the former, and the same equations will serve for both if we adopt the analytical artifice of changing the *sign* of μ. Thus for θ_2, by (4_1),

$$1-\frac{a}{b}\sin^3\theta_2=-\mu\frac{a}{b}\sin^2\theta_2(2\cos\theta_2+\mu\sin\theta_2). \qquad (4_2)$$

Ex. IV. A rectangular block lies on a rough horizontal plane, and is acted on by a horizontal force whose line of action is midway between two of the vertical sides. Find the magnitude of the force when just sufficient to produce motion, and whether the motion will be of the nature of *sliding* or *overturning*.

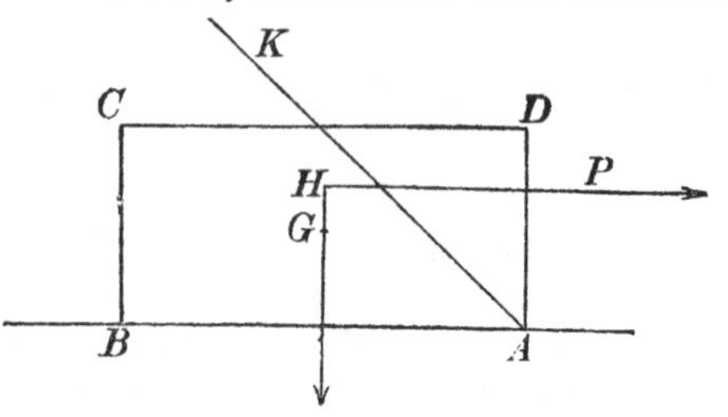

If the force P tends to overturn the body, it is evident that it will turn about the edge A, and therefore the pressure, R, of the plane and the friction, S, act at that edge. Our statical conditions are, of course,

$$R=W$$
$$S=P$$
$$Wb=Pa$$

where b is half the length of the solid, and a the distance of P from the plane. From these we have $S=\frac{b}{a}W$.

Now S cannot exceed μR, whence we must not have $\frac{b}{a}$ greater than μ, if it is to be possible to upset the body by a horizontal force in the line given for P.

A simple geometrical construction enables us to solve this and similar problems, and will be seen at once to be merely a graphic representation of the above process. Thus if we produce the directions of the applied force, and of the weight, to meet in H, and make at A the angle BAK whose co-tangent is the co-efficient of friction: there will be a tendency to upset, or not, according as H is above, or below, AK.

Ex. V. A mass, such as a gate, is supported by two rings, A and B, which pass loosely round a rough vertical post. In equilibrium, it is obvious that at A the part of the ring nearest the mass, and at B the farthest from it, will be in contact with the post. The pressures exerted on the rings, R and S, will evidently have the directions AC, CB, indicated in the diagram. If no other force besides gravity act on the mass, the line of action of its weight, W, must pass through the point C (§ 584). And it is obvious that, however small be the co-efficient of friction, provided there be friction at all, equilibrium is always possible if the distance of the centre of gravity from the post be great enough compared with the distance between the rings.

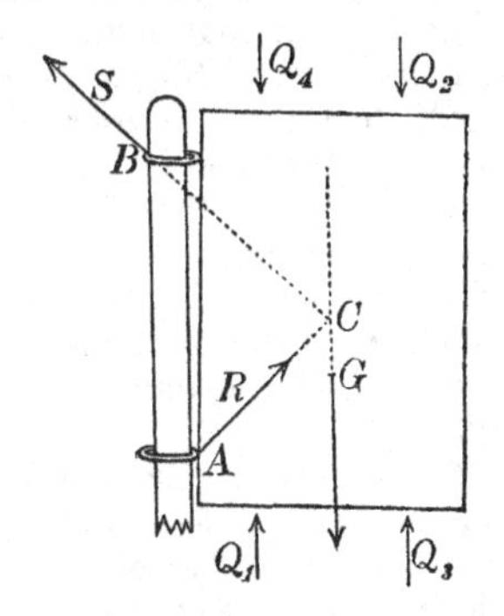

When the mass is just about to slide down, the full amount of friction is called into play, and the angles which R and S make with the horizon are each equal to the angle of repose. If we draw AC, BC according to this condition, then for equilibrium the centre of gravity G must not lie between the post and the vertical line through the point C thus determined. If, as in the figure, G lies in the vertical line through C, then a force applied upwards at Q_1, or downwards at Q_2, will remove the tendency to fall; but a force applied upwards at Q_3, or downwards at Q_4, will produce sliding at once.

A similar investigation is easily applied to the jamming of a sliding piece or drawer, and to the determination of the proper point of application of a force to move it. This we leave to the student.

As an illustration of the *use* of friction, let us consider a cord wound round a rough cylinder, and on the point of sliding.

Neglecting the weight of the cord, which is small in practice compared with the other forces; and considering a small portion AB of the cord, such that the tangents at its extremities include a very small angle θ; let T be the tension at one end,

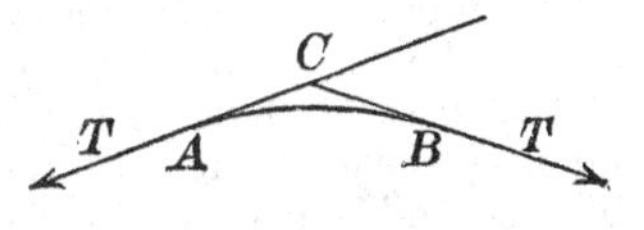

T' at the other, p the pressure of the rope on the cylinder per unit of length.

Then $p.AB = 2T\sin\frac{\theta}{2} = T\theta$ approximately. Also $\mu p.AB = T' - T$ when the rope is just about to slip, i. e.

$$\mu T\theta = T' - T,$$

or

$$T' = (1 + \mu\theta)T.$$

Hence, for equal small deflections, θ, of the rope, the tension increases in the geometrical ratio $(1 + \mu\theta) : 1$; and thus by a common theorem (compound interest payable every instant) we have $T = \epsilon^{\mu\alpha}T_0$, if T, T_0 be the tensions at the ends of a cord wrapped on a cylinder, when the external angle between the directions of the free ends is α. [ϵ is the base of Napier's Logarithms.] We thus obtain the singular result, that the *dimensions* of the cylinder have no influence on the increase of tension by friction, provided the cord is perfectly flexible.

593. Having thus briefly considered the equilibrium of a rigid body, we propose, before entering upon the subject of deformation of elastic solids, to consider certain intermediate cases, in each of which a particular assumption is made the basis of the investigation—thereby avoiding a very considerable amount of analytical difficulties.

594. Very excellent examples of this kind are furnished by the statics of a flexible and inextensible cord or chain, fixed at both ends, and subject to the action of any forces. The curve in which the chain hangs in any case may be called a *Catenary*, although the term is usually restricted to the case of a uniform chain acted on by gravity only.

595. We may consider separately the conditions of equilibrium of each element; or we may apply the general condition (§ 257) that the whole potential energy is a minimum, in the case of any conservative system of forces; or, especially when gravity is the only external force, we may consider the equilibrium of a *finite* portion of the chain treated for the time as a rigid body (§ 584).

596. The first of these methods gives immediately the three following equations of equilibrium, for the catenary in general:—

(1) The rate of variation of the tension per unit of length along the cord is equal to the tangential component of the applied force, per unit of length.

(2) The plane of curvature of the cord contains the normal component of the applied force, and the centre of curvature is on the opposite side of the arc from that towards which this force acts.

(3) The amount of the curvature is equal to the normal component of the applied force per unit of length at any point divided by the tension of the cord at the same point.

The first of these is simply the equation of equilibrium of an infinitely small element of the cord relatively to tangential motion. The second and third express that the component of the resultant of the tensions at the two ends of an infinitely small arc, along the normal through its middle point, is directly opposed and is equal to the normal applied force, and is equal to the whole amount of it on the arc. For the plane of the tangent lines in which those tensions act is (§ 12) the plane of curvature. And if θ be the angle between them (or the infinitely small angle by which the angle between their positive directions falls short of π), and T the arithmetical mean of their magnitudes, the component of their resultant along the line bisecting the angle between their positive directions is $2T\sin\frac{1}{2}\theta$, rigorously: or $T\theta$, since θ is infinitely small. Hence $T\theta = N\delta s$ if δs be the length of the arc, and $N\delta s$ the whole amount of normal force applied to it. But (§ 9) $\theta = \dfrac{\delta s}{\rho}$ if ρ be the radius of curvature; and therefore

$$\frac{1}{\rho} = \frac{N}{T},$$

which is the equation stated in words (3) above.

597. From (1) of § 596, we see that if the applied forces on any particle of the cord constitute a conservative system, and if any equal infinitely small lengths of the string experience the same force and in the same direction when brought into any one position by motion of the string, the difference of the tensions of the cord at any two points of it when hanging in equilibrium, is equal to the difference of the potential (§ 504) of the forces between the positions occupied by these points. Hence, whatever the position where the potential is reckoned zero, the tension of the string at any point is equal to the potential at the position occupied by it, with a constant added.

598. From § 596 it follows immediately that if a material particle of unit mass be carried along any catenary with a velocity, $\dot{s}$, equal to T, the numerical measure of the tension at any point, the force upon it by which this is done is in the same direction as the resultant of the applied force on the catenary at this point, and is equal to the amount of this force per unit of length, multiplied by T. For denoting by S the tangential, and (as before) by N the normal component of the applied force per unit of length at any point P of the catenary, we have, by § 596 (1), S for the rate of variation of $\dot{s}$ per unit length, and therefore $S\dot{s}$ for its variation per unit of time. That is to say,

$$\ddot{s} = S\dot{s} = ST,$$

or (§ 225) the tangential component force on the moving particle is equal to ST. Again, by § 596 (3),

$$NT = \frac{T^2}{\rho} = \frac{\dot{s}^2}{\rho},$$

or the centrifugal force of the moving particle in the circle of curvature of its path, that is to say, the normal component of the force on it, is equal to NT. And lastly, by (2) this force is in the same direction as N. We see therefore that the direction of the whole force on the moving particle is the same as that of the resultant of S and N; and its magnitude is T times the magnitude of this resultant.

599. Thus we see how, from the more familiar problems of the kinetics of a particle, we may immediately derive curious cases of catenaries. For instance: a particle under the influence of a constant force in parallel lines moves in a parabola with its axis vertical, with velocity at each point equal to that generated by the force acting through a space equal to its distance from the directrix. Hence, if z denote this distance, and f the constant force,

$$T=\sqrt{2fz}$$

in the allied parabolic catenary; and the force on the catenary is parallel to the axis, and is equal in amount per unit of length, to

$$\frac{f}{\sqrt{2fz}} \text{ or } \sqrt{\frac{f}{2z}}.$$

Hence if the force on the catenary be that of gravity, it must have its axis vertical (its vertex downwards of course for stable equilibrium) and its mass per unit length at any point must be inversely as the square root of the distance of this point above the directrix. From this it follows that the whole weight of any arc of it is proportional to its horizontal projection.

600. Or, if the question be, to find what force towards a given fixed point, will cause a cord to hang in any given plane curve with this point in its plane; it may be answered immediately from the solution of the corresponding problem in 'central forces.'

601. When a perfectly flexible string is stretched over a smooth surface, and acted on by no other force throughout its length than the resistance of this surface, it will, when in stable equilibrium, lie along a line of minimum length on the surface, between any two of its points. For (§ 584) its equilibrium can be neither disturbed nor rendered unstable by placing staples over it, through which it is free to slip, at any two points where it rests on the surface: and for the intermediate part the energy criterion of stable equilibrium is that just stated.

There being no tangential force on the string in this case, and the normal force upon it being along the normal to the surface, its osculating plane (§ 596) must cut the surface everywhere at right angles. These considerations, easily translated into pure geometry, establish the fundamental property of the geodetic lines on any surface. The analytical investigations of the question, when adapted to the case of a chain of *not* given length, stretched between two given points on a given smooth surface, constitute the direct analytical demonstration of this property.

In this case it is obvious that the tension of the string is the same at every point, and the pressure of the surface upon it is [§ 596 (3)] at each point proportional to the curvature of the string.

602. No real surface being perfectly smooth, a cord or chain may rest upon it when stretched over so great a length of a geodetic on a convex rigid body as to be not of minimum length between its extreme points: but practically, as in tying a cord round a ball, for permanent security it is necessary, by staples or otherwise, to constrain it from lateral slipping at successive points near enough to one another to make each free portion a true minimum on the surface.

603. A very important practical case is supplied by the consideration of a rope wound round a rough cylinder. We may suppose it to lie in a plane perpendicular to the axis, as we thus simplify the question very considerably without sensibly injuring the utility of the solution. To simplify still further, we shall suppose that no forces act on the rope but tensions and the reaction of the cylinder. In practice this is equivalent to the supposition that the tensions and reactions are very large compared with the weight of the rope or chain; which, however, is inadmissible in some important cases, especially such as occur in the application of the principle to brakes for laying submarine cables, to dynamometers, and to windlasses (or capstans with horizontal axes).

By § 592 we have $$T = T_0 \epsilon^{\mu\theta},$$

showing that, for equal successive amounts of integral curvature (§ 14), the tension of the rope augments in *geometrical* progression. To give an idea of the magnitudes involved, suppose $\mu = \cdot 5$, $\theta = \pi$, then

$$T = T_0 \epsilon^{\cdot 5\pi} = 481\, T_0 \text{ roughly.}$$

Hence if the rope be wound three times round the post or cylinder the ratio of the tensions of its ends, when motion is about to commence, is

$$5^6 : 1 \text{ or about } 15{,}000 : 1.$$

Thus we see how, by the aid of friction, one man may easily check the motion of the largest vessel, by the simple expedient of coiling a rope a few times round a post. This application of friction is of great importance in many other applications, especially to dynamometers (§§ 389, 390).

604. With the aid of the preceding investigations, the student may easily work out for himself the solution of the general problem of a cord under the action of any forces, and constrained by a rough surface; it is not of sufficient importance or interest to find a place here.

605. An elongated body of elastic material, which for brevity we shall generally call a *wire*, bent or twisted to any degree, subject only to the condition that the radius of curvature and the reciprocal of the twist are everywhere very great in comparison with the greatest transverse dimension, presents a case in which, as we shall see, the solution of the general equations for the equilibrium of an elastic solid is either obtainable in finite terms, or is reducible to comparatively easy questions agreeing in mathematical conditions with some of the most elementary problems of hydrokinetics, electricity, and thermal conduction. And it is only for the determination of certain constants depending on the section of the wire and the elastic quality of its substance, which measure its flexural and torsional rigidity, that the solutions of these problems are required. When the constants of flexure and torsion are known, as we shall now suppose them to be, whether from theoretical calculation or experiment, the investigation of the form and twist of any length of the wire, under the influence of any forces which do not produce a violation of the condition stated above, becomes a subject of mathematical analysis involving only such principles and formulae as those that constitute the theory of curvature (§§ 9-15) and twist in geometry or kinematics.

606. Before entering on the general theory of elastic solids, we shall therefore, according to the plan proposed in § 593, examine the dynamic properties and investigate the conditions of equilibrium of a perfectly elastic wire, without admitting any other condition or limitation of the circumstances than what is stated in § 605, and without assuming any special quality of isotropy, or of crystalline, fibrous or laminated structure in the substance.

607. Besides showing how the constants of flexural and torsional rigidity are to be determined theoretically from the form of the transverse section of the wire, and the proper data as to the elastic qualities of its substance, the complete theory simply indicates that, provided the conditional limit of deformation is not exceeded, the following laws will be obeyed by the wire under stress:—

Let the whole mutual action between the parts of the wire on the two sides of the cross section at any point (being of course the action of the matter infinitely near this plane on one side, upon the matter infinitely near it on the other side), be reduced to a single force through any point of the section and a single couple. Then—

I. The twist and curvature of the wire in the neighbourhood of this section are independent of the force, and depend solely on the couple.

II. The curvatures and rates of twist producible by any several couples separately, constitute, if geometrically compounded, the curvature and rate of twist which are actually produced by a mutual action equal to the resultant of those couples.

608. It may be added, although not necessary for our present purpose, that there is one determinate point in the cross section such that if it be chosen as the point to which the forces are transferred, a higher order of approximation is obtained for the fulfilment of these laws than if any other point of the section be taken. That point, which in the case of a wire of substance uniform through its cross section is the centre of inertia of the area of the section, we shall generally call the elastic centre, or the centre of elasticity, of the section. It has also the following important property:—The line of elastic centres, or, as we shall call it, the elastic central line, remains sensibly unchanged in length to whatever stress within our conditional limits (§ 605) the wire be subjected. The elongation or contraction produced by the neglected resultant force, if this is in such a direction as to produce any, will cause the line of *rigorously no elongation* to deviate only infinitesimally from the elastic central line, in any part of the wire finitely curved. It will, however, clearly cause there to be no line of *rigorously unchanged length*, in any straight part of the wire: but as the whole elongation would be infinitesimal in comparison with the effective actions with which we are concerned, this case constitutes no exception to the preceding statement.

609. In the most important practical cases, as we shall see later, those namely in which the substance is either 'isotropic,' which is sensibly the case with common metallic wires, or has an axis of elastic symmetry along the length of the piece, one of the three normal axes of torsion and flexure coincides with the length of the wire, and the two others are perpendicular to it; the first being an axis of pure torsion, and the two others axes of pure flexure. Thus opposing couples round the axis of the wire twist it simply without bending it; and opposing couples in either of the two principal planes of flexure, bend it into a circle. The unbent straight line of the wire, and the circular arcs into which it is bent by couples in the two principal planes of flexure, are what the three principal spirals of the general problem become in this case.

610. In the more particular case in which two principal rigidities against flexure are equal, every plane through the length of the wire is a principal plane of flexure, and the rigidity against flexure is equal in all. This is clearly the case with a common round wire, or rod, or with one of square section. It can be shown to be the case for a rod of isotropic material and of any form of normal section which is 'kinetically symmetrical' (§ 239) round all axes in its plane through its centre of inertia.

611. In this case, if one end of the rod or wire be held fixed, and a couple be applied in any plane to the other end, a uniform spiral form will be produced round an axis perpendicular to the plane of the couple. The lines of the substance parallel to the axis of the spiral are not, however, parallel to their original positions, as in each of the three principal spirals of the general problem: and lines traced along the surface of the wire parallel to its length when straight, become as it were secondary spirals, circling round the main spiral formed by the central line of the deformed wire, instead of being all spirals of equal step, as in each one of the principal spirals of the general problem. Lastly, in the present case, if we suppose the normal section of the wire to be circular, and trace uniform spirals along its surface when deformed in the manner supposed (two of which, for instance, are the lines along which it is touched by the inscribed and the circumscribed cylinder), these lines do not become straight, but become spirals laid on as it were round the wire, when it is allowed to take its natural straight and untwisted condition.

612. A wire of equal flexibility in all directions may clearly be held in any specified spiral form, and twisted to any stated degree, by a determinate force and couple applied at one end, the other end being held fixed. The direction of the force must be parallel to the axis of the spiral, and, with the couple, must constitute a system of which this line is (§ 579) the *central axis:* since otherwise there could not be the same system of balancing forces in every normal section of the spiral. All this may be seen clearly by supposing the wire to be first brought by any means to the specified condition of strain; then to have rigid planes rigidly attached to its two ends perpendicular to its axis, and these planes to be rigidly connected by a bar lying in this line. The spiral wire now left to itself cannot but be in equilibrium: although if it be too long (according to its form and degree of twist) the equilibrium may be unstable. The force along the central axis, and the couple, are to be determined by the condition that, when the force is transferred after Poinsot's manner to the elastic centre of any normal section, they give two couples together equivalent to the elastic couples of flexure and torsion.

613. A wire of equal flexibility in all directions may be held in any stated spiral form by a simple force along its axis between rigid pieces rigidly attached to its two ends, provided that, along with its spiral form a certain degree of twist be given to it. The force is determined by the condition that its moment round the perpendicular through any point of the spiral to its osculating plane at that point, must be equal and opposite to the elastic unbending couple. The degree of twist is that due (by the simple equation of torsion) to the moment of the force thus determined, round the tangent at any point of the spiral. The direction of the force being, according to the preceding condition, such as to press together the ends of the spiral, the direction of the twist in the wire is opposite to that of the tortuosity (§ 13) of its central curve.

614. The principles with which we have just been occupied are immediately applicable to the theory of spiral springs; and we shall therefore make a short digression on this curious and important practical subject before completing our investigation of elastic curves.

A common spiral spring consists of a uniform wire shaped permanently to have, when unstrained, the form of a regular helix, with the principal axes of flexure and torsion everywhere similarly situated relatively to the curve. When used in the proper manner, it is acted on, through arms or plates rigidly attached to its ends, by forces such that its form as altered by them is still a regular helix. This condition is obviously fulfilled if (one terminal being held fixed) an infinitely small force and infinitely small couple be applied to the other terminal along the axis, and in a plane perpendicular to it, and if the force and couple be increased to any degree, and always kept along and in the plane perpendicular to the axis of the altered spiral. It would, however, introduce useless complication to work out the details of the problem except for the case (§ 609) in which one of the principal axes coincides with the tangent to the central line, and is therefore an axis of pure torsion, as spiral springs in practice always belong to this case. On the other hand, a very interesting complication occurs if we suppose (what is easily realized in practice, though to be avoided if merely a good spring is desired) the normal section of the wire to be of such a figure, and so situated relatively to the spiral, that the planes of greatest and least flexural rigidity are oblique to the tangent plane of the cylinder. Such a spring when acted on in the regular manner at its ends must experience a certain degree of turning through its whole length round its elastic central curve in order that the flexural couple developed may be, as we shall immediately see it must be, precisely in the osculating plane of the altered spiral. All that is interesting in this very curious effect is illustrated later in full detail (§ 624 of our larger work) in the case of an open circular arc altered by a couple in its own plane, into a circular arc of greater or less radius; and for brevity and simplicity we shall confine the detailed investigation of spiral springs on which we now enter, to the cases in which either the wire is of equal flexural rigidity in all directions, or the two principal planes of (greatest and least or least and greatest) flexural rigidity coincide respectively with the tangent plane to the cylinder, and the normal plane touching the central curve of the wire, at any point.

615. The axial force, on the movable terminal of the spring, transferred according to Poinsot to any point in the elastic central curve, gives a couple in the plane through that point and the axis of the spiral. The resultant of this and the couple which we suppose applied to the terminal in the plane perpendicular to the axis of the spiral is the effective bending and twisting couple: and as it is in a plane perpendicular to the tangent plane to the cylinder, the component of it to which bending is due must be also perpendicular to this plane, and therefore is in the osculating plane of the spiral. This component couple therefore simply maintains a curvature different from the natural curvature of the wire, and the other, that is, the couple in the plane normal to the central curve, pure torsion. The equations of equilibrium merely express this in mathematical language.

616. The potential energy of the strained spring is

$$\tfrac{1}{2}[B(\varpi-\varpi_0)^2+A\tau^2]l,$$

if A denote the torsional rigidity, B the flexural rigidity in the plane of curvature, ϖ and ϖ_0 the strained and unstrained curvatures, and τ

the torsion of the wire in the strained condition, the torsion being reckoned as zero in the unstrained condition. The axial force, and the couple, required to hold the spring to any given length reckoned along the axis of the spiral, and to any given angle between planes through its ends and the axes, are of course (§ 244) equal to the rates of variation of the potential energy, per unit of variation of these co-ordinates respectively. It must be carefully remarked, however, that, if the terminal rigidly attached to one end of the spring be held fast, so as to fix the tangent at this end, and the motion of the other terminal be so regulated as to keep the figure of the intermediate spring always truly spiral, this motion will be somewhat complicated; as the radius of the cylinder, the inclination of the axis of the spiral to the fixed direction of the tangent at the fixed end, and the position of the point in the axis in which it is cut by the plane perpendicular to it through the fixed end of the spring, all vary as the spring changes in figure. The *effective components* of any infinitely small motion of the movable terminal are its component translation along, and rotation round, the instantaneous position of the axis of the spiral [two degrees of freedom], along with which it will generally have an infinitely small translation in some direction and rotation round some line, each perpendicular to this axis, and determined from the two degrees of arbitrary motion, by the condition that the curve remains a true spiral.

617. In the practical use of spiral springs, this condition is not rigorously fulfilled: but, instead, one of two plans is generally followed:—(1) Force, without any couple, is applied pulling out or pressing together two definite points of the two terminals, each as nearly as may be in the axis of the unstrained spiral; or (2) One terminal being held fixed, the other is allowed to slide, without any turning, in a fixed direction, being as nearly as may be the direction of the axis of the spiral when unstrained. The preceding investigation is applicable to the infinitely small displacement in either case: the couple being put equal to zero for case (1), and the instantaneous rotatory motion round the axis of the spiral equal to zero for case (2).

618. In a spiral spring of infinitely small inclination to the plane perpendicular to its axis, the displacement produced in the movable terminal by a force applied to it in the axis of the spiral is a simple rectilineal translation in the direction of the axis, and is equal to the length of the circular arc through which an equal force carries one end of a rigid arm or crank equal in length to the radius of the cylinder, attached perpendicularly to one end of the wire of the spring supposed straightened and held with the other end absolutely fixed, and the end which bears the crank, free to turn in a collar. This statement is due to J. Thomson[1], who showed that in pulling out a spiral spring of infinitely small inclination the action exercised and

[1] *Camb. and Dub. Math. Jour.* 1848.

the elastic quality used are the same as in a torsion-balance with the same wire straightened (§ 386). This theory is, as he proved experimentally, sufficiently approximate for most practical applications; spiral springs, as commonly made and used, being of very small inclination. There is no difficulty in finding the requisite correction, for the actual inclination in any case. The fundamental principle that spiral springs act chiefly by torsion seems to have been first discovered by Binet in 1814[1].

619. Returning to the case of a uniform wire straight and untwisted (that is, cylindrical or prismatic) when free from stress; let us suppose one end to be held fixed in a given direction, and no other force from without to influence it except that of a rigid frame attached to its other end acted on by a force, R, in a given line, AB, and a couple, G, in a plane perpendicular to this line. The form and twist it will have when in equilibrium are determined by the condition that the torsion and flexure at any point, P, of its length are those due to the couple G compounded with the couple obtained by bringing R to P.

620. Kirchhoff has made a very remarkable comparison between the static problem of bending and twisting a wire, and the kinetic problem of the rotation of a rigid body. We can give here but one instance, the simplest of all—the *Elastic Curve* of James Bernoulli, and the common pendulum. A uniform straight wire, either equally flexible in all planes through its length, or having its directions of maximum and minimum flexural rigidity in two planes through its whole length, is acted on by a force and couple in one of these planes, applied either directly to one end, or by means of an arm rigidly attached to it, the other end being held fast. The force and couple may, of course (§ 568), be reduced to a single force, the extreme case of a couple being mathematically included as an infinitely small force at an infinitely great distance. To avoid any restriction of the problem, we must suppose this force applied to an arm rigidly attached to the wire, although in any case in which the line of the force cuts the wire, the force may be applied directly at the point of intersection, without altering the circumstances of the wire between this point and the fixed end. The wire will, in these circumstances, be bent into a curve lying throughout in the plane through its fixed end and the line of the force, and (§ 609) its curvatures at different points will, as was first shown by James Bernoulli, be simply as their distances from this line. The curve fulfilling this condition has clearly just two independent parameters, of which one is conveniently regarded as the mean proportional, a, between the radius of curvature at any point and its distance from the line of force, and the other, the maximum distance, b, of the wire from the line of force. By choosing any value for each of these parameters it is easy to trace the corresponding curve with a very high approximation to accuracy, by commencing with a small circular arc touching at one

[1] St. Venant, *Comptes Rendus*, Sept. 1864.

extremity a straight line at the given maximum distance from the line of force, and continuing by small circular arcs, with the proper increasing radii, according to the diminishing distances of their middle points from the line of force. The annexed diagrams are, however, not so drawn, but are simply traced from the forms actually assumed by a flat steel spring, of small enough breadth not to be much disturbed by tortuosity in the cases in which different parts of it cross one another. The mode of application of the force is sufficiently explained by the indications in the diagram.

621. As we choose particularly the common pendulum for the corresponding kinetic problem, the force acting on the rigid body in the comparison must be that of gravity in the vertical through its centre of gravity. It is convenient, accordingly, not to take *unity* as the velocity for the point of comparison along the bent wire, but the velocity which gravity would generate in a body falling through a height equal to half the constant, a, of § 620: and this constant, a, will then be the length of the isochronous simple pendulum. Thus if an elastic curve be held with its line of force vertical, and if a point, P, be moved along it with a constant velocity equal to $\sqrt{ga}$, (a denoting the mean proportional between the radius of curvature at any point, and its distance from the line of force,) the tangent at P will keep always parallel to a simple pendulum, of length a, placed at any instant parallel to it, and projected with the same angular velocity. Diagrams 1 to 5, correspond to *vibrations* of the pendulum. Diagram 6 corresponds to the case in which the pendulum would just reach its position of unstable equilibrium in an infinite time. Diagram 7 corresponds to cases in which the pendulum flies round continuously in one direction, with periodically increasing and diminishing velocity. The extreme case, of the circular elastic curve, corresponds to an infinitely long pendulum flying round with finite angular velocity, which of course experiences only infinitely small variation in the course of the revolution. A conclusion worthy of remark is, that the rectification of the elastic curve is the same analytical problem as finding the time occupied by a pendulum in describing any given angle.

622. For the simple and important case of a naturally straight wire, acted on by a distribution of force, but not of couple, through its length, the condition fulfilled at a perfectly free end, acted on by neither force nor couple, is that the curvature is zero at the end, and its rate of variation from zero, per unit of length from the end, is, at the end, zero. In other words, the curvatures at points infinitely near the end are as the squares of their distances from the end in general (or, as some higher power of these distances, in singular cases). The same statements hold for the *change* of curvature produced by the stress, if the unstrained wire is not straight, but the other circumstances the same as those just specified.

623. As a very simple example of the equilibrium of a wire subject to forces through its length, let us suppose the natural form to

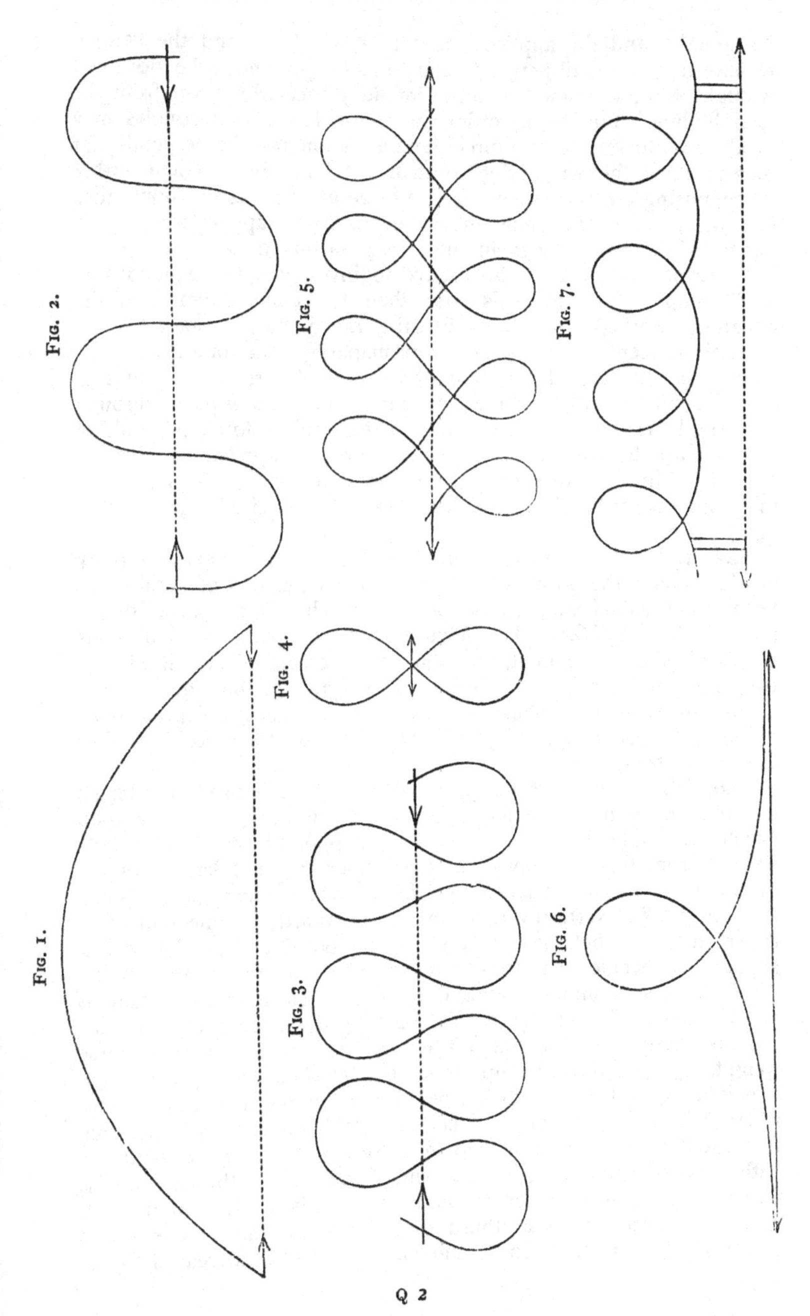
Fig. 1.
Fig. 2.
Fig. 3.
Fig. 4.
Fig. 5.
Fig. 6.
Fig. 7.

be straight, and the applied forces to be in lines, and the couples to have their axes, all perpendicular to its length, and to be not great enough to produce more than an infinitely small deviation from the straight line. Further, in order that these forces and couples may produce no torsion, let the three flexure-torsion axes be perpendicular to and along the wire. But we shall not limit the problem further by supposing the section of the wire to be uniform, as we should thus exclude some of the most important practical applications, as to beams of balances, levers in machinery, beams in architecture and engineering. It is more instructive to investigate the equations of equilibrium directly for this case than to deduce them from the equations worked out above for the much more comprehensive general problem. The particular principle for the present case is simply that the rate of variation of the rate of variation, per unit of length along the wire, of the bending couple in any plane through the length, is equal, at any point, to the applied force per unit of length, with the simple rate of variation of the applied couple subtracted. This, together with the direct equations (§ 609) between the component bending couples, gives the required equations of equilibrium.

624. If the directions of maximum and minimum flexural rigidity lie throughout the wire in two planes, the equations of equilibrium become simplified when these planes are chosen as planes of reference, XOY, XOZ. The flexure in either plane then depends simply on the forces in it, and thus the problem divides itself into two quite independent problems of integrating the equations of flexure in the two principal planes, and so finding the projections of the curve on two fixed planes agreeing with their position when the rod is straight.

625. When a uniform bar, beam, or plank is balanced on a single trestle at its middle, the droop of its ends is only $\frac{3}{5}$ of the droop which its middle has when the bar is supported on trestles at its ends. From this it follows that the former is $\frac{3}{8}$ and latter $\frac{5}{8}$ of the droop or elevation produced by a force equal to half the weight of the bar, applied vertically downwards or upwards to one end of it, if the middle is held fast in a horizontal position. For let us first suppose the whole to rest on a trestle under its middle, and let two trestles be placed under its ends and gradually raised till the pressure is entirely taken off from the middle. During this operation the middle remains fixed and horizontal, while a force increasing to half the weight, applied vertically upwards on each end, raises it through a height equal to the sum of the droops in the two cases above referred to. This result is of course proved directly by comparing the absolute values of the droop in those two cases as found above, with the deflection from the tangent at the end of the cord in the elastic curve, § 620, which is cut by the cords at right angles. It may be stated otherwise thus: the droop of the middle of a uniform beam resting on trestles at its ends is increased in the ratio of 5 to 13 by laying a mass equal in weight to itself on its middle: and, if the beam is hung by its middle, the droop of the ends is increased in the ratio of 3 to 11 by hanging on each of them a mass equal to half the weight of the beam.

626. The important practical problem of finding the distribution of the weight of a solid on points supporting it, when more than two of these are in one vertical plane, or when there are more than three altogether, which (§ 588) is indeterminate[1] if the solid is perfectly rigid, may be completely solved for a uniform elastic beam, naturally straight, resting on three or more points in rigorously fixed positions all nearly in one horizontal line, by means of the preceding results.

If there are i points of support, the $i-1$ parts of the rod between them in order and the two end parts will form $i+1$ curves expressed by distinct algebraic equations, each involving four arbitrary constants. For determining these constants we have $4i+4$ equations in all, expressing the following conditions:—

I. The ordinates of the inner ends of the projecting parts of the rod, and of the two ends of each intermediate part, are respectively equal to the given ordinates of the corresponding points of support [$2i$ equations].

II. The curves on the two sides of each support have coincident tangents and equal curvatures at the point of transition from one to the other [$2i$ equations].

III. The curvature and its rate of variation per unit of length along the rod, vanish at each end [4 equations].

Thus the equation of each part of the curve is completely determined: and by means of it we find the shearing force in any normal section. The difference between these in the neighbouring portions of the rod on the two sides of a point of support, is of course equal to the pressure on this point.

627. The solution for the case of this problem in which two of the points of support are at the ends, and the third midway between them either exactly in the line joining them, or at any given very small distance above or below it, is found at once, without analytical work, from the particular results stated in § 625. Thus if we suppose the beam, after being first supported wholly by trestles at its ends, to be gradually pressed up by a trestle under its middle, it will bear a force simply proportional to the space through which it is raised from the zero point, until all the weight is taken off the ends, and borne by the middle. The whole distance through which the middle rises during this process is, as we found, $\frac{gw}{B}\cdot\frac{8l^4}{16.24}$; and this whole elevation is $\frac{8}{5}$ of the droop of the middle in the

[1] It need scarcely be remarked that indeterminateness does not exist in nature. How it may occur in the problems of abstract dynamics, and is obviated by taking something more of the properties of matter into account, is instructively illustrated by the circumstances referred to in the text.

first position. If therefore, for instance, the middle trestle be fixed exactly in the line joining those under the ends, it will bear $\frac{5}{8}$ of the whole weight, and leave $\frac{3}{16}$ to be borne by each end. And if the middle trestle be lowered from the line joining the end ones by $\frac{7}{15}$ of the space through which it would have to be lowered to relieve itself of all pressure, it will bear just $\frac{1}{3}$ of the whole weight, and leave the other two-thirds to be equally borne by the two ends.

628. A wire of equal flexibility in all directions, and straight when freed from stress, offers, when bent and twisted in any manner whatever, not the slightest resistance to being turned round its elastic central curve, as its conditions of equilibrium are in no way affected by turning the whole wire thus equally throughout its length. The useful application of this principle, to the maintenance of equal angular motion in two bodies rotating round different axes, is rendered somewhat difficult in practice by the necessity of a perfect attachment and adjustment of each end of the wire, so as to have the tangent to its elastic central curve exactly in line with the axis of rotation. But if this condition is rigorously fulfilled, and the wire is of exactly equal flexibility in every direction, and exactly straight when free from stress, it will give, against any constant resistance, an accurately uniform motion from one to another of two bodies rotating round axes which may be inclined to one another at any angle, and need not be in one plane. If they are in one plane, if there is no resistance to the rotatory motion, and if the action of gravity on the wire is insensible, it will take some of the varieties of form (§ 620) of the plane elastic curve of James Bernoulli. But however much it is altered from this, whether by the axis not being in one plane, or by the torsion accompanying the transmission of a couple from one shaft to the other, and necessarily, when the axes are in one plane, twisting the wire out of it, or by gravity, the elastic central curve will remain at rest, the wire in every normal section rotating round it with uniform angular velocity, equal to that of each of the two bodies which it connects. Under Properties of Matter, we shall see, as indeed may be judged at once from the performances of the vibrating spring of a chronometer for twenty years, that imperfection in the elasticity of a metal wire does not exist to any such degree as to prevent the practical application of this principle, even in mechanism required to be durable.

It is right to remark, however, that if the rotation be too rapid, the equilibrium of the wire rotating round its unchanged elastic central curve may become unstable, as is immediately discovered by experiments (leading to very curious phenomena), when, as is often done in illustrating the kinetics of ordinary rotation, a rigid body is hung by a steel wire, the upper end of which is kept turning rapidily.

629. The definitions and investigations regarding strain, §§ 135-161, constitute a kinematical introduction to the theory of elastic solids. We must now, in commencing the elementary dynamics of the subject, consider the forces called into play through the interior of a solid when brought into a condition of strain. We adopt, from Rankine[1], the term *stress* to designate such forces, as distingushed from strain defined (§ 135) to express the merely geometrical idea of a change of volume or figure.

630. When through any space in a body under the action of force, the mutual force between the portions of matter on the two sides of any plane area is equal and parallel to the mutual force across any equal, similar, and parallel plane area, the stress is said to be homogeneous through that space. In other words, the stress experienced by the matter is homogeneous through any space if all equal, similar, and similarly turned portions of matter within this space are similarly and equally influenced by force.

631. To be able to find the distribution of force over the surface of any portion of matter homogeneously stressed, we must know the direction, and the amount per unit area, of the force across a plane area cutting through it in any direction. Now if we know this for any three planes, in three different directions, we can find it for a plane in any direction as we see in a moment by considering what is necessary for the equilibrium of a tetrahedron of the substance. The resultant force on one of its sides must be equal and opposite to the resultant of the forces on the three others, which is known if these sides are parallel to the three planes for each of which the force is given.

632. Hence the stress, in a body homogeneously stressed, is completely specified when the direction, and the amount per unit area, of the force on each of three distinct planes is given. It is, in the analytical treatment of the subject, generally convenient to take these planes of reference at right angles to one another. But we should immediately fall into error did we not remark that the specification here indicated consists not of nine but in reality only of six, independent elements. For if the equilibrating forces on the six faces of a cube be each resolved into three components parallel to its three edges, OX, OY, OZ, we have in all 18 forces; of which each pair acting perpendicularly on a pair of opposite faces, being equal and directly opposed, balance one another. The twelve tangential components that remain constitute three pairs of couples having their axes in the direction of the three edges, each of which must separately be in equilibrium. The diagram shows the pair of equilibrating couples having OY for axis; from the consideration of which we infer that the

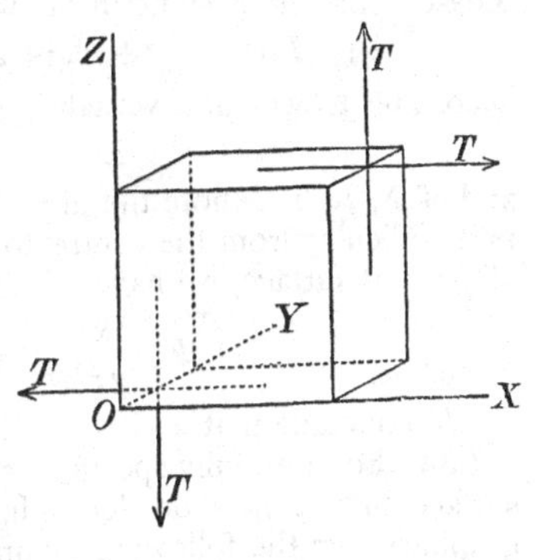

[1] *Cambridge and Dublin Mathematical Journal*, 1850.

forces on the faces (zy), parallel to OZ, are equal to the forces on the faces (yx), parallel to OX. Similarly, we see that the forces on the faces (yx), parallel to OY, are equal to those on the faces (xz), parallel to OZ; and that the forces on (xz), parallel to OX, are equal to those on (zy), parallel to OY.

633. Thus, any three rectangular planes of reference being chosen, we may take six elements thus, to specify a stress: P, Q, R the normal components of the forces on these planes; and S, T, U the tangential components, respectively perpendicular to OX, of the forces on the two planes meeting in OX, perpendicular to OY, of the forces on the planes meeting in OY, and perpendicular to OZ, of the forces on the planes meeting in OZ; each of the six forces being reckoned, per unit of area. A normal component will be reckoned as positive when it is a traction tending to separate the portions of matter on the two sides of its plane. P, Q, R are sometimes called simple longitudinal stresses, and S, T, U simple shearing stresses.

From these data, to find in the manner explained in § 631, the force on any plane, specified by l, m, n, the direction-cosines of its normal; let such a plane cut OX, OY, OZ in the three points X, Y, Z. Then, if the area XYZ be denoted for a moment by A, the areas YOZ, ZOX, XOY, being its projections on the three rectangular planes, will be respectively equal to Al, Am, An. Hence, for the equilibrium of the tetrahedron of matter bounded by those four triangles, we have, if F, G, H denote the components of the force experienced by the first of them, XYZ, per unit of its area,

$$F.A = P.lA + U.mA + T.nA,$$

and the two symmetrical equations for the components parallel to OY and OZ. Hence, dividing by A, we conclude

$$\left.\begin{aligned} F &= Pl + Um + Tn \\ G &= Ul + Qm + Sn \\ H &= Tl + Sm + Rn \end{aligned}\right\}. \tag{1}$$

These expressions stand in the well-known relation to the ellipsoid

$$Px^2 + Qy^2 + Rz^2 + 2(Syz + Tzx + Uxy) = 1, \tag{2}$$

according to which, if we take

$$x = lr, y = mr, z = nr,$$

and if λ, μ, ν denote the direction-cosines and p the length of the perpendicular from the centre to the tangent plane at (x, y, z) of the ellipsoidal surface, we have

$$F = \frac{\lambda}{pr}, \quad G = \frac{\mu}{pr}, \quad H = \frac{\nu}{pr}.$$

We conclude that

634. For any fully specified state of stress in a solid, a quadratic surface may always be determined, which shall represent the stress graphically in the following manner:—

To find the direction and the amount per unit area, of the force acting across any plane in the solid, draw a line perpendicular to this plane from the centre of the quadratic to its surface. The required force will be equal to the reciprocal of the product of the length of this line into the perpendicular from the centre to the tangent plane at the point of intersection, and will be perpendicular to the latter plane.

635. From this it follows that for any stress whatever there are three determinate planes at right angles to one another such that the force acting in the solid across each of them is precisely perpendicular to it. These planes are called the principal or normal planes of the stress; the forces upon them, per unit area,—its principal or normal tractions; and the lines perpendicular to them,—its principal or normal axes, or simply its axes. The three principal semi-diameters of the quadratic surface are equal to the reciprocals of the square roots of the normal tractions. If, however, in any case each of the three normal tractions is negative, it will be convenient to reckon them rather as *positive pressures;* the reciprocals of the square roots of which will be the semi-axes of a real stress-ellipsoid representing the distribution of force in the manner explained above, with pressure substituted throughout for traction.

636. When the three normal tractions are all of one sign, the stress-quadratic is an ellipsoid; the cases of an ellipsoid of revolution and a sphere being included, as those in which two, or all three, are equal. When one of the three is negative and the two others positive, the surface is a hyperboloid of one sheet. When one of the normal tractions is positive and the two others negative, the surface is a hyperboloid of two sheets.

637. When one of the three principal tractions vanishes, while the other two are finite, the stress-quadratic becomes a cylinder, circular, elliptic, or hyperbolic, according as the other two are equal, unequal of one sign, or of contrary signs. When two of the three vanish, the quadratic becomes two planes; and the stress in this case is (§ 633) called a simple longitudinal stress. The theory of principal planes, and normal tractions just stated (§ 635), is then equivalent to saying that any stress whatever may be regarded as made up of three simple longitudinal stresses in three rectangular directions. The geometrical interpretations are obvious in all these cases.

638. The composition of stresses is of course to be effected by adding the component tractions thus:—If $(P_1, Q_1, R_1, S_1, T_1, U_1)$, $(P_2, Q_2, R_2, S_2, T_2, U_2)$, etc., denote, according to § 633, any given set of stresses acting simultaneously in a substance, their joint effect is the same as that of a simple resultant stress of which the specification in corresponding terms is $(\Sigma P, \Sigma Q, \Sigma R, \Sigma S, \Sigma T, \Sigma U)$.

639. Each of the statements that have now been made (§§ 630, 638) regarding stresses, is applicable to *infinitely small* strains, if for traction perpendicular to any plane, reckoned per unit of its area, we substitute *elongation,* in the lines of the traction, reckoned per unit of length; and for *half the tangential traction* parallel to any direction, *shear* in the same direction, reckoned in the manner explained in § 154. The student will find it a useful exercise to study in detail this transference of each one of those statements, and to justify it by modifying in the proper manner the results of § 150, 151, 152, 153, 154, 161, to adapt them to infinitely small strains. It must be remarked that the strain-quadratic thus formed according to the rule of § 634, which may have any of the varieties of character mentioned in §§ 636, 637, is not the same as the strain-ellipsoid of § 141, which is always essentially an ellipsoid, and which, for an infinitely small strain, differs infinitely little from a sphere.

The comparison of § 151, with the result of § 632 regarding tangential tractions is particularly interesting and important.

640. The following tabular synopsis of the meaning of the elements constituting the corresponding rectangular specifications of a strain and stress explained in preceding sections, will be found convenient:—

Components of the strain.	Components of the stress.	Planes; of which relative motion, or across which force is reckoned.	Direction of relative motion or of force.
e	P	yz	x
f	Q	zx	y
g	R	xy	z
a	S	$\begin{cases} yx \\ zx \end{cases}$	y z
b	T	$\begin{cases} zy \\ xy \end{cases}$	z x
c	U	$\begin{cases} xz \\ yz \end{cases}$	x y

641. If a unit cube of matter under any stress (P, Q, R, S, T, U) experience the infinitely small simple longitudinal strain e alone, the work done on it will be Pe; since, of the component forces, P, U, T parallel to OX, U and T do no work in virtue of this strain. Similarly, Qf, Rg are the works done if, the same stress acting, the simple longitudinal strains f or g are experienced, either alone. Again, if the cube experiences a simple shear, a, whether we regard it (§ 151) as a differential sliding of the planes yx, parallel to y, or of the planes zx, parallel to z, we see that the work done is Sa: and similarly, Tb if the strain is simply a shear b, parallel to OZ, of planes zy, or parallel to OX, of planes xy: and Uc if the strain is a shear c, parallel to OX, of planes xz, or parallel to OY, of planes yz. Hence the whole work done by the stress (P, Q, R, S, T, U) on a unit cube taking the strain (e, f, g, a, b, c), is

$$Pe + Qf + Rg + Sa + Tb + Uc. \qquad (3)$$

It is to be remarked that, inasmuch as the action called a stress is a system of forces which balance one another if the portion of matter experiencing it is rigid, it cannot do any work when the matter moves in any way without change of shape: and therefore no amount of translation or rotation of the cube taking place along with the strain can render the amount of work done different from that just found.

If the side of the cube be of any length p, instead of unity, each force will be p^2 times, and each relative displacement p times, and, therefore, the work done p^3 times the respective amounts reckoned above. Hence a body of any shape, and of cubic content C, subjected throughout to a uniform stress (P, Q, R, S, T, U) while taking uniformly throughout a strain (e, f, g, a, b, c), experiences an amount of work equal to

$$(Pe+Qf+Rg+Sa+Tb+Uc)C. \tag{4}$$

It is to be remarked that this is necessarily equal to the work done on the bounding surface of the body by forces applied to it from without. For the work done on any portion of matter within the body is simply that done on its surface by the matter touching it all round, as no force acts at a distance from without on the interior substance. Hence if we imagine the whole body divided into any number of parts, each of any shape, the sum of the work done on all these parts is, by the disappearance of equal positive and negative terms expressing the portions of the work done on each part by the contiguous parts on all its sides, and spent by these other parts in this action, reduced to the integral amount of work done by force from without applied all round the outer surface.

642. If, now, we suppose the body to yield to a stress (P, Q, R, S, T, U), and to oppose this stress only with its innate resistance to change of shape, the differential equation of work done will [by (4) with de, df, etc., substituted for e, f, etc.] be

$$dw=Pde+Qdf+Rdg+Sda+Tdb+Udc. \tag{5}$$

If w denote the whole amount of work done per unit of volume in any part of the body while the substance in this part experiences a strain (e, f, g, a, b, c) from some initial state regarded as a state of no strain. This equation, as we shall see later, under Properties of Matter, expresses the work done in a natural fluid, by distorting stress (or difference of pressure in different directions) working against its innate viscosity; and w is then, according to Joule's discovery, the dynamic value of the heat generated in the process. The equation may also be applied to express the work done in straining an imperfectly elastic solid, or an elastic solid of which the temperature varies during the process. In all such applications the stress will depend partly on the speed of the straining motion, or on the varying temperature, and not at all, or not solely, on the state of strain at any moment, and the system will not be dynamically conservative.

643. *Definition.*—A perfectly elastic body is a body which, when brought to any one state of strain, requires at all times the same stress to hold it in this state; however long it be kept strained, or however rapidly its state be altered from any other strain, or from no strain, to the strain in question. Here, according to our plan

(§§ 396, 401) for Abstract Dynamics, we ignore variation of temperature in the body. If, however, we add a condition of absolutely no variation of temperature, or of recurrence to one specified temperature after changes of strain, we have a definition of that property of perfect elasticity towards which highly elastic bodies in nature approximate; and which is rigorously fulfilled by all fluids, and may be so by some real solids, as homogeneous crystals. But inasmuch as the elastic reaction of every kind of body against strain varies with varying temperature, and (a thermodynamic consequence of this, as we shall see later) any increase or diminution of strain in an elastic body is necessarily [1] accompanied by a change of temperature; even a perfectly elastic body could not, in passing through different strains, act as a rigorously conservative system, but on the contrary, must give rise to dissipation of energy in consequence of the conduction or radiation of heat induced by these changes of temperature.

But by making the changes of strain quickly enough to prevent any sensible equalization of temperature by conduction or radiation (as, for instance, Stokes has shown, is done in sound of musical notes travelling through air); or by making them slowly enough to allow the temperature to be maintained sensibly constant [2] by proper appliances; any highly elastic, or perfectly elastic body in nature may be got to act very nearly as a conservative system.

644. In nature, therefore, the integral amount, w, of work defined as above, is for a perfectly elastic body, independent (§ 246) of the series of configurations, or states of strain, through which it may have been brought from the first to the second of the specified conditions, provided it has not been allowed to change sensibly in temperature during the process.

When the whole amount of strain is infinitely small, and the stress-components are therefore all altered in the same ratio as the strain-components if these are altered all in any one ratio; w must be a homogeneous quadratic function of the six variables e, f, g, a, b, c, which, if we denote by (e, e), $(f,f)\ldots(e,f)\ldots$ constants depending on the quality of the substance and on the directions chosen for the axes of co-ordinates, we may write as follows:—

$$\begin{aligned} w=\tfrac{1}{2}\{&(e,e)e^2+(f,f)f^2+(g,g)g^2+(a,a)a^2+(b,b)b^2+(c,c)c^2\\ &+2(e,f)ef+2(e,g)eg+2(e,a)ea+2(e,b)eb+2(e,c)ec\\ &\qquad+2(f,g)fg+2(f,a)fa+2(f,b)fb+2(f,c)fc\\ &\qquad\qquad+2(g,a)ga+2(g,b)gb+2(g,c)gc\\ &\qquad\qquad\qquad+2(a,b)ab+2(a,c)ac\\ &\qquad\qquad\qquad\qquad+2(b,c)bc\} \end{aligned}$$

The 21 co-efficients (e, e), $(f,f)\ldots(b, c)$, in this expression constitute the 21 'co-efficients of elasticity,' which Green first showed to be proper and essential for a complete theory of the dynamics of an elastic solid subjected to infinitely small strains. The only condition

[1] 'On the Thermoelastic and Thermomagnetic Properties of Matter' (W. Thomson). *Quarterly Journal of Mathematics*, April 1857.

[2] *Ibid.*

that can be theoretically imposed upon these co-efficients is that they must not permit w to become negative for any values, positive or negative, of the strain-components $e, f, \ldots c$. Under Properties of Matter, we shall see that a false theory (Boscovich's), falsely worked out by mathematicians, has led to relations among the co-efficients of elasticity which experiment has proved to be false.

645. The average stress, due to elasticity of the solid, when strained from its natural condition to that of strain (e, f, g, a, b, c) is (as from the assumed applicability of the principle of superposition we see it must be) just half the stress required to keep it in this state of strain.

646. A body is called homogeneous when any two equal, similar parts of it, with corresponding lines parallel and turned towards the same parts, are undistinguishable from one another by any difference in quality. The perfect fulfilment of this condition without any limit as to the smallness of the parts, though conceivable, is not generally regarded as probable for any of the real solids or fluids known to us, however seemingly homogeneous. It is, we believe, held by all naturalists that there is a *molecular structure*, according to which, in *compound* bodies such as water, ice, rock-crystal, etc., the constituent substances lie side by side, or arranged in groups of finite dimensions, and even in bodies called *simple* (i.e. not known to be chemically resolvable into other substances) there is no ultimate homogeneousness. In other words, the prevailing belief is that every kind of matter with which we are acquainted has a more or less *coarse-grained* texture, whether having visible molecules, as great masses of solid stone or brick-building, or natural granite or sandstone rocks; or, molecules too small to be visible or directly measurable by us (but *not infinitely small*)[1] in seemingly homogeneous metals, or continuous crystals, or liquids, or gases. We must of course return to this subject under Properties of Matter; and in the meantime need only say that the definition of *homogeneousness* may be applied practically on a very large scale to masses of building or coarse-grained conglomerate rock, or on a more moderate scale to blocks of common sandstone, or on a very small scale to seemingly homogeneous metals[2]; or on a scale of extreme, undiscovered fineness, to vitreous bodies, continuous crystals, solidified gums, as India rubber, gum-arabic, etc., and fluids.

647. The substance of a homogeneous solid is called *isotropic* when a spherical portion of it, tested by any physical agency, exhibits no difference in quality however it is turned. Or, which amounts to the same, a cubical portion cut from any position in an isotropic body exhibits the same qualities relatively to each pair of parallel faces. Or two equal and similar portions cut from *any* positions

[1] Probably not *undiscoverably* small, although of dimensions not yet known to us.

[2] Which, however, we know, as recently proved by Deville and Van Troost, are porous enough at high temperatures to allow very free percolation of gases.

in the body, not subject to the condition of parallelism (§ 646), are undistinguishable from one another. A substance which is not isotropic, but exhibits differences of quality in different directions, is called *aeolotropic*.

648. An individual body, or the substance of a homogeneous solid, may be isotropic in one quality or class of qualities, but aeolotropic in others.

Thus in abstract dynamics a rigid body, or a group of bodies rigidly connected, contained within and rigidly attached to a rigid spherical surface, is kinetically symmetrical (§ 239) if its centre of inertia is at the centre of the sphere, and if its moments of inertia are equal round all diameters. It is also isotropic relatively to gravitation if it is centrobaric (§ 542), so that the centre of figure is not merely a centre of inertia, but a true centre of gravity. Or a transparent substance may transmit light at different velocities in different directions through it (that is, be *doubly refracting*), and yet a cube of it may (and generally does in natural crystals) absorb the same part of a beam of white light transmitted across it perpendicularly to any of its three pairs of faces. Or (as a crystal which exhibits *dichroism*) it may be aeolotropic relatively to the latter, or to either, optic quality, and yet it may conduct heat equally in all directions.

649. The remarks of § 646 relative to homogeneousness in the aggregate, and the supposed ultimately heterogeneous texture of all substances however seemingly homogeneous, indicate corresponding limitations and non-rigorous practical interpretations of isotropy.

650. To be elastically isotropic, we see first that a spherical or cubical portion of any solid, if subjected to uniform normal pressure (positive or negative) all round, must, in yielding, experience no deformation: and therefore must be equally compressed (or dilated) in all directions. But, further, a cube cut from any position in it, and acted on by *tangential* or distorting stress (§ 633) in planes parallel to two pairs of its sides, must experience simple deformation, or shear (§ 150), in the same direction, unaccompanied by condensation or dilatation[1], and the same in amount for all the three ways in which a stress may be thus applied to any one cube, and for different cubes taken from any different positions in the solid.

651. Hence the elastic quality of a perfectly elastic, homogeneous, isotropic solid is fully defined by two elements;—its resistance to compression, and its resistance to distortion. The amount of uniform pressure in all directions, per unit area of its surface, required to produce a stated very small compression, measures the first of

[1] It must be remembered that the changes of figure and volume we are concerned with are so small that the principle of superposition is applicable; so that if any distorting stress produced a condensation, an opposite distorting stress would produce a dilatation, which is a violation of the isotropic condition. But it is possible that a distorting stress may produce, in a truly isotropic solid, condensation or dilatation in proportion to the square of its value: and it is probable that such effects may be sensible in India rubber, or cork, or other bodies susceptible of great deformations or compressions, with persistent elasticity.

these, and the amount of the distorting stress required to produce a stated amount of distortion measures the second. The numerical measure of the first is the compressing pressure divided by the diminution of the bulk of a portion of the substance which, when uncompressed, occupies the unit volume. It is sometimes called the *elasticity of volume*, or the *resistance to compression*. Its reciprocal, or the amount of compression on unit of volume divided by the compressing pressure, or, as we may conveniently say, the compression per unit of volume, per unit of compressing pressure, is commonly called the *compressibility*. The second, or resistance to change of shape, is measured by the tangential stress (reckoned as in § 633) divided by the amount of the distortion or shear (§ 154) which it produces, and is called the *rigidity* of the substance, or its *elasticity of figure*.

652. From § 148 it follows that a strain compounded of a simple extension in one set of parallels, and a simple contraction of equal amount in any other set perpendicular to those, is the same as a simple shear in either of the two sets of planes cutting the two sets of parallels at 45°. And the numerical measure (§ 154) of this shear, or simple distortion, is equal to *double* the amount of the elongation or contraction (each measured, of course, per unit of length). Similarly, we see (§ 639) that a longitudinal traction (or negative pressure) parallel to one line, and an equal longitudinal positive pressure parallel to any line at right angles to it, is equivalent to a distorting stress of tangential tractions (§ 632) parallel to the planes which cut those lines at 45°. And the numerical measure of this distorting stress, being (§ 633) the amount of the tangential traction in either set of planes, is equal to the amount of the positive or negative normal pressure, *not doubled.*

653. Since then any stress whatever may be made up of simple longitudinal stresses, it follows that, to find the relation between any stress and the strain produced by it, we have only to find the strain produced by a single longitudinal stress, which we may do at once thus:—A simple longitudinal stress, P, is equivalent to a uniform dilating tension $\frac{1}{3}P$ in all directions, compounded with two distorting stresses, each equal to $\frac{1}{3}P$, and having a common axis in the line of the given longitudinal stress, and their other two axes any two lines at right angles to one another and to it. The diagram, drawn in a plane through one of these latter lines, and the former, sufficiently indicates the synthesis; the only forces not shown being those perpendicular to its plane.

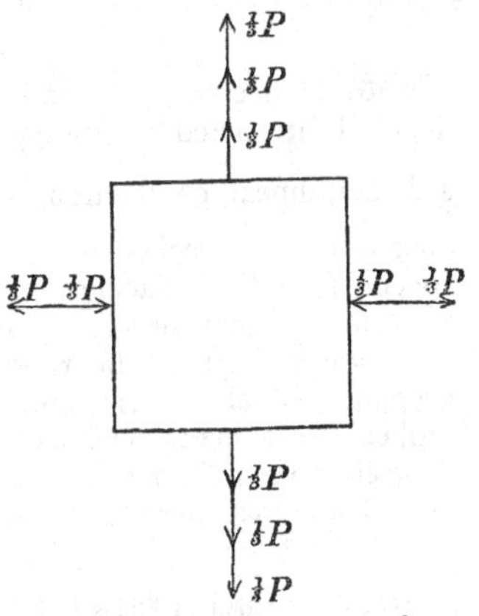

Hence if n denote the *rigidity*, and k the *resistance to dilatation* [being the same as the reciprocal of the compressibility (§ 651)], the effect will be an equal dilatation in all directions, amounting, per unit of volume, to

$$\frac{\frac{1}{3}P}{k} \qquad (1)$$

compounded with two equal distortions, each amounting to

$$\frac{\frac{1}{3}P}{n} \qquad (2)$$

and having (§ 650) their axes in the directions just stated as those of the distorting stresses.

654. The dilatation and two shears thus determined may be conveniently reduced to simple longitudinal strains by still following the indications of § 652, thus:—

The two shears together constitute an elongation amounting to $\frac{\frac{1}{3}P}{n}$ in the direction of the given force, P, and equal contraction amounting to $\frac{\frac{1}{6}P}{n}$ in all directions perpendicular to it. And the cubic dilatation $\frac{\frac{1}{3}P}{k}$ implies a lineal dilatation, equal in all directions, amounting to $\frac{\frac{1}{9}P}{k}$. On the whole, therefore, we have

$$\left.\begin{array}{l}\text{linear elongation} = P\left(\frac{1}{3n} + \frac{1}{9k}\right), \text{ in the direction of the applied stress, and} \\ \text{linear contraction} = P\left(\frac{1}{6n} - \frac{1}{9k}\right), \text{ in all directions perpendicular to the applied stress.}\end{array}\right\} \qquad (3)$$

655. Hence when the ends of a column, bar, or wire, of isotropic material, are acted on by equal and opposite forces, it experiences a lateral lineal contraction, equal to $\frac{3k-2n}{2(3k+n)}$ of the longitudinal dilatation, each reckoned as usual per unit of lineal measure. One specimen of the fallacious mathematics above referred to (§ 644), is a celebrated conclusion of Navier's and Poisson's that this ratio is $\frac{1}{4}$, which requires the rigidity to be $\frac{3}{5}$ of the resistance to compression, for all solids: and which was first shown to be false by Stokes[1] from many obvious observations, proving enormous discrepancies from it in many well-known bodies, and rendering it most improbable that there is any approach to a constancy of ratio between

[1] 'On the Friction of Fluids in Motion, and the Equilibrium and Motion of Elastic Solids.' *Trans. Camb. Phil. Jour.*, April 1845. See also *Camb. and Dub. Math. Jour.*, March 1848.

rigidity and resistance to compression in any class of solids. Thus clear elastic jellies, and India rubber, present familiar specimens of isotropic homogeneous solids, which, while differing very much from one another in rigidity ('stiffness'), are probably all of very nearly the same compressibility as water. This being $\frac{1}{308000}$ per pound per square inch; the resistance to compression, measured by its reciprocal, or, as we may read it, '308000 lbs. per square inch,' is obviously many hundred times the absolute amount of the rigidity of the stiffest of those substances. A column of any of them, therefore, when pressed together or pulled out, within its limits of elasticity, by balancing forces applied to its ends (or an India rubber band when pulled out), experiences no sensible change of volume, though a very sensible change of length. Hence the proportionate extension or contraction of any transverse diameter must be sensibly equal to $\frac{1}{2}$ the longitudinal contraction or extension: and for all ordinary stresses, such substances may be practically regarded as incompressible elastic solids. Stokes gave reasons for believing that metals also have in general greater resistance to compression, in proportion to their rigidities, than according to the fallacious theory, although for them the discrepancy is very much less than for the gelatinous bodies. This probable conclusion was soon experimentally demonstrated by Werthiem, who found the ratio of lateral to longitudinal change of lineal dimensions, in columns acted on solely by longitudinal force, to be about $\frac{1}{3}$ for glass or brass; and by Kirchhoff, who, by a very well-devised experimental method, found ·387 as the value of that ratio for brass, and ·294 for iron. For copper we find that it probably lies between ·226 and ·441, by recent experiments[1] of our own, measuring the torsional and longitudinal rigidities (§§ 609, 657) of a copper wire.

656. All these results indicate rigidity *less* in proportion to the compressibility than according to Navier's and Poisson's theory. And it has been supposed by many naturalists, who have seen the necessity of abandoning that theory as inapplicable to ordinary solids, that it may be regarded as the proper theory for an ideal *perfect solid*, and as indicating an amount of rigidity not quite reached in any real substance, but approached to in some of the most rigid of natural solids (as, for instance, iron). But it is scarcely possible to hold a piece of cork in the hand without preceiving the fallaciousness of this last attempt to maintain a theory which never had any good foundation. By careful measurements on columns of cork of various forms (among them, cylindrical pieces cut in the ordinary way for bottles) before and after compressing them longitudinally in a Brahmah's press, we have found that the change of lateral dimensions is insensible both with small longitudinal contractions and return dilatations, within the limits of elasticity, and with such enormous longitudinal contractions as to $\frac{1}{6}$ or $\frac{1}{8}$ of the original length. It is thus proved decisively that cork is much more rigid, while metals,

[1] 'On the Elasticity and Viscosity of Metals' (W. Thomson), *Proc. R. S.*, May 1865.

glass, and gelatinous bodies are all less rigid, in proportion to resistance to compression than the supposed 'perfect solid'; and the utter worthlessness of the theory is experimentally demonstrated.

657. The modulus of elasticity of a bar, wire, fibre, thin filament, band, or cord of any material (of which the substance need not be isotropic, nor even homogeneous within one normal section), as a bar of glass or wood, a metal wire, a natural fibre, an India rubber band, or a common thread, cord, or tape, is a term introduced by Dr. Thomas Young to designate what we also sometimes call its *longitudinal rigidity:* that is, the quotient obtained by dividing the simple longitudinal force required to produce any infinitesimal elongation or contraction by the amount of this elongation or contraction reckoned as always per unit of length.

658. Instead of reckoning the modulus in units of weight, it is sometimes convenient to express it in terms of the weight of the unit length of the rod, wire, or thread. The modulus thus reckoned, or, as it is called by some writers, the length of the modulus, is of course found by dividing the weight-modulus by the weight of the unit length. It is useful in many applications of the theory of elasticity; as, for instance, in this result, which will be proved later:—the velocity of transmission of longitudinal vibrations (as of sound) along a bar or cord, is equal to the velocity acquired by a body in falling from a height equal to half the length of the modulus[1].

659. The *specific modulus of elasticity of an isotropic substance*, or, as it is most often called, simply *the modulus of elasticity of the substance*, is the modulus of elasticity of a bar of it having some definitely specified sectional area. If this be such that the weight of unit length is unity, the *modulus of the substance* will be the same as the length of the modulus of any bar of it; a system of reckoning which, as we have seen, has some advantages in application. It is, however, more usual to choose a common unit of area as the sectional area of the bar referred to in the definition. There must also be a definite understanding as to the unit in terms of which the force is measured, which may be either the *absolute unit* (§ 188): or the gravitation unit for a specified locality; that is (§ 191), the weight in that locality of the unit of mass. Experimenters hitherto have stated their results in terms of the gravitation unit, each for his own locality; the accuracy hitherto attained being scarcely in any cases sufficient to

[1] It is to be understood that the vibrations in question are so much spread out through the *length* of the body, that inertia does not sensibly influence the transverse contractions and dilatations which (unless the substance have in this respect the peculiar character presented by cork, § 656) take place along with them. Also, under thermodynamics, we shall see that changes of temperature produced by the varying strains cause changes of stress which, in ordinary solids, render the velocity of transmission of longitudinal vibrations sensibly greater than that calculated by the rule stated in the text, if we use the *static modulus* as understood from the definition there given; and we shall learn to take into account the thermal effect by using a definite *static modulus*, or *kinetic modulus*, according to the circumstances of any case that may occur.

require corrections for the different forces of gravity in the different places of observation.

660. The most useful and generally convenient specification of the modulus of elasticity of a substance is in grammes-weight per square centimetre. This has only to be divided by the specific gravity of the substance to give the *length of the modulus*. British measures, however, being still unhappily sometimes used in practical and even in high scientific statements, we may have occasion to refer to reckonings of the modulus in pounds per square inch or per square foot, or to length of the modulus in feet.

661. The reckoning most commonly adopted in British treatises on mechanics and practical statements is pounds per square inch. The modulus thus stated must be divided by the weight of 12 cubic inches of the solid, or by the product of its specific gravity into 4337[1], to find the length of the modulus, in feet.

To reduce from pounds per square inch to grammes per square centimetre, multiply by 70·31, or divide by ·014223. French engineers generally state their results in kilogrammes per square millimetre, and so bring them to more convenient numbers, being $\frac{1}{100000}$ of the inconveniently large numbers expressing moduli in grammes-weight per square centimetre.

662. The same statements as to units, reducing factors, and nominal designations, are applicable to the resistance to compression of any elastic solid or fluid, and to the rigidity (§ 651) of an isotropic body; or, in general, to any one of the 21 co-efficients in the expressions for terms in stresses of strains, or to the reciprocal of any one of the 21 co-efficients in the expressions for strains in terms of stresses, as well as to the modulus defined by Young.

663. In §§ 652, 653 we examined the effect of a simple longitudinal stress, in producing elongation in its own direction, and contraction

[1] This decimal being the weight in pounds of 12 cubic inches of water. The one great advantage of the French metrical system is, that the mass of the unit volume (1 cubic centimetre) of water at its temperature of maximum density (3°·945 C.) is unity (1 gramme) to a sufficient degree of approximation for almost all practical purposes. Thus, according to this system, the density of a body and its specific gravity mean one and the same thing; whereas on the British no-system the density is expressed by a number found by multiplying the specific gravity by one number or another, according to the choice of a cubic inch, cubic foot, cubic yard, or cubic mile that is made for the unit of volume; and the grain, scruple, gunmaker's drachm, apothecary's drachm, ounce Troy, ounce avoirdupois, pound Troy, pound avoirdupois, stone (Imperial, Ayrshire, Lanarkshire, Dumbartonshire), stone for hay, stone for corn, quarter (of a hundred-weight), quarter (of corn), hundredweight, or ton, that is chosen for unit of mass. It is a remarkable phenomenon, belonging rather to moral and social than to physical science, that a people tending naturally to be regulated by common sense should voluntarily condemn themselves, as the British have so long done, to unnecessary hard labour in every action of common business or scientific work related to measurement, from which all the other nations of Europe have emancipated themselves. We have been informed, through the kindness of Professor W. H. Miller, of Cambridge, that he concludes, from a very trustworthy comparison of standards by Kupffer, of St. Petersburgh, that the weight of a cubic decimetre of water at temperature of maximum density is 1000·013 grammes.

in lines perpendicular to it. With stresses substituted for strains, and strains for stresses, we may apply the same process to investigate the longitudinal and lateral tractions required to produce a simple longitudinal strain (that is, an elongation in one direction, with no change of dimensions perpendicular to it) in a rod or solid of any shape.

Thus a simple longitudinal strain e is equivalent to a cubic dilatation e without change of figure (or linear dilatation $\frac{1}{3}e$ equal in all directions), and two distortions consisting each of dilatation $\frac{1}{3}e$ in the given direction, and contraction $\frac{1}{3}e$ in each of two directions perpendicular to it and to one another. To produce the cubic dilatation, e, alone requires (§ 651) a normal traction ke equal in all directions. And, to produce either of the distortions simply, since the measure (§ 154) of each is $\frac{2}{3}e$, requires a distorting stress equal to $n\times\frac{2}{3}e$, which consists of tangential tractions each equal to this amount, positive (or drawing outwards) in the line of the given elongation, and negative (or pressing inwards) in the perpendicular direction. Thus we have in all

$$\left.\begin{array}{l}\text{normal traction}=(k+\tfrac{4}{3}n)\,e, \text{ in the direction of the given strain, and}\\ \text{normal traction}=(k-\tfrac{2}{3}n)\,e, \text{ in every direction perpendicular to the given strain.}\end{array}\right\}\quad(4)$$

664. If now we suppose any possible infinitely small strain (e, f, g, a, b, c), according to the specification of § 640, to be given to a body, the stress (P, Q, R, S, T, U) required to maintain it will be expressed by the following formulae, obtained by successive applications of § 663 (4) to the components e, f, g separately, and of § 651 to a, b, c:—

$$\left.\begin{array}{l}S=na,\ T=nb,\ U=nc,\\ P=\mathfrak{A}e+\mathfrak{B}(f+g),\\ Q=\mathfrak{A}f+\mathfrak{B}(g+e),\\ R=\mathfrak{A}g+\mathfrak{B}(e+f),\\ \text{where}\quad \left.\begin{array}{l}\mathfrak{A}=k+\dfrac{4n}{3},\\ \mathfrak{B}=k-\dfrac{2n}{3},\end{array}\right\}\ n=\tfrac{1}{2}(\mathfrak{A}-\mathfrak{B}).\end{array}\right\}\quad(5)$$

665. Similarly, by § 651 and § 654 (3), we have

$$\left.\begin{array}{l}a=\dfrac{1}{n}S,\ b=\dfrac{1}{n}T,\ c=\dfrac{1}{n}U,\\ Me=\{P-\sigma(Q+R)\},\\ Mf=\{Q-\sigma(R+P)\},\\ Mg=\{R-\sigma(P+Q)\},\\ \text{where}\quad M=\dfrac{9nk}{3k+n}\\ \text{and}\quad \sigma=\dfrac{3k-2n}{2(3k+n)}=\tfrac{1}{2}\dfrac{M}{n}-1,\end{array}\right\}\quad(6)$$

as the formulae expressing the strain (e, f, g, a, b, c) in terms of the stress (P, Q, R, S, T, U). They are of course merely the algebraic inversions of (5); and they might have been found by solving these for e, f, g, a, b, c, regarded as the unknown quantities. M is here introduced to denote Young's modulus.

666. To express the equation of energy for an isotropic substance, we may take the general formula,

$$w = \tfrac{1}{2}(Pe + Qf + Rg + Sa + Tb + Uc),$$

and eliminate from it P, Q, etc., by (5) of § 664, or, again, e, f, etc., by (6) of § 665, we thus find

$$\left.\begin{aligned} 2w &= \left(k + \frac{4n}{3}\right)(e^2 + f^2 + g^2) + 2\left(k - \frac{2n}{3}\right)(fg + ge + ef) + n(a^2 + b^2 + c^2) \\ &= \tfrac{1}{3}\left\{\left(\frac{1}{n} + \frac{1}{3k}\right)(P^2 + Q^2 + R^2) - 2\left(\frac{1}{2n} - \frac{1}{3k}\right)(QR + RP + PQ)\right\} + \frac{1}{n}(S^2 + T^2 - U^2). \end{aligned}\right\} \quad (7)$$

667. The mathematical theory of the equilibrium of an elastic solid presents the following general problems:—

A solid of any given shape, when undisturbed, is acted on in its substance by force distributed through it in any given manner, and displacements are arbitrarily produced, or forces arbitrarily applied, over its bounding surface. It is required to find the displacement of every point of its substance.

This problem has been thoroughly solved for a shell of homogeneous isotropic substance bounded by surfaces which, when undisturbed, are spherical and concentric; but not hitherto for a body of any other shape. The limitations under which solutions have been obtained for other cases (thin plates and rods), leading, as we have seen, to important practical results, have been stated above (§ 605). To demonstrate the laws (§ 607) which were taken in anticipation will also be one of our applications of the general equations for interior equilibrium of an elastic solid, which we now proceed to investigate.

668. Any portion in the interior of an elastic solid may be regarded as becoming perfectly rigid (§ 584) without disturbing the equilibrium either of itself or of the matter round it. Hence the traction exerted by the matter all round it, regarded as a distribution of force applied to its surface, must, with the applied forces acting on the substance of the portion considered, fulfil the conditions of equilibrium of forces acting on a rigid body. This statement, applied to an infinitely small rectangular parallelepiped of the body, gives the general differential equations of internal equilibrium of an elastic solid. It is to be remarked that *three* equations suffice; the conditions of equilibrium for the *couples* being secured by the relation established above (§ 632) among the six pairs of tangential component tractions on the six faces of the figure.

669. One of the most beautiful applications of the general equations of internal equilibrium of an elastic solid hitherto made is that of M. de St. Venant to 'the torsion of prisms[1].' To one end of a long straight prismatic rod, wire, or solid or hollow cylinder of any form, a given couple is applied in a plane perpendicular to the length, while the other end is held fast: it is required to find the degree of twist produced, and the distribution of strain and stress throughout the prism. The conditions to be satisfied here are that the resultant action between the substance on the two sides of any normal section is a couple in the normal plane, equal to the given couple. Our work for solving the problem will be much simplified by first establishing the following preliminary propositions:—

670. Let a solid (whether aeolotropic or isotropic) be so acted on by force applied from without to its boundary, that throughout its interior there is no normal traction on any plane parallel or perpendicular to a given plane, XOY, which implies, of course, that there is no distorting stress with axes in or parallel to this plane, and that the whole stress at any point of the solid is a simple distorting stress of tangential forces in some direction in the plane parallel to XOY, and in the plane perpendicular to this direction. Then—

(1) The interior distorting stress must be equal, and similarly directed, in all parts of the solid lying in any line perpendicular to the plane XOY.

(2) It being premised that the traction at every point of any surface perpendicular to the plane XOY is, by hypothesis, a distribution of force in lines perpendicular to this plane; the integral amount of it on any closed prismatic or cylindrical surface perpendicular to XOY, and bounded by planes parallel to it, is zero.

(3) The matter within the prismatic surface and terminal planes of (2) being supposed for a moment (§ 584) to be rigid, the distribution of tractions referred to in (2) constitutes a couple whose moment, divided by the distance between those terminal planes, is equal to the resultant force of the tractions on the area of either, and whose plane is parallel to the lines of these resultant forces. In other words, the moment of the distribution of forces over the prismatic surface referred to in (2) round any line (OY or OX) in the plane XOY, is equal to the sum of the components (T or S), perpendicular to the same line, of the traction in either of the terminal planes multiplied by the distance between these planes.

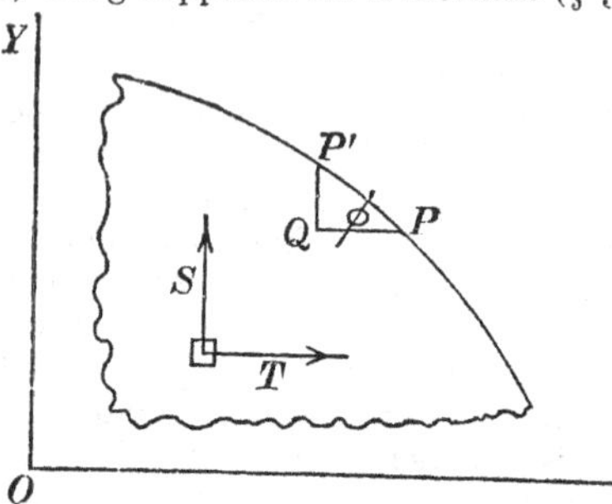

[1] *Mémoires des Savants Étrangers*, 1855. 'De la Torsion des Prismes, avec des Considérations sur leur Flexion,' etc.

To prove (1) consider for a moment as rigid (§ 584) an infinitesimal prism, AB (of sectional area ω), perpendicular to XOY, and having plane ends, A, B, parallel to it. There being no forces on its sides (or cylindrical boundary) perpendicular to its length, its equilibrium so far as motion in the direction of any line (OX), perpendicular to its length, requires that the components of the tractions on its ends be equal and in opposite directions. Hence, in the notation § 633, the distorting-stress components, T, must be equal at A and B; and so must the stress components S, for the same reason.

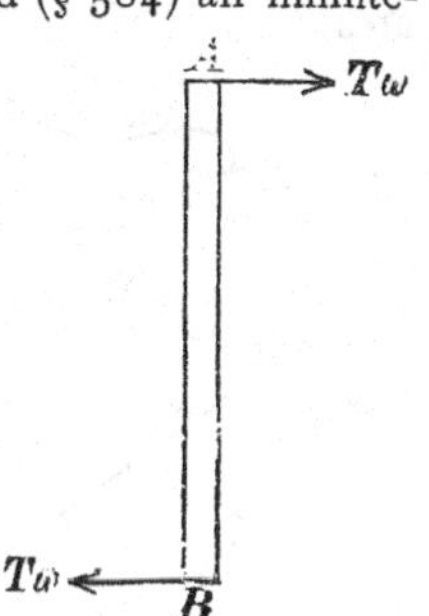

To prove (2) and (3) we have only to remark that they are required for the equilibrium of the rigid prism referred to in (3).

671. For a solid or hollow circular cylinder, the solution of § 669 (given first, we believe, by Coulomb) obviously is that each circular normal section remains unchanged in its own dimensions, figure, and internal arrangement (so that every straight line of its particles remains a straight line of unchanged length), but is turned round the axis of the cylinder through such an angle as to give a uniform *rate of twist* equal to the applied couple divided by the product of the moment of inertia of the circular area (whether annular or complete to the centre) into the rigidity of the substance.

672. Similarly, we see that if a cylinder or prism of any shape be compelled to take exactly the state of strain above specified (§ 671) with the line through the centres of inertia of the normal sections, taken instead of the axis of the cylinder, the mutual action between the parts of it on the two sides of any normal section will be a couple of which the moment will be expressed by the same formula, that is, the product of the rigidity, into the rate of twist, into the moment of inertia of the section round its centre of inertia.

673. But for any other shape of prism than a solid or symmetrical hollow circular cylinder, the supposed state of strain will require, besides the terminal opposed couples, force parallel to the length of the prism, distributed over the prismatic boundary, in proportion to the distance along the tangent, from each point of the surface, to the point in which this line is cut by a perpendicular to it from the centre of inertia of the normal section. To prove this let a normal section of the prism be represented in the annexed diagram (page 248). Let PK represent the shear at any point, P, close to the prismatic boundary, be resolved into PN and PT respectively along the normal and tangent. The whole shear, PK, being equal to τr, its component, PN, is equal to $\tau r \sin \omega$ or $\tau.PE$. The corresponding component of the required stress is $n\tau.PE$, and involves (§ 632) equal forces in the plane of the diagram, and in the plane through

TP perpendicular to it, each amounting to $n\tau.PE$ per unit of area.

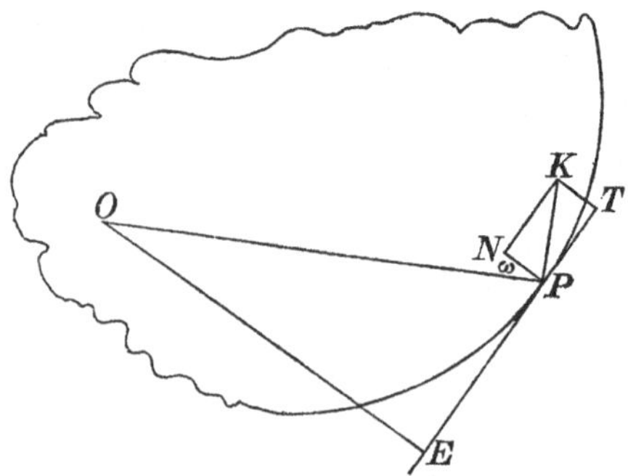

An application of force equal and opposite to the distribution thus found over the prismatic boundary, would of course alone produce in the prism, otherwise free, a state of strain which, compounded with that supposed above, would give the state of strain actually produced by the sole application of balancing couples to the two ends. The result, it is easily seen (and it will be proved below), consists of an increased twist, together with a warping of naturally plane normal sections, by infinitesimal displacements perpendicular to themselves, into certain surfaces of anticlastic curvature, with equal opposite curvatures in the principal sections (§ 122) through every point. This theory is due to St. Venant, who not only pointed out the falsity of the supposition admitted by several previous writers, that Coulomb's law holds for other forms of prism than the solid or hollow circular cylinder, but discovered fully the nature of the requisite correction, reduced the determination of it to a problem of pure mathematics, worked out the solution for a great variety of important and curious cases, compared the results with observation in a manner satisfactory and interesting to the naturalist, and gave conclusions of great value to the practical engineer.

674. We take advantage of the identity of mathematical conditions in St. Venant's torsion problem, and a hydrokinetic problem first solved a few years earlier by Stokes [1], to give the following statement, which will be found very useful in estimating deficiencies in torsional rigidity below the amount calculated from the fallacious extension of Coulomb's law:—

675. Conceive a liquid of density n completely filling a closed infinitely light prismatic box of the same shape within as the given elastic prism and of length unity, and let a couple be applied to the box in a plane perpendicular to its length. The *effective* moment of inertia of the liquid [2] will be equal to the correction by which the torsional rigidity of the elastic prism calculated by the false extension of Coulomb's law, must be diminished to give the true torsional rigidity.

Further, the actual *shear* of the solid, in any infinitely thin plate of

[1] 'On some cases of Fluid Motion,' *Cambridge Philosophical Transactions*, 1843.

[2] That is the moment of inertia of a rigid solid which, as will be proved in Vol. II., may be fixed within the box, if the liquid be removed, to make its motions the same as they are with the liquid in it.

it between two normal sections, will at each point be, when reckoned as a differential sliding (§ 151) parallel to their planes, equal to and in the same direction as the velocity of the liquid relatively to the containing box.

676. St. Venant's treatise abounds in beautiful and instructive graphical illustrations of his results, from which we select the following:—

(1) *Elliptic cylinder.* The plain and dotted curvilineal arcs are 'contour lines' (*coupes topographiques*) of the section as warped by

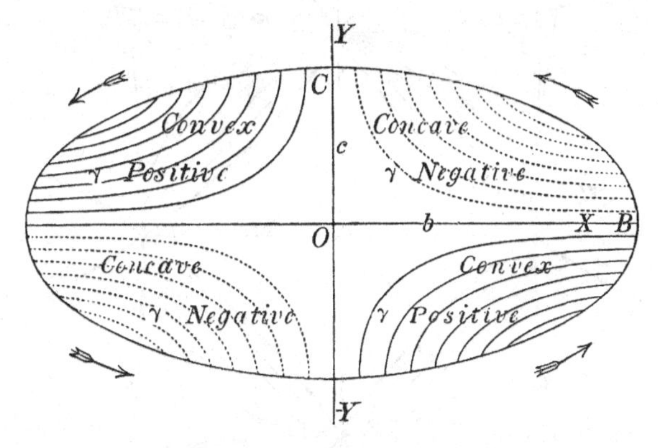

torsion; that is to say, lines in which it is cut by a series of parallel planes, each perpendicular to the axis. These lines are equilateral hyperbolas in this case. The arrows indicate the direction of rotation in the part of the prism *above* the plane of the diagram.

(2) *Equilateral triangular prism.*—The contour lines are shown

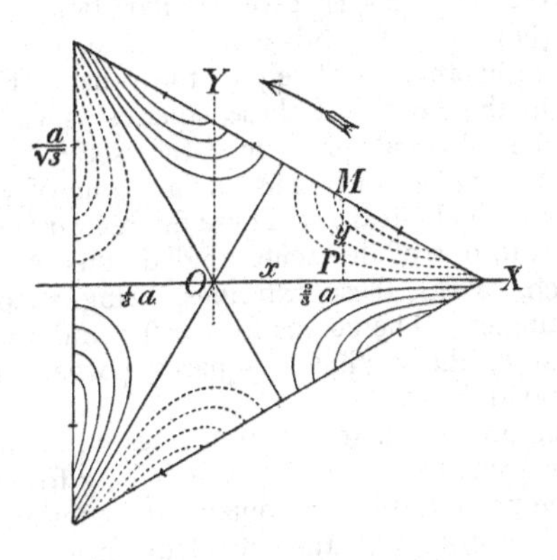

as in case (1); the dotted curves being those where the warped section falls *below* the plane of the diagram, the direction of rotation of the part of the prism above the plane being indicated by the bent arrow.

(3) This diagram shows a series of lines given by St. Venant, and more or less resembling squares. Their common equation containing only one constant a. It is remarkable that the values

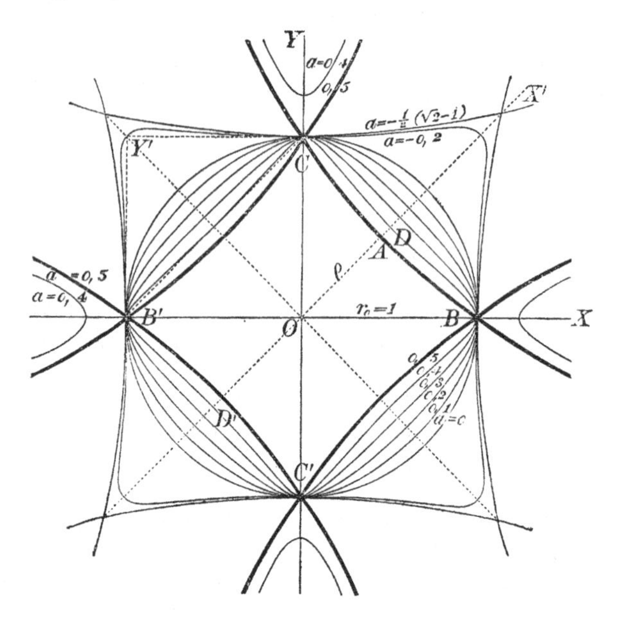

$a=0{\cdot}5$ and $a=-\frac{1}{2}(\sqrt{2}-1)$ give similar but not equal curvilineal squares (hollow sides and acute angles), one of them turned through half a right angle relatively to the other. Everything in the diagram outside the larger of these squares is to be cut away as irrelevant to the physical problem; the series of closed curves remaining exhibits figures of prisms, for any one of which the torsion problem is solved algebraically. These figures vary continuously from a circle, inwards to one of the acute-angled squares, and outwards to the other: each, except these extremes, being a continuous closed curve with no angles. The curves for $a=0{\cdot}4$ and $a=-0{\cdot}2$ approach remarkably near to the rectilineal squares, partially indicated in the diagram by dotted lines.

(4) This diagram shows the contour lines, in all respects as in the cases (1) and (2) for the case of a prism having for section the figure indicated. The portions of curve outside the continuous closed curve are merely indications of mathematical extensions irrelevant to the physical problem.

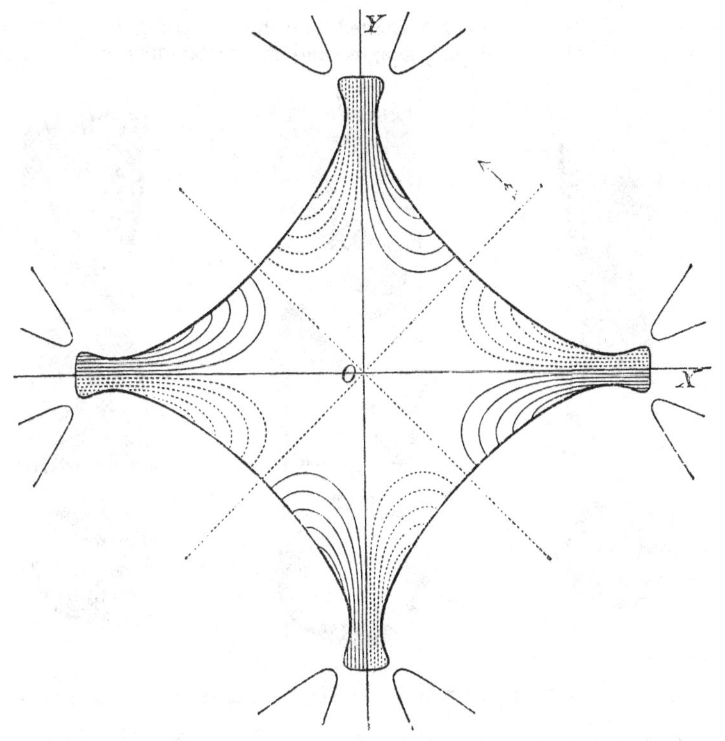

(5) This shows, as in the other cases, the contour lines for the warped section of a square prism under torsion.

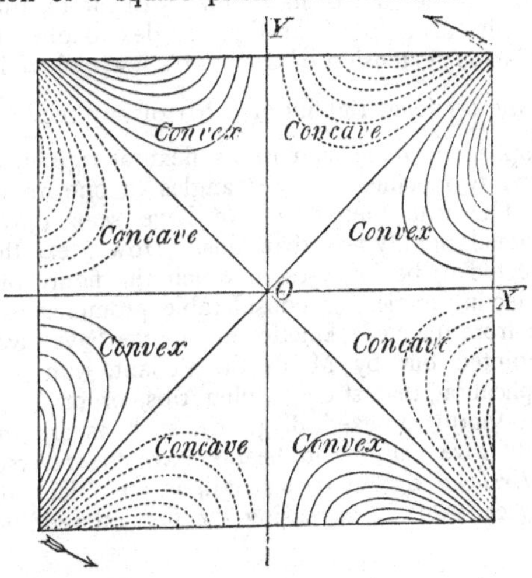

(6), (7), (8). These are shaded drawings, showing the appearances presented by elliptic, square, and flat rectangular bars under

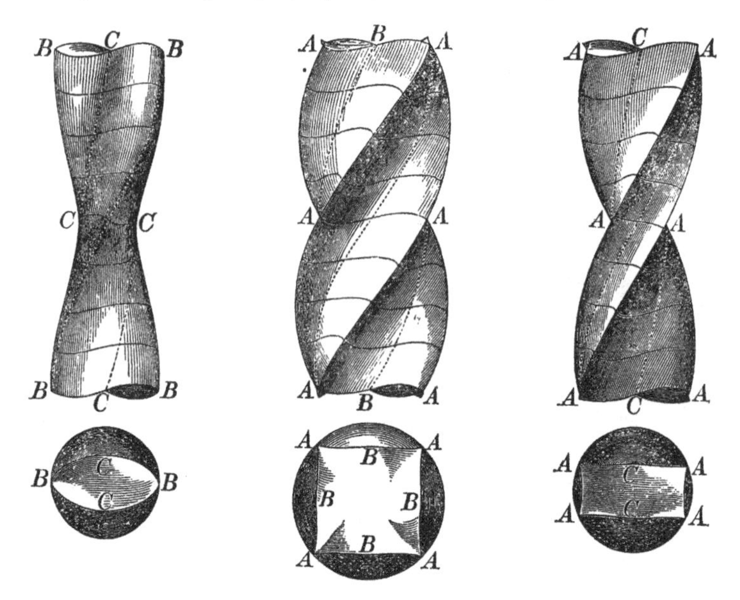

exaggerated torsion, as may be realized with such a substance as India rubber.

677. Inasmuch as the moment of inertia of a plane area about an axis through its centre of inertia perpendicular to its plane is obviously equal to the sum of its moments of inertia round any two axes through the same point, at right angles to one another in its plane, the fallacious extension of Coulomb's law, referred to in § 673, would make the torsional rigidity of a bar of any section equal to $\frac{n}{M}$ (§ 665) multiplied into the sum of its flexural rigidities (see below, § 679) in any two planes at right angles to one another through its length. The true theory, as we have seen (§ 675), always gives a torsional rigidity less than this. How great the deficiency may be expected to be in cases in which the figure of the section presents projecting angles, or considerable prominences (which may be imagined from the hydrokinetic analogy we have given in § 675), has been pointed out by M. de St. Venant, with the important practical application, that strengthening ribs, or projections (see, for instance, the fourth annexed diagram), such as are introduced in engineering to give stiffness to beams, have the reverse of a good effect when *torsional* rigidity or strength is an object, although they are truly of great value in increasing the flexural rigidity, and giving strength to bear ordinary strains, which are always more or less flexural. With remarkable ingenuity and mathematical skill he has drawn beautiful illustrations of this important practical principle from his algebraic and transcendental solutions. Thus for an equilateral

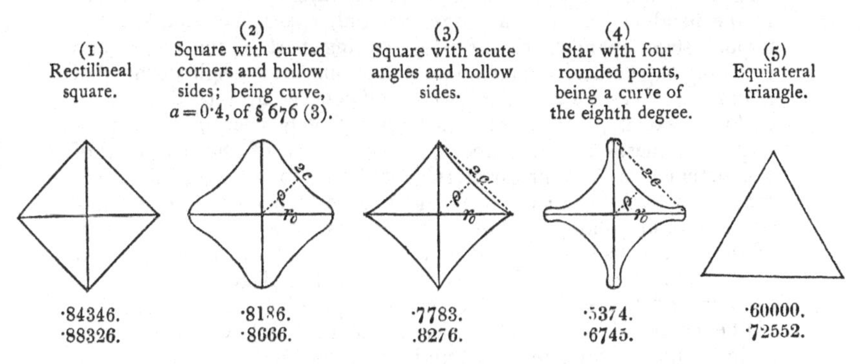

triangle, and for the rectilineal and three curvilineal squares shown in the annexed diagram, he finds for the torsional rigidities the values stated. The number immediately below the diagram indicates in each case the fraction which the true torsional rigidity is of the old fallacious estimate (§ 673); the latter being the product of the rigidity of the substance into the moment of inertia of the cross section round an axis perpendicular to its plane through its centre of inertia. The second number indicates in each case the fraction which the torsional rigidity is of that of a solid circular cylinder of the same sectional area.

678. M. de St. Venant also calls attention to a conclusion from his solutions which to many may be startling, that in his simpler cases the places of greatest distortion are those points of the boundary which are nearest to the axis of the twisted prism in each case, and the places of least distortion those farthest from it. Thus in the elliptic cylinder the substance is most strained at the ends of the smaller principal diameter, and least at the ends of the greater. In the equilateral triangular and square prisms there are longitudinal lines of maximum strain through the middles of the sides. In the oblong rectangular prism there are two lines of greater maximum strain through the middles of the broader pair of sides, and two lines of less maximum strain through the middles of the narrow sides. The strain is, as we may judge from (§ 675) the hydrokinetic analogy, excessively small, but not evanescent, in the projecting ribs of a prism of the figure shown in (4) § 677. It is quite evanescent infinitely near the angle, in the triangular and rectangular prisms, and in each other case as (3) of § 677, in which there is a finite angle, whether acute or obtuse, projecting outwards. This reminds us of a general remark we have to make, although consideration of space may oblige us to leave it without formal proof. A solid of any elastic substance, isotropic or aeolotropic, bounded by any surfaces presenting projecting edges or angles, or re-entrant angles or edges, however obtuse, cannot experience any finite stress or strain in the neighbourhood of a *projecting* angle (trihedral, polyhedral, or conical); in the neighbourhood of an edge, can only experience simple longitudinal stress parallel to the neighbouring part of the edge; and generally experiences infinite stress and strain in the neighbourhood of a *re-entrant* edge or angle; when influenced by any distribution of force, exclusive of surface tractions infinitely near the angles or edges in question. An important application of the last part of this statement is the practical rule, well known in mechanics, that every re-entering edge or angle ought to be rounded to prevent risk of rupture, in solid pieces designed to bear stress. An illustration of these principles is afforded by the complete mathematical solution of the torsion problem for prisms of fan-shaped sections, such as the annexed figures. In the cases corresponding to figures (4), (5), (6) below, the distortion at the centre of the circle vanishes in (4), is finite and determinate in (5), and infinite in (6).

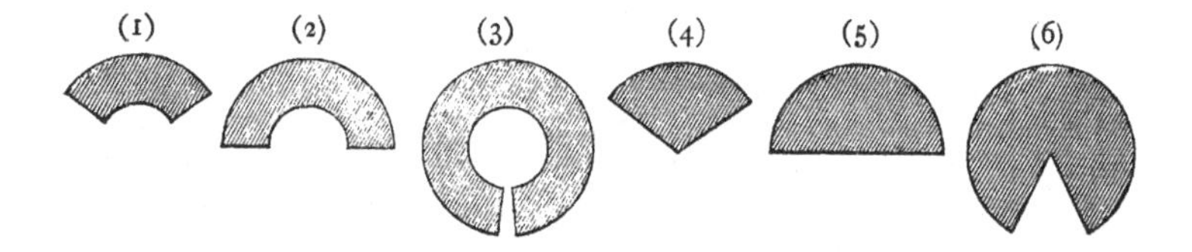

679. Hence in a rod of isotropic substance the principal axes of flexure (§ 609) coincide with the principal axes of inertia of the area of the normal section; and the corresponding flexural rigidities are the moments of inertia of this area round these axes multiplied by Young's modulus. Analytical investigation leads to the following results, due to St. Venant. Imagine the whole rod divided, parallel to its length, into infinitesimal filaments (prisms when the rod is straight). Each of these contracts or swells laterally with sensibly the same freedom as if it were separated from the rest of the substance, and becomes elongated or shortened in a straight line to the same extent as it is really elongated or shortened in the circular arc which it becomes in the bent rod. The distortion of the cross section by which these changes of lateral dimensions are necessarily accompanied is illustrated in the annexed diagram, in which either the whole normal section of a rectangular beam, or a rectangular area in the normal section of a beam of any figure, is represented in its strained and unstrained figures, with the central point O common to the two. The flexure is in planes perpendicular to YOY, and concave upwards (or towards X); G the centre of curvature, being in the direction indicated, but too far to be included in the diagram. The straight sides AC, BD, and all straight lines parallel to them, of the unstrained rectangular area become concentric arcs of circles concave in the

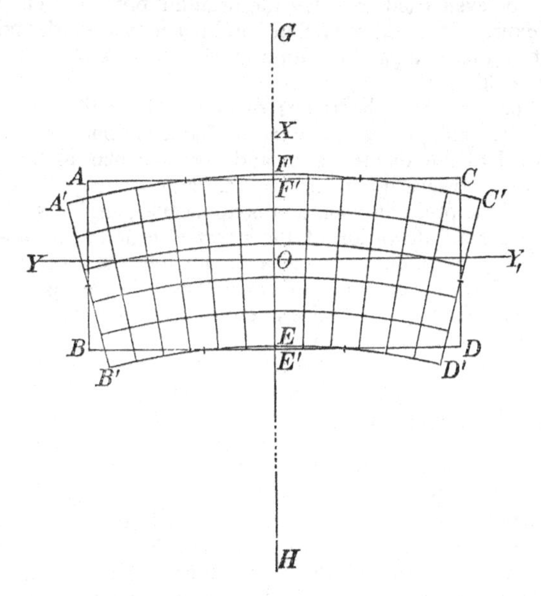

opposite direction, their centre of curvature, H, being for rods of gelatinous substance, or of glass or metal, from 2 to 4 times as far from O on one side as G is on the other. Thus the originally plane sides AC, BD of a rectangular bar become anticlastic surfaces, of curvatures $\frac{1}{\rho}$ and $\frac{-\sigma}{\rho}$, in the two principal sections. A flat rectangular, or a square, rod of India rubber [for which σ amounts (§ 655) to very nearly $\frac{1}{2}$, and which is susceptible of very great amounts of strain without utter loss of corresponding elastic action], exhibits this phenomenon remarkably well.

680. The conditional limitation (§ 605) of the curvature to being very small in comparison with that of a circle of radius equal to the greatest diameter of the normal section (not obviously necessary, and indeed not generally known to be necessary, we believe, when the greatest diameter is perpendicular to the plane of curvature) now receives its full explanation. For unless the *breadth*, AC, of the bar (or diameter perpendicular to the plane of flexure) be very small in comparison with the mean proportional between the radius, OH, and the thickness, AB, the distances from OY to the corners A', C' would fall short of the half thickness, OE, and the distances to B', D' would exceed it by differences comparable with its own amount. This would give rise to sensibly less and greater shortenings and stretchings in the filaments towards the corners, and so vitiate the solution. Unhappily mathematicians have not hitherto succeeded in solving, possibly not even tried to solve, the beautiful problem thus presented by the flexure of a broad very thin band (such as a watch spring) into a circle of radius comparable with a third proportional to its thickness and its breadth.

681. But, provided the radius of curvature of the flexure is not only a large multiple of the *greatest* diameter, but also of a third proportional to the diameters in and perpendicular to the plane of flexure; then however great may be the ratio of the greatest diameter to the least, the preceding solution is applicable; and it is remarkable that the necessary distortion of the normal section (illustrated in the diagram of § 679) does not sensibly impede the free lateral contractions and expansions in the filaments, even in the case of a broad thin lamina (whether of precisely rectangular section, or of unequal thicknesses in different parts).

682. In our sections on hydrostatics, the problem of finding the deformation produced in a spheroid of incompressible liquid by a given disturbing force will be solved; and then we shall consider the application of the preceding methods to an elastic solid sphere in their bearing on the theory of the tides and the rigidity of the earth. This proposed application, however, reminds us of a general remark of great practical importance, with which we shall leave elastic solids for the present. Considering different elastic solids of similar substance and similar shapes, we see that if by forces applied to them in any way they are similarly strained, the surface tractions in or across similarly situated elements of surface, whether of their boundaries or of surfaces imagined as cutting through their substances, must be equal, reckoned as usual per unit of area. Hence; the force across, or in, any such surface, being resolved into components parallel to any directions; the whole amounts of each such component for similar surfaces of the different bodies are in proportion to the *squares* of their lineal dimensions. Hence, if equilibrated similarly under the action of gravity, or of their kinetic reactions (§ 230) against equal accelerations (§ 32), the greater body would be more strained than the less; as the amounts of gravity or of kinetic reaction of similar portions of them are as the *cubes* of their linear dimensions. Definitively, the strains at similarly situated points of the bodies will be in simple proportion to their linear dimensions, and the displacements will be as the squares of these lines, provided that there is no strain in any part of any of them too great to allow the principle of superposition to hold with sufficient exactness, and that no part is turned through more than a very small angle relatively to any other part. To illustrate by a single example, let us consider a uniform long, thin, round rod held horizontally by its middle. Let its substance be homogeneous, of density ρ, and Young's modulus, M; and let its length, l, be p times its diameter. Then (as the moment of inertia of a circular area of radius r round a diameter is $\frac{1}{4}\pi r^4$) the flexural rigidity of the rod will (§ 679) be $\frac{M}{4}\pi\left(\frac{l}{2p}\right)^4$. This gives us for the curvature at the middle of the rod the elongation and contraction where greatest, that is, at the highest and lowest points of the normal section through the middle point; and the droop of the ends; the following expressions

$$\frac{2p^2\rho}{M};\ \frac{pl\rho}{M};\ \text{and}\ \frac{p^2l^2\rho}{8M}.$$

Thus, for a rod whose length is 200 times its diameter, if its substance be iron or steel, for which $\rho=7{\cdot}75$, and $\mathrm{M}=194\times10^7$ grammes per square centimetre, the maximum elongation and contraction (being at the top and bottom of the middle section where it is held) are each equal to $8\times10^{-6}\times l$, and the droop of its ends $2\times10^{-5}\times l^2$. Thus a steel or iron wire, ten centimetres long, and half a millimetre in diameter, held horizontally by its middle, would experience only ·000008 of maximum elongation and contraction, and only ·002 of a centimetre of droop in its ends: a round steel rod, of half a centimetre in diameter, and one metre long, would experience ·00008 of maximum elongation and contraction, and ·2 of a centimetre of droop: a round steel rod, of ten centimetres diameter, and twenty metres long, must be of remarkable temper (see *Properties of Matter*) to bear being held by the middle without taking a very sensible permanent set: and it is probable that no temper of steel is high enough in a round shaft forty metres long, if only two decimetres in diameter, to allow it to be held by its middle without either bending it to some great angle, and beyond all appearance of elasticity, or breaking it.

683. In passing from the dynamics of perfectly elastic solids to abstract hydrodynamics, or the dynamics of perfect fluids, it is convenient and instructive to anticipate slightly some of the views as to intermediate properties observed in real solids and fluids, which, according to the general plan proposed (§ 402) for our work, will be examined with more detail under *Properties of Matter*.

By induction from a great variety of observed phenomena, we are compelled to conclude that no change of volume or of shape can be produced in any kind of matter without dissipation of energy (§ 247); so that if in any case there is a return to the primitive configuration, some amount (however small) of work is always required to compensate the energy dissipated away, and restore the body to the same physical and the same palpably kinetic condition as that in which it was given. We have seen (§ 643), by anticipating something of thermodynamic principles, how such dissipation is inevitable, even in dealing with the *absolutely perfect* elasticity of volume presented by every fluid, and possibly by some solids, as, for instance, homogeneous crystals. But in metals, glass, porcelain, natural stones, wood, India-rubber, homogeneous jelly, silk fibre, ivory, etc., a distinct *frictional resistance*[1] against every change of shape is, as we shall see later, under *Properties of Matter*, demonstrated by many experiments, and is found to depend on the speed with which the change of shape is made. A very remarkable and obvious proof of frictional resistance to change of shape in ordinary solids, is afforded by the gradual, more or less rapid, subsidence of vibrations of elastic solids; marvellously rapid in India-rubber, and even in homogeneous jelly; less rapid in glass and metal springs, but still demonstrably, much more rapid than can be accounted for by the resistance of the air. This molecular friction in elastic solids may be properly called *viscosity of solids*, because, as being an internal resistance to change of shape depending on the rapidity of the change, it must be classed with fluid molecular friction, which by general consent is called *viscosity of fluids*. But, at the same time, we feel bound to remark that the word viscosity, as used hitherto by the best writers, when solids or heterogeneous semisolid-semifluid masses are referred to, has not been distinctly applied to molecular friction, especially not to the molecular friction of a highly elastic solid within its limits of high elasticity, but has rather been employed to designate a property of slow, continual yielding through very great, or altogether unlimited, extent of change of shape, under the action of continued stress. It is in this sense that Forbes, for instance, has used the word in stating that 'Viscous Theory of Glacial Motion' which he demonstrated by his grand observations on glaciers. As, however, he, and many other writers after him, have used the words plasticity and plastic, both with reference to homogeneous solids (such as wax or pitch, even though also brittle; soft metals; etc.), and to heterogeneous semisolid-semifluid masses (as mud, moist earth, mortar, glacial ice, etc.), to designate the property[2], common to all those cases, of experiencing, under continued stress either quite continued and unlimited change of shape, or gradually very great change at a diminishing (asymptotic) rate through infinite time; and as the use of the term *plasticity* implies no more than does *viscosity*, any physical theory or explanation of the property, the word viscosity is without inconvenience left available for the definition we have given of it above.

684. A *perfect fluid*, or (as we shall call it) a fluid, is an unrealizable conception, like a rigid, or a smooth, body: it is defined as a body incapable of resisting a change of shape: and therefore incapable of experiencing distorting or tangential stress (§ 640). Hence its pressure on any surface, whether of a solid or of a contiguous portion of

[1] See *Proceedings of the Royal Society*, May 1865, 'On the Viscosity and Elasticity of Metals' (W. Thomson).

[2] Some confusion of ideas might have been avoided on the part of writers who have professedly objected to Forbes' theory while really objecting only (and we believe groundlessly) to his usage of the word viscosity, if they had paused to consider that no one physical explanation can hold for those several cases; and that Forbes' theory is merely the proof by observation that glaciers have the property that mud (heterogeneous), mortar (heterogeneous), pitch (homogeneous), water (homogeneous), all have of changing shape indefinitely and continuously under the action of continued stress.

the fluid, is at every point perpendicular to the surface. In equilibrium, all common liquids and gaseous fluids fulfil the definition. But there is finite resistance, of the nature of friction, opposing change of shape at a finite rate; and, therefore, while a fluid is changing shape, it exerts tangential force on every surface other than normal planes of the stress (§ 635) required to keep this change of shape going on. Hence; although the hydrostatical results, to which we immediately proceed, are verified in practice; in treating of hydrokinetics, in a subsequent chapter, we shall be obliged to introduce the consideration of fluid friction, except in cases where the circumstances are such as to render its effects insensible.

685. With reference to a fluid the *pressure at any point in any direction* is an expression used to denote the average pressure per unit of area on a plane surface imagined as containing the point, and perpendicular to the direction in question, when the area of that surface is indefinitely diminished.

686. At any point in a fluid at rest the pressure is the same in all directions: and, if no external forces act, the pressure is the same at every. point. For the proof of these and most of the following propositions, we imagine, according to § 584, a definite portion of the fluid to become solid, without changing its mass, form, cr dimensions.

Suppose the fluid to be contained in a closed vessel, the pressure within depending on the pressure exerted on it by the vessel, and not on any external force such as gravity.

687. The resultant of the fluid pressures on the elements of any portion of a spherical surface must, like each of its components, pass through the centre of the sphere. Hence, if we suppose (§ 584) a portion of the fluid in the form of a plano-convex lens to be solidified, the resultant pressure on the plane side must pass through the centre of the sphere; and, therefore, being perpendicular to the plane, must pass through the centre of the circular area. From this it is obvious that the pressure is the same at all points of any plane in the fluid. Hence the resultant pressure on any plane surface passes through its centre of inertia.

Next, imagine a triangular prism of the fluid, with ends perpendicular to its faces, to be solidified. The resultant pressures on its ends act in the line joining the centres of inertia of their areas, and are equal since the resultant pressures on the sides are in directions perpendicular to this line. Hence the pressure is the same in all parallel planes.

But the centres of inertia of the three faces, and the resultant pressures applied there, lie in a triangular section parallel to the ends. The pressures act at the middle points of the sides of this triangle, and perpendicularly to them, so that their directions meet in a point. And, as they are in equilibrium, they must be proportional to the respective sides of the triangle; that is, to the breadths, or areas, of the faces of the prism. Thus the resultant pressures on the faces must be proportional to the areas of the faces, and therefore the pressure is equal in any two planes which meet.

Collecting our results, we see that the pressure is the same at all points, and in all directions, throughout the fluid mass.

688. Hence if a force be applied at the centre of inertia of each face of a polyhedron, with magnitude proportional to the area of the face, the polyhedron will be in equilibrium. For we may suppose the polyhedron to be a solidified portion of the fluid. The resultant pressure on each face will then be proportional to its area, and will act at its centre of inertia; which, in this case, is the *Centre of Pressure*.

689. Another proof of the equality of pressure throughout a mass of fluid, uninfluenced by other external force than the pressure of the containing vessel, is easily furnished by the energy criterion of equilibrium, § 254; but, to avoid complication, we will consider the fluid to be incompressible. Suppose a number of pistons fitted into cylinders inserted in the sides of the closed vessel containing the fluid. Then, if A be the area of one of these pistons, p the average pressure on it, x the distance through which it is pressed, in or out; the energy criterion is that no work shall be done on the whole, i. e. that

$$A_1p_1x_1 + A_2p_2x_2 + \ldots = \Sigma(Apx) = 0,$$

as much work being restored by the pistons which are forced out, as is done by those forced in. Also, since the fluid is incompressible, it must have gained as much space by forcing out some of the pistons as it lost by the intrusion of the others. This gives

$$A_1x_1 + A_2x_2 + \ldots = \Sigma(Ax) = 0.$$

The last is the only condition to which x_1, x_2, etc., in the first equation, are subject; and therefore the first can only be satisfied if

$$p_1 = p_2 = p_3 = \text{etc.},$$

that is, if the pressure be the same on each piston. Upon this property depends the action of Bramah's *Hydrostatic Press*.

If the fluid be compressible, the work expended in compressing it from volume V to $V - \delta V$, at mean pressure p, is $p\delta V$.

If in this case we *assume* the pressure to be the same throughout, we obtain a result consistent with the energy criterion.

The work done on the fluid is $\Sigma\ (Apx)$, that is, in consequence of the assumption, $p\Sigma(Ax)$.

But this is equal to $$p\delta V,$$
for, evidently, $$\Sigma(Ax) = \delta V.$$

690. When forces, such as gravity, act from external matter upon the substance of the fluid, either in proportion to the density of its own substance in its different parts, or in proportion to the density of electricity, or of magnetic polarity, or of any other conceivable accidental property of it, the pressure will still be the same in all directions at any one point, but will now vary continuously from point to point. For the preceding demonstration (§ 687) may still be applied by simply taking the dimensions of the prism small enough; since the pressures are as the squares of its linear dimensions, and the effects of the applied forces such as gravity, as the cubes.

691. When forces act on the whole fluid, surfaces of equal pressure, if they exist, must be at every point perpendicular to the direction of the resultant force. For, any prism of the fluid so situated that the whole pressures on its ends are equal must experience from the applied forces no component in the direction of its length; and, therefore, if the prism be so small that from point to point of it the direction of the resultant of the applied forces does not vary sensibly, this direction must be perpendicular to the length of the prism. From this it follows that whatever be the physical origin, and the law, of the system of forces acting on the fluid, and whether it be conservative or non-conservative, the fluid cannot be in equilibrium unless the lines of force possess the geometrical property of being at right angles to a series of surfaces.

692. Again, considering two surfaces of equal pressure infinitely near one another, let the fluid between them be divided into columns of equal transverse section, and having their lengths perpendicular to the surfaces. The difference of pressure on the two ends being the same for each column, the resultant applied forces on the fluid masses composing them must be equal. Comparing this with § 506, we see that if the applied forces constitute a conservative system, the density of matter, or electricity, or whatever property of the substance they depend on, must be equal throughout the layer under consideration. This is the celebrated hydrostatic proposition that *in a fluid at rest, surfaces of equal pressure are also surfaces of equal density and of equal potential.*

693. Hence when gravity is the only external force considered, surfaces of equal pressure and equal density are (when of moderate extent) horizontal planes. On this depends the action of levels, siphons, barometers, etc.; also the separation of liquids of different densities (which do not mix or combine chemically) into horizontal strata, etc., etc. The free surface of a liquid is exposed to the pressure of the atmosphere simply; and therefore, when in equilibrium, must be a surface of equal pressure, and consequently level. In extensive sheets of water, such as the American lakes, differences of atmospheric pressure, even in moderately calm weather, often produce considerable deviations from a truly level surface.

694. The rate of increase of pressure per unit of length in the direction of the resultant force, is equal to the intensity of the force reckoned per unit of volume of the fluid. Let F be the resultant force per unit of volume in one of the columns of § 692; p and p' the pressures at the ends of the column, l its length, S its section. We have, for the equilibrium of the column,

$$(p'-p)\, S = SlF.$$

Hence the rate of increase of pressure per unit of length is F.

If the applied forces belong to a conservative system, for which V and V' are the values of the potential at the ends of the column, we have (§ 504)

$$V'-V=-lF\rho,$$

where ρ is the density of the fluid. This gives

$$p'-p=-\rho(V'-V),$$

or

$$dp=-\rho dV.$$

Hence in the case of gravity as the only impressed force the rate of increase of pressure per unit of depth in the fluid is ρ, in gravitation measure (usually employed in hydrostatics). In kinetic or absolute measure (§ 189) it is $g\rho$.

If the fluid be a gas, such as air, and be kept at a constant temperature, we have $\rho=cp$, where c denotes a constant, the reciprocal of H, the 'height of the homogeneous atmosphere,' defined (§ 695) below. Hence, in a calm atmosphere of uniform temperature we have

$$\frac{dp}{p}=-cdV;$$

and from this, by integration,

$$p=p_0\epsilon^{-cV};$$

where p_0 is the pressure at any particular level (the sea-level, for instance) where we choose to reckon the potential as zero.

When the differences of level considered are infinitely small in comparison with the earth's radius, as we may practically regard them, in measuring the heights of mountains, or of a balloon, by the barometer, the force of gravity is constant, and therefore differences of potential (force being reckoned in units of weight) are simply equal to differences of level. Hence if x denote height of the level of pressure p above that of p_0, we have, in the preceding formulae, $V=x$, and therefore

$$p=p_0\epsilon^{-cx}; \text{ that is,}$$

695. If the air be at a constant temperature, the pressure diminishes in geometrical progression as the height increases in arithmetical progression. This theorem is due to Halley. Without formal mathematics we see the truth of it by remarking that differences of pressure are (§ 694) equal to differences of level multiplied by the density of the fluid, or by the proper mean density when the density differs sensibly between the two stations. But the density, when the temperature is constant, varies in simple proportion to the pressure, according to Boyle's law. Hence differences of pressure between pairs of stations differing equally in level are proportional to the proper mean values of the whole pressure, which is the well-known compound interest law. The rate of diminution of pressure per unit of length upwards in proportion to the whole pressure at any point, is of course equal to the reciprocal of the height above that point that the atmosphere must have, if of constant density, to give that pressure by its weight. The height thus defined is commonly called 'the height of the homogeneous atmosphere,' a very convenient conventional expression. It is equal to the product of the volume occupied by the unit mass of the gas at any pressure into the value of that pressure reckoned per unit of area, in terms of the weight of the unit of mass. If we denote it by H, the exponential expression of the law is

$$p=p_0\epsilon^{-\frac{z}{H}},$$

which agrees with the final formula of § 694.

The value of H for dry atmospheric air, at the freezing temperature, according to Regnault, is, in the latitude of Paris, 799,020 centimetres, or 26,215 feet. Being inversely as the force of gravity in different latitudes (§ 187), it is 798,533 centimetres, or 26,199 feet, in the latitude of Edinburgh and Glasgow.

696. It is both necessary and sufficient for the equilibrium of an incompressible fluid completely filling a rigid closed vessel, and influenced only by a conservative system of forces, that its density be uniform over every equipotential surface, that is to say, every surface cutting the lines of force at right angles. If, however, the boundary, or any part of the boundary, of the fluid mass considered, be not rigid; whether it be of flexible solid matter (as a membrane, or a thin sheet of elastic solid), or whether it be a mere geometrical boundary, on the other side of which there is another fluid, or *nothing* [a case which, without believing in vacuum as a reality, we may admit in abstract dynamics (§ 391)], a farther condition is necessary to secure that the pressure from without shall fulfil the hydrostatic equation at every point of the boundary. In the case of a bounding membrane, this condition must be fulfilled either through pressure artificially applied from without, or through the interior elastic forces of the matter of the membrane. In the case of another fluid of different density touching it on the other side of the boundary, all round or over some part of it, with no separating membrane, the condition of equilibrium of a heterogeneous fluid is to be fulfilled relatively to the whole fluid mass made up of the two; which shows that at the boundary the pressure must be constant and equal to that of the fluid on the other side. Thus water, oil, mercury, or any other liquid, in an open vessel, with its free surface exposed to the air, requires for equilibrium simply that this surface be level.

697. Recurring to the consideration of a finite mass of fluid completely filling a rigid closed vessel, we see, from what precedes, that, if homogeneous and incompressible, it cannot be disturbed from equilibrium by any conservative system of forces; but we do not require the analytical investigation to prove this, as we should have 'the perpetual motion' if it were denied, which would violate the hypothesis that the system of forces is conservative. On the other hand, a non-conservative system of forces cannot, under any circumstances, equilibrate a fluid which is either uniform in density throughout, or of homogeneous substance, rendered heterogeneous in density only through difference of pressure. But if the forces, though not conservative, be such that through every point of the space occupied by the fluid a surface may be drawn which shall cut at right angles all the lines of force it meets, a heterogeneous fluid will rest in equilibrium under their influence, provided (§ 692) its density, from point to point of every one of these orthogonal surfaces, varies inversely as the product of the resultant force into the thickness of the infinitely thin layer of space between that surface and another of the orthogonal surfaces infinitely near it on either side. (Compare § 506).

698. If we imagine all the fluid to become rigid except an infinitely thin closed tabular portion lying in a surface of equal density, and if the fluid in this tabular circuit be moved any length along the tube and left at rest, it will remain in equilibrium in the new position, all positions of it in the tube being indifferent because of its homogeneousness. Hence the work (positive or negative) done by the force (X, Y, Z) on any portion of the fluid in any displacement along the tube is balanced by the work (negative or positive) done on the remainder of the fluid in the tube. Hence a single particle, acted on only by X, Y, Z, while moving round the circuit, that is moving along any closed curve on a surface of equal density, has, at the end of one complete circuit, done just as much work against the force in some parts of its course, as the forces have done on it in the remainder of the circuit.

699. The following imaginary example, and its realization in a subsequent section (§ 701), show a curiously interesting practical application of the theory of fluid equilibrium under extraordinary circumstances, generally regarded as a merely abstract analytical theory, practically useless and quite unnatural, 'because forces in nature follow the conservative law.'

700. Let the lines of force be circles, with their centres all in one line, and their planes perpendicular to it. They are cut at right angles by planes through this axis; and therefore a fluid may be in equilibrium under such a system of forces. The system will not be conservative if the intensity of the force be according to any other law than inverse proportionality to distance from this axial line; and the fluid, to be in equilibrium, must be heterogeneous, and be so distributed as to vary in density from point to point of every plane through the axis, inversely as the product of the force into the distance from the axis. But from one such plane to another it may be either uniform in density, or may vary arbitrarily. To particularize farther, we may suppose the force to be in direct simple proportion to the distance from the axis. Then the fluid will be in equilibrium if its density varies from point to point of every plane through the axis, inversely as the square of that distance. If we still farther particularize by making the force uniform all round each circular line of force, the distribution of force becomes precisely that of the kinetic reactions of the parts of a rigid body against accelerated rotation. The fluid pressure will (§ 691) be equal over each plane through the axis. And in one such plane, which we may imagine carried round the axis in the direction of the force, the fluid pressure will increase in simple proportion to the angle at a rate per unit angle (§ 55) equal to the product of the density at unit distance into the force at unit distance. Hence it must be remarked, that if any closed line (or circuit) can be drawn round the axis, without leaving the fluid, there cannot be equilibrium without a firm partition cutting every such circuit, and maintaining the difference of pressures on the two sides of it, corresponding to the angle 2π. Thus, if the axis pass through the fluid in any part, there must be a partition extending from this part of the axis continuously to the outer bounding surface of the fluid. Or if the bounding surface of the whole fluid be annular (like a hollow anchor-ring, or of any irregular shape), in other words, if the fluid fills a tubular circuit; and the axis (A) pass through the aperture of the ring (without passing into the fluid); there must be a firm partition (CD) extending somewhere continuously across the channel, or passage, or tube, to stop the circulation of the fluid round it; otherwise there could not be equilibrium with the supposed forces in action. If we further suppose the density of the fluid to be uniform round each of the circular lines of force in the system we have so far considered (so that the density shall be equal over every circular cylinder having the line of their centres for its axis, and shall vary from one such cylindrical surface to another, inversely as the squares of their radii), we may, without disturbing the equilibrium, impose any conservative system of force in lines perpendicular to the axis; that is (§ 506), any system of force in this direction, with intensity varying as some function of the distance. If this function be the simple distance, the superimposed system of force agrees precisely with the reactions against curvature, that is to say, the centrifugal forces, of the parts of a rotating rigid body.

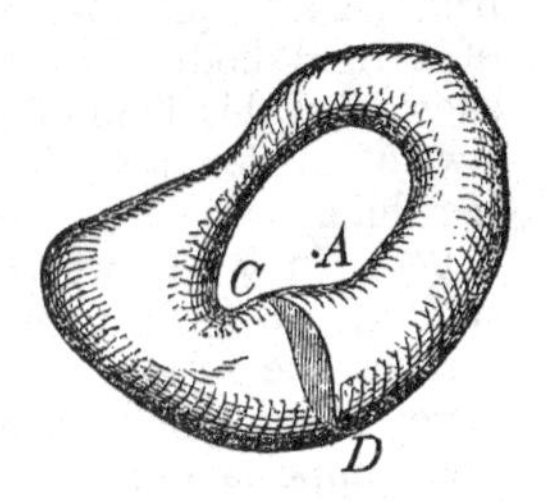

701. Thus we arrive at the remarkable conclusion, that if a rigid closed box be completely filled with incompressible heterogeneous fluid, of density varying inversely as the square of the distance from a certain line, and if the box be movable round this line as a fixed axis, and be urged in any way by forces applied to its outside, the fluid will remain in equilibrium relatively to the box; that is to say, will move round with the box as if the whole were one rigid body, and will come to rest with the box if the box be brought again to rest: provided always the preceding condition as to partitions be fulfilled if the axis pass through the fluid, or be surrounded by continuous lines of fluid. For, in starting from rest, *if* the fluid moves like a rigid solid, we have reactions against acceleration, tangential to the circles of motion, and equal in amount to $\dot{\omega}r$ per unit of mass of the fluid at distance r from the axis, $\dot{\omega}$ being the rate of acceleration (§ 57) of the angular velocity; and (see Vol. II.) we have, in the direction perpendicular to the axis outwards, reaction against curvature of path, that is to say, 'centrifugal force,' equal to $\omega^2 r$ per unit of mass of the fluid. Hence the equilibrium which we have demonstrated in the preceding section, for the fluid supposed at rest, and arbitrarily influenced by two systems of force (the circular non-conservative and the radical conservative system) agreeing in law with these forces of kinetic reaction, proves for us now the D'Alembert (§ 230) equilibrium condition for the motion of the whole fluid as of a rigid body experiencing accelerated rotation: that is to say, shows that this kind of motion fulfils for the actual circumstances the laws of motion, and, therefore, that it is *the* motion actually taken by the fluid.

702. In § 688 we considered the resultant pressure on a plane surface, when the pressure is uniform. We may now consider (briefly) the resultant pressure on a plane area when the pressure varies from point to point, confining our attention to a case of great importance;—that in which gravity is the only applied force, and the fluid is a nearly incompressible liquid such as water. In this case the determination of the position of the Centre of Pressure is very simple; and the whole pressure is the same as if the plane area were turned about its centre of inertia into a horizontal position.

The pressure at any point at a depth z in the liquid may be expressed by

$$p = \rho z + p_0,$$

where ρ is the (constant) density of the liquid, and p_0 the (atmospheric) pressure at the free surface, reckoned in units of weight per unit of area.

Let the axis of x be taken as the intersection of the plane of the immersed plate with the free surface of the liquid, and that of y perpendicular to it and in the plane of the plate. Let a be the inclination of the plate to the vertical. Let also A be the area of the portion of the plate considered, and $\bar{x}$, $\bar{y}$, the co-ordinates of its centre of inertia.

Then the whole pressure is

$$\iint p\,dx\,dy = \iint (p_0 + \rho y \cos a)\,dx\,dy$$
$$= Ap_0 + A\rho\bar{y}\cos a.$$

The moment of the pressure about the axis of x is

$$\iint py\,dx\,dy = Ap_0\bar{y} + Ak^2\rho\cos a,$$

k being the radius of gyration of the plane area about the axis of x.

For the moment about y we have

$$\iint px\,dx\,dy = Ap_0\bar{x} + \rho\cos a \iint xy\,dx\,dy.$$

The first terms of these three expressions merely give us again the results of § 688; we may therefore omit them. This will be equivalent to introducing a stratum of additional liquid above the free surface such as to produce an equivalent to the atmospheric pressure.

If the origin be now shifted to the upper surface of this stratum we have

$$\text{Pressure} = A\rho\bar{y}\cos\alpha,$$

$$\text{Moment about } Ox = Ak^2\rho\cos\alpha,$$

$$\text{Distance of centre of pressure from axis of } x = \frac{k^2}{\bar{y}}.$$

But if k_1 be the radius of gyration of the plane area about a horizontal axis in its plane, and passing through its centre of inertia, we have

$$k^2 = k_1^2 + \bar{y}^2.$$

Hence the distance, measured parallel to the axis of y, of the centre of pressure from the centre of inertia is

$$\frac{k_1^2}{\bar{y}};$$

and, as we might expect, diminishes as the plane area is more and more submerged. If the plane area be turned about the line through its centre of inertia parallel to the axis of x, this distance varies as the cosine of its inclination to the vertical; supposing, of course, that by the rotation neither more nor less of the plane area is submerged.

703. A body, wholly or partially immersed in any fluid influenced by gravity, loses, through fluid pressure, in apparent weight an amount equal to the weight of the fluid displaced. For if the body were removed, and its place filled with fluid homogeneous with the surrounding fluid, there would be equilibrium, even if this fluid be supposed to become rigid. And the resultant of the fluid pressure upon it is therefore a single force equal to its weight, and in the vertical line through its centre of gravity. But the fluid pressure on the originally immersed body was the same all over as on the solidified portion of fluid by which for a moment we have imagined it replaced, and therefore must have the same resultant. This proposition is of great use in Hydrometry, the determination of specific gravity, etc., etc.

704. The following lemma, while in itself interesting, is of great use in enabling us to simplify the succeeding investigations regarding the stability of equilibrium of floating bodies:—

Let a homogeneous solid, the weight of unit of volume of which we suppose to be unity, be cut by a horizontal plane in $XYX'Y'$. Let O be the centre of inertia, and let XX', YY' be the principal axes, of this area.

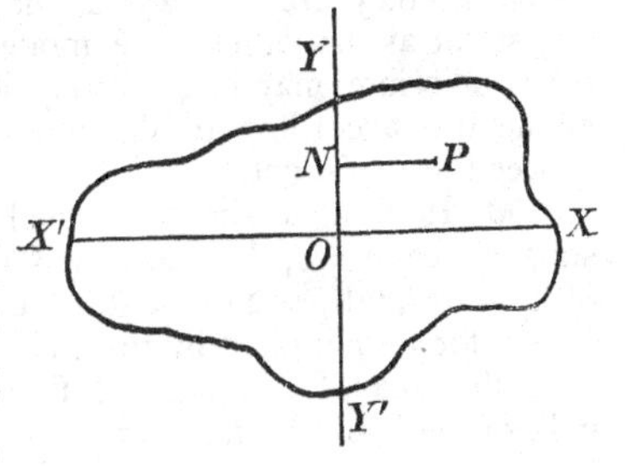

Let there be a second plane section of the solid, through YY', inclined to the first at an infinitely small angle, θ. Then (1) the volumes of the two wedges cut from the solid by these sections are equal; (2) their centres of inertia lie in one plane perpendicular to YY'; and (3) the moment of the weight of each of these, round YY', is equal to the moment of inertia about it of the corresponding portion of the area multiplied by θ.

Take OX, OY as axes, and let θ be the angle of the wedge: the thickness of the wedge at any point P, (x, y), is θx, and the volume of a right prismatic portion whose base is the elementary area $dxdy$ at P is

$$\theta x dx dy.$$

Now let [ ] and ( ) be employed to distinguish integrations extended over the portions of area to the right and left of the axis of y respectively, while integrals over the whole area have no such distinguishing mark. Let a and a' be these areas, v and v' the volumes of the wedges; $(\bar{x}, \bar{y})$, $(\bar{x}', \bar{y}')$ the co-ordinates of their centres of inertia. Then

$$v = \theta\,[\textstyle\iint x dx dy] = a\bar{x}$$

$$-v' = \theta\,(\textstyle\iint x dx dy) = a'\bar{x}',$$

whence $v - v' = \theta \iint x dx dy = 0$ since O is the centre of inertia. Hence

$$v = v', \text{ which is (1).}$$

Again, taking moments about XX',

$$v\bar{y} = \theta\,[\textstyle\iint xy dx dy],$$

and

$$-v'\bar{y}' = \theta\,(\textstyle\iint xy dx dy).$$

Hence

$$v\bar{y} - v'\bar{y}' = \theta \textstyle\iint xy dx dy.$$

But for a principal axis $\Sigma xy dm$ vanishes. Hence $v\bar{y} - v'\bar{y}' = 0$, whence, since $v = v'$, we have

$$\bar{y} = \bar{y}', \text{ which proves (2).}$$

And (3) is merely a statement in words of the obvious equation

$$[\textstyle\iint x.x\theta dx dy] = \theta\,[\textstyle\iint x^2.dx dy].$$

705. If a positive amount of work is required to produce any possible infinitely small displacement of a body from a position of equilibrium, the equilibrium in this position is stable (§ 256). To apply this test to the case of a floating body, we may remark, first, that any possible infinitely small displacement may (§§ 30, 106) be conveniently regarded as compounded of two horizontal displacements in lines at right angles to one another, one vertical displacement, and three rotations round rectangular axes through any chosen point. If one of these axes be vertical, then three of the component displacements, viz. the two horizontal displacements and the rotation about the vertical axis, require no work (positive or negative), and therefore, so far as they are concerned, the equilibrium is essentially neutral. But so far as the other three modes of displacement are concerned, the equilibrium may be positively stable, or may be unstable, or may be neutral, according to the fulfilment of conditions which we now proceed to investigate.

706. If, first, a simple vertical displacement, downwards, let us suppose, be made, the work is done against an increasing resultant of upward fluid pressure, and is of course equal to the mean increase of this force multiplied by the whole space. If this space be denoted by z, the area of the plane of flotation by A, and the weight of unit bulk of the liquid by w, the increased bulk of immersion is clearly Az, and therefore the increase of the resultant of fluid pressure is wAz, and is in a line vertically upward through the centre of gravity of A. The mean force against which the work is done is therefore $\frac{1}{2}wAz$, as this is a case in which work is done against a force increasing from zero in simple proportion to the space. Hence the work done is $\frac{1}{2}wAz^2$. We see, therefore, that so far as vertical displacements alone are concerned, the equilibrium is necessarily stable, unless the body is wholly immersed, when the area of the plane of flotation vanishes, and the equilibrium is neutral.

707. The lemma of § 704 suggests that we should take, as the two horizontal axes of rotation, the principal axes of the plane of flotation. Considering then rotation through an infinitely small angle

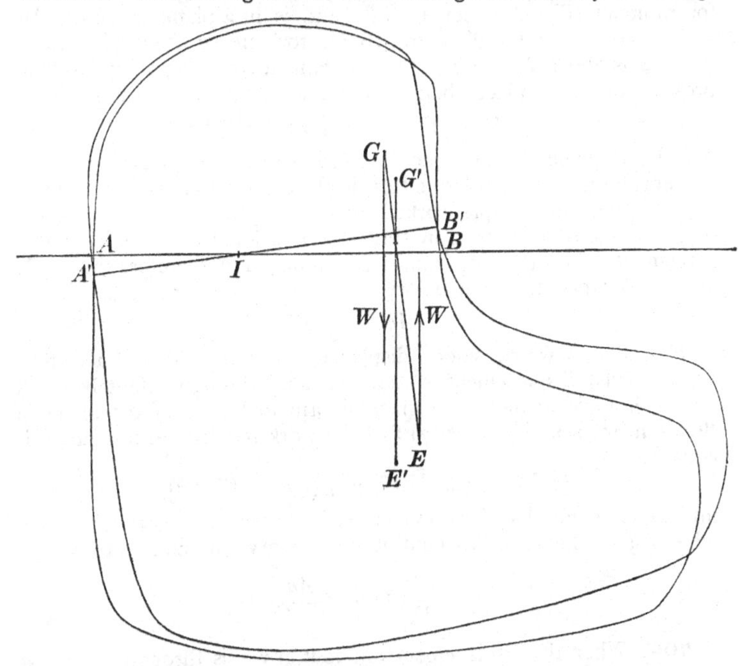

θ round one of these, let G and E be the displaced centres of gravity of the solid, and of the portion of its volume which was immersed when it was floating in equilibrium, and G', E' the positions which they then had; all projected on the plane of the diagram which we suppose to be through I the centre of inertia of the plane of flotation. The resultant action of gravity on the displaced body is W, its weight, acting downwards through G; and that of the fluid pressure on it is W upwards through E corrected by the amount (upwards) due to the additional immersion of the wedge AIA', and the amount (downwards) due to the extruded wedge $B'IB$. Hence the whole action of gravity and fluid pressure on the displaced body is the couple of forces up and down in verticals through G and E, and the correction due to the wedges. This correction consists of a force vertically upwards through the centre of gravity of $A'IA$, and downwards through that of BIB'. These forces are equal [§ 704 (1)], and therefore constitute a couple which [704 (2)] has the axis of the displacement for its axis, and which [§ 704 (3)] has its moment equal to $\theta w k^2 A$ if A be the area of the plane of flotation, and k its radius of gyration (§ 235) round the principal axis in question. But since GE, which was vertical ($G'E'$) in the position of equilibrium, is inclined at the infinitely small angle θ to the vertical in the displaced body, the couple of forces W in the verticals through G and E has for moment $Wh\theta$, if h denote GE; and is in a plane perpendicular to the axis, and in the direction tending to increase the displacement, when G is above E. Hence the resultant action of gravity and fluid pressure on the displaced body is a couple whose moment is

$$(wAk^2 - Wh)\theta, \text{ or } w(Ak^2 - Vh)\theta,$$

if V be the volume immersed. It follows that when $Ak^2 > Vh$ the equilibrium is stable, so far as this displacement alone is concerned.

Also, since the couple worked against in producing the displacement increases from zero in simple proportion to the angle of displacement, its mean value is half the above; and therefore the whole amount of work done is equal to

$$\tfrac{1}{2}w(Ak^2 - Vh)\theta^2.$$

708. If now we consider a displacement compounded of a vertical (downwards) displacement z, and rotations through infinitely small angles θ, θ' round the two horizontal principal axes of the plane of flotation, we see (§§ 706, 707) that the work required to produce it is equal to

$$\tfrac{1}{2}w[Az^2 + (Ak^2 - Vh)\theta^2 + (Ak'^2 - Vh)\theta'^2],$$

and we conclude that, for complete stability with reference to all possible displacements of this kind, it is necessary and sufficient that

$$h < \frac{Ak^2}{V}, \text{ and } < \frac{Ak'^2}{V}.$$

709. When the displacement is about any axis through the centre of inertia of the plane of flotation, the resultant of fluid pressures is equal to the weight of the body; but it is only when the axis is a principal axis of the plane of flotation that this resultant is in the plane of displacement. In such a case the point of intersection of the resultant with the line originally vertical, and through the centre of gravity of the body, is called the *Metacentre*. And it is obvious, from the above investigation, that for either of these planes of displacement the condition of stable equilibrium is that the metacentre shall be *above* the centre of gravity.

710. We shall conclude with the consideration of one case of the equilibrium of a revolving mass of fluid subject only to the gravitation of its parts, which admits of a very simple synthetical solution, without any restriction to approximate sphericity; and for which the following remarkable theorem was discovered by Newton and Maclaurin :—

711. An oblate ellipsoid of revolution, of any given eccentricity, is a figure of equilibrium of a mass of homogeneous incompressible fluid, rotating about an axis with determinate angular velocity, and subject to no forces but those of gravitation among its parts.

The angular velocity for a given eccentricity is independent of the bulk of the fluid, and proportional to the square root of its density.

712. The proof of this proposition is easily obtained from the results already deduced with respect to the attraction of an ellipsoid and the properties of the free surface of a fluid.

We know, § 538, that if APB be a meridian section of a homogeneous oblate spheroid, AC the polar axis, CB an equatorial radius, and P any point on the surface, the attraction of the spheroid may be resolved into two parts; one, Pp, perpendicular to the polar axis, and varying as the ordinate PM; the other, Ps, parallel to the polar axis, and varying as PN. These components are not equal when MP and PN are equal, else the resultant attraction at all points in the surface would pass through C; whereas we know that it is in some such direction as Pf, cutting the radius BC *between* B and C, but at a point nearer to C than n the foot of the normal at P. Let then

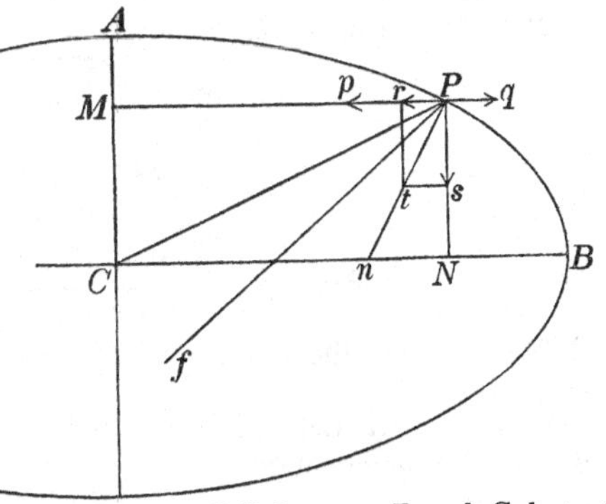

$$Pp = a.PM,$$
$$\text{and } Ps = \beta.PN,$$

where a and β are known constants, depending merely on the density (ρ), and eccentricity (e), of the spheroid.

Also, we know by geometry that $Nn = (1 - e^2)\, CN$.

Hence; to find the magnitude of a force Pq perpendicular to the axis of the spheroid, which, when compounded with the attraction, will bring the resultant force into the normal Pn: make $pr = Pq$, and we must have

$$\frac{Pr}{Ps} = \frac{Nn}{PN} = (1 - e^2)\frac{CN}{PN} = (1 - e^2)\frac{\beta Pp}{aPs}.$$

Hence $$Pr = (1 - e^2)\frac{\beta}{a} Pp,$$

$$Pp - Pq = (1 - e^2)\frac{\beta}{a}Pp,$$

or

$$Pq = \left(1 - (1 - e^2)\frac{\beta}{a}\right)Pp$$

$$= (a - (1 - e^2)\beta)PM.$$

Now if the spheroid were to rotate with angular velocity ω about AC, the centrifugal force, §§ 39, 42, 225, would be in the direction Pq, and would amount to

$$\omega^2 PM.$$

Hence, if we make

$$\omega^2 = a - (1 - e^2)\beta,$$

the whole force on P, that is, the resultant of the attraction and centrifugal force, will be in the direction of the normal to the surface, which is the condition for the free surface of a mass of fluid in equilibrium.

Now, (§ 522 of our larger work)

$$a = 2\pi\rho\left(\frac{\sqrt{1-e^2}}{e^3}\sin^{-1}e - \frac{1-e^2}{e^2}\right),$$

$$\beta = 4\pi\rho\left(\frac{1}{e^2} - \frac{\sqrt{1-e^2}}{e^2}\sin^{-1}e\right).$$

Hence

$$\omega^2 = 2\pi\rho\left\{\frac{(3-2e^2)\sqrt{1-e^2}}{e^2}\sin^{-1}e - 3\frac{1-e^2}{e^2}\right\}. \qquad (1)$$

This determines the angular velocity, and proves it to be proportional to $\sqrt{\rho}$.

713. If, after Laplace, we introduce instead of e a quantity ϵ defined by the equation

$$\left.\begin{aligned} 1 - e^2 &= \frac{1}{1+\epsilon^2}, \\ \text{or} \quad \epsilon &= \frac{e}{\sqrt{1-e^2}} = \tan(\sin^{-1}e), \end{aligned}\right\} \qquad (2)$$

the expression (1) for ω^2 is much simplified, and

$$\frac{\omega^2}{2\pi\rho} = \frac{3+\epsilon^2}{\epsilon^2}\tan^{-1}\epsilon - \frac{3}{\epsilon^2}. \qquad (3)$$

When e, and therefore also ϵ, is small, this formula is most easily calculated from

$$\frac{\omega^2}{2\pi\rho} = \tfrac{4}{15}\epsilon^2 - \tfrac{8}{35}\epsilon^4 + \text{etc.} \qquad (4)$$

of which the first term is sufficient when we deal with spheroids so little oblate as the earth.

The following table has been calculated by means of these simplified formulae. The last figure in each of the four last columns is given to the nearest unit. The two last columns will be explained a few sections later:—

i.	ii.	iii.	iv.	v.
e.	$\frac{1}{\epsilon}$.	$\frac{\omega^2}{2\pi\rho}$	$\frac{2\pi}{\omega}$ when $\rho = 3{\cdot}68 \times 10^{-7}$.	$(1+\epsilon^2)^{\frac{2}{3}} \frac{\omega^2}{2\pi\rho}$
0·1	9·950	0·0027	79,966	0·0027
·2	4·899	·0107	39,397	·0110
·3	3·180	·0243	26,495	·0258
·4	2·291	·0436	19,780	·0490
·5	1·732	·0690	15,730	·0836
·6	1·333	·1007	13,022	·1356
·7	1·020	·1387	11,096	·2172
·8	0·750	·1816	9,697	·3588
·9	·4843	·2203	8,804	·6665
·91	·4556	·2225	8,759	·7198
·92	·4260	·2241	8,729	·7813
·93	·3952	·2247	8,718	·8533
·94	·3629	·2239	8,732	·9393
·95	·3287	·2213	8,783	1·045
·96	·2917	·2160	8,891	1·179
·97	·2506	·2063	9,098	1·359
98	·2030	·1890	9,504	1·627
·99	·1425	·1551	10,490	2·113
1·00	0·0000	0.0000	∞	∞

From this we see that the value of $\frac{\omega^2}{2\pi\rho}$ increases gradually from zero to a maximum as the eccentricity e rises from zero to about 0·93, and then (more quickly) falls to zero as the eccentricity rises from 0·93 to unity. The values of the other quantities corresponding to this maximum are given in the table.

714. If the angular velocity exceed the value calculated from

$$\frac{\omega^2}{2\pi\rho} = 0{\cdot}2247, \qquad (5)$$

when for ρ is substituted the density of the liquid, equilibrium is impossible in the form of an ellipsoid of revolution. If the angular velocity fall short of this limit there are always two ellipsoids of revolution which satisfy the conditions of equilibrium. In one of these the eccentricity is greater than 0·93, in the other less.

715. It may be useful, for special applications, to indicate briefly how ρ is measured in these formulae. In the definitions of §§ 476, 477, on which the attraction formulae are based, unit mass is defined as exerting unit force on unit mass at unit distance; and unit volume-density is that of a body which has unit mass in unit volume. Hence, with the foot as our linear unit, we have for the earth's attraction on a particle of unit mass at its surface

$$\frac{\frac{4}{3}\pi\sigma R^3}{R^2} = \tfrac{4}{3}\pi\sigma R = 32{\cdot}2\,;$$

where R is the radius of the earth (supposed spherical) in feet, and σ its mean density, expressed in terms of the unit just defined.

Taking 20,900,000 feet as the value of R, we have

$$\sigma = 0{\cdot}000000368 = 3{\cdot}68 \times 10^{-7}. \qquad (6)$$

As the mean density of the earth is somewhere about 5·5 times that of water, the density of water in terms of our present unit is

$$\frac{3{\cdot}68}{5{\cdot}5} 10^{-7} = 6{\cdot}7 \times 10^{-8}.$$

716. The fourth column of the table above gives the time of rotation in seconds, corresponding to each value of the eccentricity, ρ being assumed equal to the mean density of the earth. For a mass of water these numbers must be multiplied by $\sqrt{5{\cdot}5}$; as the time of rotation to give the same figure is inversely as the square root of the density.

For a homogeneous liquid mass, of the earth's mean density, rotating in $23^h\ 46^m\ 4^s$ we find $e = 0{\cdot}093$, which corresponds to an ellipticity of about $\frac{1}{230}$.

717. An interesting form of this problem, also discussed by Laplace, is that in which the moment of momentum and the mass of the fluid are given, not the angular velocity; and it is required to find what is the eccentricity of the corresponding ellipsoid of revolution, the result proving that there can be but one.

It is evident that a mass of any ordinary liquid (not a *perfect fluid*, § 684), if left to itself in any state of motion, must preserve unchanged its moment of momentum, § 202. But the viscosity, or internal friction, § 684, will, if the mass remain continuous, ultimately destroy all relative motion among its parts; so that it will ultimately rotate as a rigid solid. If the final form be an ellipsoid of revolution, we can easily show that there is a single definite value of its eccentricity. But, as it has not yet been discovered whether there is any other form consistent with *stable* equilibrium, we do not know that the mass will necessarily assume the form of this particular ellipsoid. Nor in fact do we know whether even the ellipsoid of rotation may not become an *unstable* form if the moment of momentum exceed some limit depending on the mass of the fluid. We shall return to this subject in Vol. II., as it affords an excellent example of that difficult and delicate question *Kinetic Stability*, § 300.

If we call a the equatorial semi-axis of the ellipsoid, e its eccentricity, and ω its angular velocity of rotation, the *given* quantities are the mass

$$M = \tfrac{4}{3}\pi\rho a^3 \sqrt{1-e^2},$$

and the moment of momentum

$$A = \tfrac{8}{15}\pi\rho\omega a^5 \sqrt{1-e^2}.$$

These equations, along with (2), determine the three quantities, a, e, and ω.

Eliminating a between the two just written, and expressing e as before in terms of ϵ, we have

$$\frac{A^2}{M^{\frac{10}{3}}} = \frac{2}{25}\left(\frac{3}{4}\right)^{\frac{1}{3}} \frac{\omega^2(1+\epsilon^2)^{\frac{2}{3}}}{(\pi\rho)^{\frac{4}{3}}}.$$

This gives

$$\frac{\omega^2}{2\pi\rho} = \frac{k}{(1+\epsilon^2)^{\frac{2}{3}}}$$

where k is a *given* multiple of $\rho^{\frac{1}{3}}$. Substituting in 771 (2) we have

$$k = (1+\epsilon^2)^{\frac{2}{3}} \left(\frac{3+\epsilon^2}{\epsilon^3} \tan^{-1}\epsilon - \frac{3}{\epsilon^2}\right).$$

Now the last column of the table in § 713 shows that the value of this function of ϵ (which vanishes with ϵ) continually increases with ϵ, and becomes infinite when ϵ is infinite. Hence there is always one, and only one, value of ϵ, and therefore of e, which satisfies the conditions of the problem.

718. All the above results might without much difficulty have been obtained analytically, by the discussion of the equations; but we have preferred, for once, to show by an actual case that numerical calculation may sometimes be of very great use.

719. No one seems yet to have attempted to solve the general problem of finding all the forms of equilibrium which a mass of homogeneous incompressible fluid rotating with uniform angular velocity may assume. Unless the velocity be so small that the figure differs but little from a sphere, the problem presents difficulties of an exceedingly formidable nature. It is therefore of some importance to know that we can by a synthetical process show that another form, besides that of the ellipsoid of revolution, may be compatible with equilibrium; viz. an ellipsoid with three unequal axes, of which the least is the axis of rotation. This curious theorem was discovered by Jacobi in 1834, and seems, simple as it is, to have been enunciated by him as a challenge to the French mathematicians[1]. For the proof we must refer to our larger work.

[1] See a Paper by Liouville, *Journal de l'École Polytechnique*, cahier xxiii., footnote to p. 290.

APPENDIX.

KINETICS.

(*a*) In the case of the *Simple Pendulum*, a heavy particle is suspended from a point by a light inextensible string. If we suppose it to be drawn aside from the vertical position of equilibrium and allowed to fall, it will oscillate in one plane about its lowest position. When the string has an inclination θ to the vertical, the weight mg of the particle may be resolved into $mg \cos \theta$ which is balanced by the tension of the string, and $mg \sin \theta$ in the direction of the tangent to the path. If l be the length of the string, the distance (along the arc) from the position of equilibrium is $l\theta$.

Now if the angle of oscillation be small (not above 3° or 4° say), the sine and the angle are nearly equal to each other. Hence the acceleration of the motion (which is rigorously $g \sin \theta$) may be written $g\theta$. Hence we have a case of motion in which the acceleration is proportional to the distance from a point in the path, that is, by § 74, *Simple Harmonic Motion*. The square of the angular velocity in the corresponding circular motion is $\frac{\text{acceleration}}{\text{displacement}} = \frac{g}{l}$, and the period of the harmonic motion is therefore $2\pi\sqrt{\frac{l}{g}}$. In the case of the pendulum, the time of an oscillation from side to side of the vertical is usually taken—and is therefore $\pi\sqrt{\frac{l}{g}}$.

(*b*) Thus the times of vibration of different pendulums are as the square roots of their lengths, for *any* arcs of vibration, provided only these be *small*.

Also the times of vibration of the same pendulum at different places are inversely as the square roots of the apparent force of gravity on a unit mass at these places.

(*c*) It was found experimentally by Newton that pendulums of the same length vibrate in equal times at the same place whatever be the material of which their bobs are formed. This would evidently not be the case unless the weight were in every case proportional to the amount of matter in the bob.

(*d*) If the simple pendulum be slightly disturbed in *any* way from its position of equilibrium, it will in general describe very nearly an ellipse about its lowest position as centre. This is easily seen from § 82.

(*e*) If the arc of vibration be considerable, the motion will not be simple harmonic, and the time of vibration will be greater than that above stated; since the acceleration being as the sine of the displacement, is in less and less ratio to the displacement as the latter is greater.

In this case, the motion for *any* disturbance is, for one revolution, approximately elliptic as before; but the ellipse slowly turns round the vertical, in the direction in which the bob moves.

(*f*) The bob may, however, be so projected as to revolve uniformly in a horizontal circle, in which case the apparatus is called a *Conical Pendulum.* Here we have $l \sin \theta$ for the radius of the circle, and the force in the direction of the radius is $T \sin \theta$, where T is the tension of the string. $T \cos \theta$ balances mg—and thus the force in the radius of the circle is $mg \tan \theta$. The square of the angular velocity in the circle is therefore $\frac{g}{l \cos \theta}$, and the time of revolution $2\pi \sqrt{\frac{l \cos \theta}{g}}$; or $2\pi \sqrt{\frac{h}{g}}$, where h is the height of the point of suspension above the plane of the circle. Thus all conical pendulums with the same height revolve in the same time.

(*g*) A rigid mass oscillating about a horizontal axis, under the action of gravity, constitutes what is called a *Compound Pendulum.*

When in the course of its motion the body is inclined at any angle θ to the position in which it hangs, when in equilibrium, it experiences from gravity, and the resistance of the supports of its axis, a couple, which is easily seen to be equal to

$$gWh \sin \theta,$$

where W is the mass and h the distance of its centre of gravity from the axis. This couple produces (§§ 232, 235) acceleration of angular velocity, calculated by dividing the moment of the couple by the moment of inertia of the body. Hence, if I denote the moment of inertia about the supporting axis, the angular acceleration is equal to

$$\frac{gW \sin \theta . h}{I}.$$

Its motion is, therefore, identical (§ (*a*)) with that of the simple pendulum of length equal to $\frac{I}{Wh}$.

If a rigid body be supported about an axis, which either passes very nearly through the centre of gravity, or is at a very great distance from this point, the length of the equivalent simple pendulum will be very great: and it is clear that some particular distance for the point of support from the centre of gravity will render the length of the corresponding simple pendulum, and, therefore, the time of vibration, least possible.

To investigate these circumstances for all axes parallel to a given line, through the centre of gravity, let k be the radius of gyration round this line, we have (§ 198),

$$I=W(k^2+h^2);$$

and, therefore, if l be the length of the isochronous simple pendulum,

$$l=\frac{h^2+k^2}{h}=\frac{(h-k)^2+2hk}{h}=2k+\frac{(h-k)^2}{h}.$$

The second term of the last of these forms vanishes when $h=k$, and is positive for all other values of h. The smallest value of l is, therefore, $2k$, and this, the shortest length of the isochronous simple pendulum, is realized when the axis of support is at the distance k from the centre of inertia.

To find at what distance h, from the centre of inertia the axis must be fixed to produce a pendulum isochronous with the simple pendulum, of given length l, we have the quadratic equation

$$h^2-hl=-k^2.$$

For the solution to be possible we have seen that l must be greater than, or at least equal to, $2k$. If $l=2k$, the roots of this equation are equal, k being their common value. For any value of l greater than $2k$, the equation has two real roots whose sum is equal to l, and product equal to k^2: hence, for any distance from the centre of inertia less than k, another distance greater than k, which is a third proportional to it and k, gives the same time of vibration; and the length of the simple pendulum corresponding to either case, is equal to the sum of the distances of the two axes from the centre of inertia. This sum is equal to the distance between them if the two axes are in one plane, through the centre of inertia, and on opposite sides of this point; and, therefore, for axes thus placed, and not equidistant from the centre of inertia, if the times of oscillation of the body when successively supported upon them are found to be equal, it may be inferred that the distance between them is equal to the length of the isochronous simple pendulum. As a simple pendulum exists only in theory, this proposition was taken advantage of by Kater for the practical determination of the force of gravity at any station.

(*h*) *A uniformly heavy and perfectly flexible cord, placed in the interior of a smooth tube in the form of any plane curve, and subject to no external forces, will exert no pressure on the tube if it have everywhere the same tension, and move with a certain definite velocity.*

For, as in § 592, the statical pressure due to the curvature of the rope per unit of length is $T\frac{\theta}{\sigma}$ (where σ is the length of the arc AB in that figure) directed inwards to the centre of curvature. Now, the element σ, whose mass is $m\sigma$, is moving in a curve whose curvature is $\frac{\theta}{\sigma}$ with velocity v (suppose). The requisite force is $\frac{mv^2\sigma\theta}{\sigma}=mv^2\theta$;

and for unit of length $mv^2\frac{\theta}{\sigma}$. Hence if $T=mv^2$ the theorem is true. If we suppose a portion of the tube to be straight, and the whole to be moving with velocity v parallel to this line, and *against* the motion of the cord, we shall have the straight part of the cord reduced to rest, and an undulation, of *any*, but *unvarying*, form and dimensions, running along it with the linear velocity $\sqrt{\frac{T}{m}}$.

Suppose the cord stretched by an appended mass of W pounds, and suppose its length l feet and its own mass w pounds. Then $T=Wg$, $lm=w$, and the velocity of the undulation is

$$\sqrt{\frac{Wlg}{w}} \text{ feet per second.}$$

(*j*) *When an incompressible liquid escapes from an orifice, the velocity is the same as would be acquired by falling from the free surface to the level of the orifice.*

For, as we may neglect (provided the vessel is large compared with the orifice) the kinetic energy of the bulk of the liquid; the kinetic energy of the escaping liquid is due to the loss of potential energy of the whole by the depression of the free surface. Thus the proposition at once.

(*k*) *The small oscillations of a liquid in a U tube follow the harmonic law.*

The tube being of uniform section S, a depression of level, x, from the mean, on one side, leads to a rise, x, on the other; and if the whole column of fluid be of length $2a$, we have the mass $2aS\rho$ disturbed through a space x, and acted on by a force $2Sxg\rho$ tending to bring it back. The time of oscillation is therefore (§ (*a*)) $2\pi\sqrt{\frac{a}{g}}$ and is the same for all liquids whatever be their densities.

Printed in the United States
By Bookmasters